NEUROCHEMISTRY IN
CLINICAL APPLICATION

ADVANCES IN EXPERIMENTAL MEDICINE AND BIOLOGY

Editorial Board:

NATHAN BACK, *State University of New York at Buffalo*
IRUN R. COHEN, *The Weizmann Institute of Science*
DAVID KRITCHEVSKY, *Wistar Institute*
ABEL LAJTHA, *N. S. Kline Institute for Psychiatric Research*
RODOLFO PAOLETTI, *University of Milan*

Recent Volumes in this Series

Volume 357
LACTOFERRIN: Structure and Function
Edited by T. William Hutchens, Sylvia V. Rumball, and Bo Lönnerdal

Volume 358
ACTIN: Biophysics, Biochemistry, and Cell Biology
Edited by James E. Estes and Paul J. Higgins

Volume 359
TAURINE IN HEALTH AND DISEASE
Edited by Ryan J. Huxtable and Dietrich Michalk

Volume 360
ARTERIAL CHEMORECEPTORS: Cell to System
Edited by Ronan G. O'Regan, Philip Nolan, Daniel S. McQueen, and David J. Paterson

Volume 361
OXYGEN TRANSPORT TO TISSUE XVI
Edited by Michael C. Hogan, Odile Mathieu-Costello, David C. Poole, and Peter D. Wagner

Volume 362
ASPARTIC PROTEINASES: Structure, Function, Biology, and Biomedical Implications
Edited by Kenji Takahashi

Volume 363
NEUROCHEMISTRY IN CLINICAL APPLICATION
Edited by Lily C. Tang and Steven J. Tang

Volume 364
DIET AND BREAST CANCER
Edited under the auspices of the American Institute for Cancer Research;
Scientific Editor: Elizabeth K. Weisburger

Volume 365
MECHANISMS OF LYMPHOCYTE ACTIVATION AND IMMUNE REGULATION V:
Molecular Basis of Signal Transduction
Edited by Sudhir Gupta, William E. Paul, Anthony DeFranco, and Roger Perlmutter

---

A Continuation Order Plan is available for this series. A continuation order will bring delivery of each new volume immediately upon publication. Volumes are billed only upon actual shipment. For further information please contact the publisher.

NEUROCHEMISTRY IN CLINICAL APPLICATION

Edited by

Lily C. Tang and
Steven J. Tang
SLCT, Inc.
Bethesda, Maryland

PLENUM PRESS • NEW YORK AND LONDON

Library of Congress Cataloging-in-Publication Data

```
Neurochemistry in clinical application / edited by Lily C. Tang and
  Steven J. Tang.
      p.   cm. -- (Advances in experimental medicine and biology ; v.
  363)
      "Proceedings of the International Neuropharmacology Symposim, held
  November 9-11, 1992 in Guangzhou, China"--T.p. verso.
      Includes bibliographical references and index.
      ISBN 0-306-44836-X
      1. Neuropharmacology--Congresses.  2. Neurochemistry--Congresses.
  I. Tang, Lily C.   II. Tang, Steven J.   III. International
  Neuropharmacology Symposium (1992 : Canton, China)   IV. Series.
      [DNLM: 1. Nervous System Diseases--drug therapy--congresses.
  2. Nervous System Diseases--physiopathology--congresses.  3. Nervous
  System--drug effects--congresses.  4. Mental Disorders--drug
  therapy--congresses.  5. Receptors, Drug--physiology--congresses.
  6. Brain Chemistry--congresses.   W1 AD559 v. 363 1994 / WL 100
  N4914 1994]
  RM315.N458 1994
  615'.78--dc20
  DNLM/DLC
  for Library of Congress                                      94-45371
                                                                   CIP
```

Proceedings of the International Neuropharmacology Symposium,
held November 9–11, 1992, in Guangzhou, China

ISBN 0-306-44836-X

© 1995 Plenum Press, New York
A Division of Plenum Publishing Corporation
233 Spring Street, New York, N. Y. 10013

10 9 8 7 6 5 4 3 2 1

All rights reserved

No part of this book may be reproduced, stored in a retrieval system, or transmitted in any form or by any means, electronic, mechanical, photocopying, microfilming, recording, or otherwise, without written permission from the Publisher

Printed in the United States of America

Soong Ching-ling (1893-1981)   Sung Yat-sen (1866-1925)

Dedicated to Dr. Sun Yat-sen

PREFACE

Neurological disorders cause untold suffering and financial burden to hundreds of thousands of people, not only to the patients, but also the relatives and society. As of today, though numerous scientists and clinicians have devoted their efforts to understand and combat these diseases, there is still no cure or satisfactory solution to the problems. Furthermore, the brain is the most essential organ of a human being. Realizing the importance of the brain, the president of the United States, George Bush, declared the 90s as the Decade of the Brain in January, 1992.

Being in neuroscience research for almost three decades, I initiated, planned and organized the first international neuropharmacology symposium. After long negotiation and fund raising, with the assistance and moral support of Dr. Abel Lajtha, director of the Center of Neurochemistry in New York, USA, we finally successfully put the program together. The Sun Yat-sen Foundation in China supported all the local expenses of the symposium and Sun Yat-sen University of Medicine in Guangzhou, China served as host organization. The symposium was held in Guangzhou, China, November 9–11, 1992, the eve of Dr. Sun Yat-sen's birthday.

Dr. Sun Yat-sen was born in Cui Heng Cun, on the outskirts of Guangzhou, China on November 12, 1866. He finished his high school education in the British and American Christian school in Honolulu, where he was exposed to Western influence. He had been long frustrated and discontented with the backwardness and corruption of the Ching Dynasty. During his years in medical training in Hong Kong he was very active in promoting reform of China.

After he graduated from medical school, he practiced in Macao and Guangzhou alongside the traditional Chinese doctors. He also started a Western and traditional Chinese medicine pharmacy. Two years later, he quit the medical field and became actively involved in the revolution to overthrow the imperial system and became the first president in the democratic China. He has since been known as the "Father of Chinese Revolution" and is well repected and honored by both Taiwan and mainland China.

While he was practicing medicine, he tried to merge the differences of the Western and traditional Chinese medicine for helping the suffering of the sick. He led the revolution and succeeded in overthrowing the imperial system with the goal of making a sudden leap to get China from backwardness to modernity. His careers, both in medicine and politics, were selfless. He attempted to absorb foreign knowledge to help China with the aim of allowing China to secure independence and equality in the world.

I, as one of Dr. Sun Yat-sen's decendants, would like to follow his idealistic dream, hoping through the International Neuropharmacology Symposium to bring better understanding between the Western and Chinese medical treatments. It is only right to dedicate this symposium to Dr. Sun Yat-sen in recognition of his vision in medicine. The symposium is also a centennial celebration of Dr. Sun Yat-sen's graduation from medical school and beginning his medical career. The objectives of this symposium are:

1. To promote an interchange of new findings in neuroscience between East and West.

2. To arrange review lectures (and chapters) on selective neurodisorders that affect a large population of the general public, presented by established senior scientists to stimulate discussion and provide younger scientists the chance to interact and hopefully provide the opportunity of collaboration between groups with related interests.

3. To integrate different perspectives for unifying understanding of neurodisorders (Parkinson's and Alzheimer's disease, psychological imbalance, depressions, strokes and epilepsy) and lessen the gap of clinical and basic sciences.

4. To promote the understanding of eastern (traditional Chinese herbs) and western approaches in seeking relief of these devastating nervous system disorders.

Lily C. Tang

CONTENTS

Perspective of Neurochemistry in Neurological Disorders 1
    Lily C. Tang

Selegiline Induces "Trophic-Like" Rescue of Dying Neurons without Mao Inhibition .. 15
    William G. Tatton, K. Ansari, W. Ju, P. T. Salo and Peter H. Yu

Aliphatic Propargylamines, a New Series of Potent Selective, Irreversible
    Non-Amphetamine-Like MAO-B Inhibitors 17
    Peter H. Yu, Bruce A. Davis, and Alan A. Boulton

Medical Treatment of Parkinson's Disease 25
    D. L. Xu

Neuropharmacological Effects of (−)-Stepholidine and Its Analogues on Brain
    Dopaminergic System 27
    Guo-Zhang Jin and Bao-Cun Sun

Neurotoxicity of MPTP and Uptake of MPPT into Dopamine and Norepinephrine
    Neurons in Mice 29
    E. H. Y. Lee and K. T. Lu

Linopirdine 47
    S. William Tam and Robert Zaczek

Differential Changes in Regional Brain Ganglioside and Neutral Glycosphingolipid
    Contents in Alzheimer's Disease 57
    N.M.K. Ng Ying Kin, L.H. Pan, J.H. Louvaris, Y. Robitaille and N.P.V. Nair

Action of Organophosphate Anticholinesterases on the Three Conformational States of
    Nicotinic Receptor 65
    Mugen Chi and Manji Sun

Purification and Characterization of a Novel Neurotoxin-Kappa Bungarotoxin 75
    Hu Ben-rong and Zhang Ze

Development of Antidepressant Drugs 77
    David T. Wong and Frank P. Bymaster

The Identification of Heterogeneity of 5-HT Receptors with [H]Rs-42358-197 97
    Erik H.F. Wong, Douglas W. Bonhaus, and Richard M. Eglen

Advances in Clinical Research on Common Mental Disorders with Computer
    Controlled Electro-Acupuncture Treatment 109
    Hechun Luo, Yunkai Jia Xiugin Feng, Xueying Zhao and Lily C. Tang

The Importance of Glutamate Receptors in Brain Ischemia 123
    Anker Jon Hansen

Biochemical Study of the Postischemic Neuronal Damage 133
    Dehua Jiang, Xue Rong, Qingyou Li and Zhongyou Wei

Effects of Ilexonin A on Circulatory Neuroregulation 143
    Luo, Rong Jing, Chen, Jie Wen, Zhou le Quan, Chen Li, Jia Ke Liang, Chen Chao
    Feng, Hu Wei An, Luo Zhuo Ling, Yang Bei Xin

Study on Cerebrovascular Disease of the Elderly in China 155
    Wang Xin-de

Cellular Physiology of Epileptogenic Phenomena 165
    Eiichi Sugaya, Aiko Sugaya, Kagemasa Kajiwara, Tadashi Tsuda, Noriyo Kubota,
    Noriyuki Yuyama, Masahiro Motoki, Tamaki Takagi Hisaaki Takagi, Tamiko
    Ookura, and Hideko Nagasawa

The Clinical Pharmacology of Antiepileptic Drugs 181
    Qu Zhi-ping

Hemodynamic Actions of Huatuo Reconstruction Pill on Anesthetized Animals 183
    Huang Shou Jian

Treatment of Affective Disorders 189
    Nobutaka Motohashi

Studies of Diagnosis and Pathogenesis of Wilson's Disease 197
    Xiuling Leung, Rong Chen, Zhoulin Liu and Yinru Zhang

Evidence for Presynaptic Damage in Myasthenia Gravis 207
    Chuan-Zhen Lu

PERSPECTIVE OF NEUROCHEMISTRY IN NEUROLOGICAL DISORDERS

Lily C. Tang

Department of Experimental Therapeutics
Walter Reed Army Institute of Research
Washington, DC 20307 and
SLCT, Inc.
P.O. Box 1634
Bethesda, MD 20827

INTRODUCTION

Degeneration of basal forebrain cholinergic neurons have been considered as one of the earliest and prominent neuropathological features of diseases of the human brain that give rise to loss of memory and dementia including Parkinson's and Alzheimer's diseases, and other neurological disorders. Genetic factors, aluminum or other toxic factors, immunological disturbances, disturb glucose metabolism, deficiency of essential nutrients, and stress are some of the etiology. The pathogenesis of the disorder involves not only the structural changes but also neurochemical disturbances and neuroendocrine dysfunction.

The most actively investigated area of biochemistry in relation to brain function in the last three decades has been that of the cerebral biogenic amines. Research on the neurotransmitters, receptors, ion channels and second messengers flourished. The currently recognized neurotransmitters are listed in Figure 1. The neurotransmitters release from the presynaptic to the cleft and act on the post synaptic membrane receptors. The receptors are coupled to the ion channels. The neurotransmitters influence the ions (Na^+, Ca^{++}, K^+) fluxes, and initiate a cascade of reactions. These cations affect the cyclase activities and second messengers. Once the reactions are disturbed, these abnormal activities give rise to various neurological problems. With the recent break through in biotechnology, new neurotransmitters, and receptors (subtypes) were discovered, purified, sequenced and characterized which made it possible to treat neurodisorders more effectively with biochemicals. This report will review the progress and potential effects of state of art in biotechnology and neurochemistry in clinical applications.

PARKINSON'S DISEASE

In late 1950's and beginning of 1960's, Dopamine (3-hydroxytyramine, 3,4-dihydroxyphenylethylamine) was discovered in both animal and human brains. Dopamine has

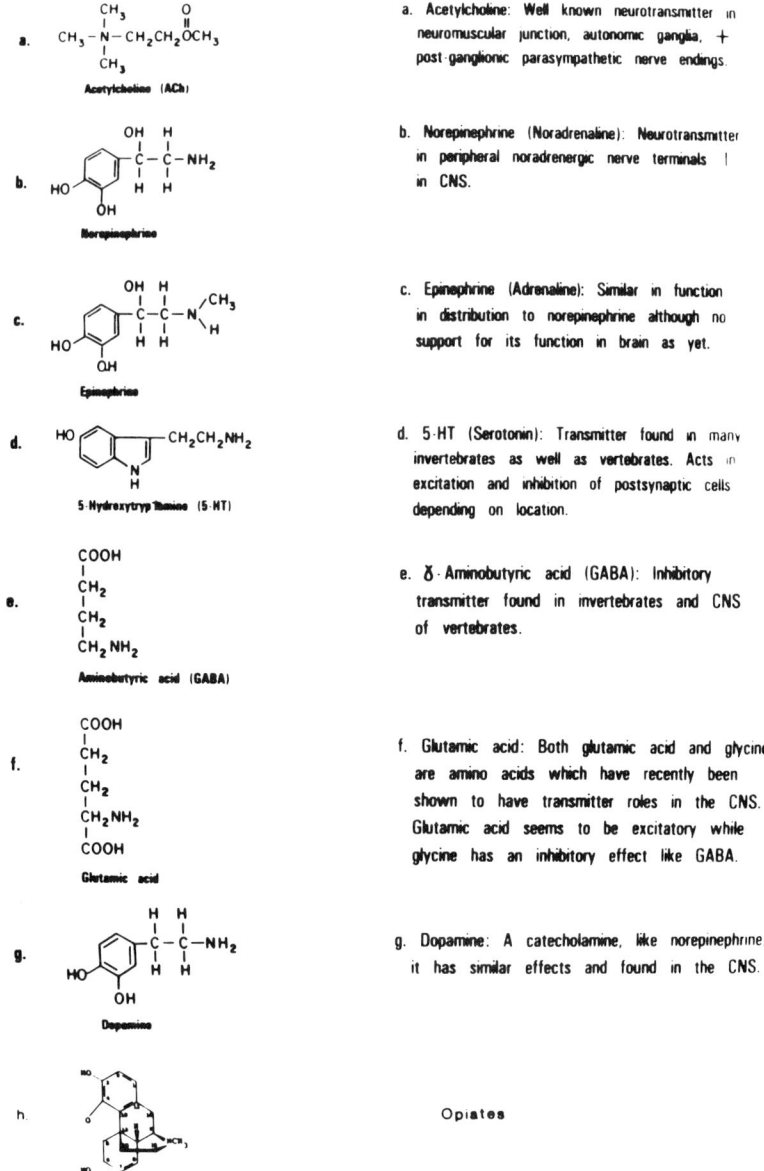

Figure 1. Major Neurotransmitters.

found to induce Parkinsonism in psychotics.[12] It diminishes the brain dopamine when administered to mice.[5] Ehringer and Hornykiewicz found that Parkinson's disease patients have low levels of dopamine and norepinephrine in the corpus striatum of their brians.[11] Since dopamine does not cross the blood brain barrier, in 1966 our group, led by the late Dr. George Cotzias, we pioneer the use of oral D,L-dopa, levodopa[15], the precursor of dopamine[16] to alleviate the symptoms of Parkinson's disease. In order to reduce the amount of L-dopa administered, we have investigated the possibility of inhibiting the peripheral dopa decarboxy-

we pioneer the use of oral D,L-dopa, levodopa[15], the precursor of dopamine[16] to alleviate the symptoms of Parkinson's disease. In order to reduce the amount of L-dopa administered, we have investigated the possibility of inhibiting the peripheral dopa decarboxylase to enhance the cerebral effects of L-dopa in mice.[17] With the positive result, we introduced the use of dopa decarboxylase inhibitor and L-dopa (sinemet or carbidopa)[18,19] to treat Parkinson's disease patients. The success on using L-dopa by a new technique of administration is the first to launch a rational biochemical attack on a complicated neurological disease and revolutionized the field of neuroscience, especially on the prospect of correlating neurochemical imbalance with neurological disorders and correcting them biochemically. The introduction of L-dopa for the treatment of Parkinson's disease is an excellent example of the development of a therapeutically useful agent by the interaction of basic and clinical research. As of to-day, L-dopa remains the drug of choice in the treatment of Parkinson's disease.[20–25] However, the effectiveness of L-dopa therapy declines on continuous use and usually accompanies by acute side effects (e.g. nausea, vomiting, orthostatic hypotension),[26] involuntary movements and on-off phenomenon.[15,27,28] Managing these complication is as challenging as the treatment of Parkinson's disease symptoms. Numerous efforts have been made to find more effective phamacologic therapies, not only to improve parkinsonism symptoms, but also to prevent disease progression. Other dopaminergic agonists, such as bromocriptine[29], pergoline[30], lisuride[31], and apomorphine[32], have been used alone or together with L-dopa. All were found beneficial but the effectiveness declined in a short period of time. Also the undesired side effects exist in all of them.

We had found that the monoamine oxidase (MAO) inhibitors can promote the entrance of amines (catecholamines), including dopamine into the brain.[33] The MAO inhibitors appear to have merit for treating Parkinson's disease and can slow down the progress of the disease.[34–36] Various laboratories have studied the enzyme extensively. William Tatton from University of Toronto will briefly describe his study on this enzyme inhibitor selegiline and its "tropic like" factor which may rescue the dying neurons. The chapter following Dr. Tatton's will be on MAO-B inhibitor analogues, their structures, functions and pharmacological implications discussed by Peter Yu.

We and others have attempted to synthesize or design agents acting on the dopaminergic system: increase dopamine synthesis (L-dopa, sinemet), enhance dopamine release (amantidin, CNS stimulants, nicotine), block dopamine re-uptake (tricyclic), and inhibit dopamine catabolism (catechol-O-methyl transferase (COMT) inhibitors,[37] MAO-B inhibitors). Other than the therapeutic agents that act on the dopaminergic system, manipulating other neurotransmitters have also been considered. The chemicals that interact with the cholinergic system (anticholinergics,[38–40] amantidine, tricyclic[41]) have been tried but presented minimal effect. Clonazepam, lisuride, 5-hydroxytryptamine[42] which have implication with serotonin were also being used with some effects. Research has been conducted on the relation of gamma-amino-butyric acid (GABA) with Parkinson's disease. Baclofen progabide has been used in combination with L-dopa.[43] There are indication that glutamate inhibitors (N-methyl-D-aspartate NMDA and AMPA antagonists) has a locomotor-stimulating effect in monamine-depleted rodents. Parkinson's disease patients has been treated with lamotigine, a MNDA-antagonist. It appears to have certain antiparkinsonism activity.[44–46]

The discovery of 1-methyl,4 phenyl,1,2,3,6-tetrahydropyridine (MPTP), a potent toxin that selectively damages substantia nigra dopaminergic neurons provided an animal model[47,48] for designing new novel therapeutic strategies and agents. These therapeutic approaches have focus on protective strategies designed to prevent or slow disease progression and promote neuronal regeneration. A study on basic mechanism of MPTP is included in this symposium.

L-dopa is a nutrient by virtue of being a large neutral amino acid. Although it does not become incorporate into protein but it shares its transport and its decarboxylation mechanism with a variety of large neutral amino acids. It must share its O-methylation with a variety of catechol derivatives; and it must share its capability to release pituitary growth hormone at least with arginine. The release of growth hormone might be useful in the treatment for other non-neurological diseases that respond to growth hormone. We have potentiated the responses of mice to L-dopa by injecting either bovine or porcine growth hormone.[49,50] The intermittent release of this hormone during treatment of L-dopa might be potentiating the intermittent cerebral effects of L-dopa. In our experiment, the patients of high protein intakes tend to cancel both the therapeutic effects and the cerebral side effects of L-dopa. Whereas low-protein diets potentiates and stabilizes the therapeutic effects.[51,52]

Recently, antioxidants have been considered to be one of the possible treatment for parkinsonism.[53] It is still in the basic investigation stage.

Improvements in surgical techniques have led to new applications of stereotactic surgery in the treatment of contralateral tremor, rigidity and even bradykinesia.[54,55] The stereotactic functional surgery is to combine the use of magnetic resonance imaging (MRI) with microelectrode recording. The tentative target points are determined by using stereotaxy assisted software system which revised the distortion of MRI images. Consequently, the accuracy and safety of the microelectrode recording will be increased.

The advance in neurosurgery techniques and biotechnology provide other novel therapeutics interventions, including implantation of polymer capsules[56] with dopaminergic drugs or cell lines that secret dopamine or trophic factors.[57] Neurosurgery is playing an increasingly important role for the future treatment of Parkinson's disease in the last 5 years.

Subthalamotomy has drastically improved bradykinesia in MPTP monkeys.[58] There were encouraging results on trial of ablative procedures on the lateral globus pallidus in human subject.[59] In the mean time, reports on autografts of adrenal medulla to the caudate can provide some benefits.[60–63] With the promise of adrenal medulla transplantation, other more effective brain grafting strategies have been considered. Though transplantation of fetal (6–10 weeks gestation) substantia nigra has shown promising results, but has caused both medical and ethical limitations.[64–72]

Currently, epidermoid growth factor (EGF), brain-derived neurotrophic factor (BDNF), other trophic molecules, such as GM1 ganglioside, and gene therapy have been investigated in animal models.[73–75] The future stragtergies of therapies for Parkinson's disease would have to consider all the alternative treatments available. The patients' syndromes should be evaluated and treated individually with combination of drugs and/or surgical maneuver.

ALZHEIMER'S DISEASE

The neurochemical findings in Alzheimer's disease (AD) have provided the basis of pharmacological investigations. The most essential dysfunction of the brain is probably in the mesolimbic acetylcholine system. Memory deficits can be demonstrated during the period of induced cortical cholinergic hypofunction.[76] The following have been found in AD patients: reduction of acetylcholinesterase (AChE), specific loss of choline acetyltransferase (CAT) activity in the cortex and the hippocampus[77–79], degeneration of ascending acetylcholine (ACh) neurons[80], and decreased ACh synthesis and choline uptake.[81,82] Basic neuroscience research has demonstrated that the brain regions known to be rich in cholinergic neuronal elements, play an important role in learning and memory.[83–85] Research has further implicated the muscarinic receptor in the hippocampus in the mechanism of short term memory.[86] Galanin, a neuropeptide isolated from the rat brain, is known to co-locate with CAT in the meganocellular neurons of the basal nucleus of Meynert, which is believed to be involved in dementing disorders. Galanin inhibits both ACh release in the hippocampus and memory acquisition. The

been to correct the neurochemicals dysfunctions. Cholinomimetic agents have been used to treat memory problems associated with AD and aging. Three approaches have been adopted: the precursor therapy, anti-AChE treatment, and muscarinic receptor agonists treatment. The precursor therapy is based on the observations that acute injections of[88,89], or dietary choline chloride[90] or lecithin [91] could produce a significant, but transient increase in brain ACh levels in rats and guinea pigs. Yet numerous clinical trials have failed to demonstrate beneficial effect with either choline or lecithin in demented or non demented aged patients.[92,93] The approaches of supplying AChE inhibitors or cholinergic agonists for improving geriatric memory have been tried. The anti-AChE physostigmine has shown moderate improvement in cognitive tasks within a very restricted dose range.[94] Beyond the restricted dose range, physostigmine can cause marked impairment in performance. Although the early attempts in using physostigmine in geriatric patients have not been successful[95–97], recently, it has been generally accepted that physostigmine can be effective in improving geriatric memory. Administering it intravenously shows the most consistent effect.[98–103] There is evidence suggesting that oral administration of physostigmine may accomplish better controlled effective dose.[104] Continuous systemic infusion of physostigmine affects the spatial learning in water maze of rats with bilateral lesions of nucleus basalis magnocellularis.[105] These results further demonstrated that anti-AChE agents can reduce memory loss by enhancing ACh transmission. The third approach uses the cholinergic drugs that directly stimulate central muscarinic receptors. There are indication degeneration of cholinergic forebrain nuclei may account for the loss of CAT activity in AD patients. If one assumes this degeneration plays a major role in the cognitive symptoms of the disease, then the most effective means to treat the deficit would be to stimulate the surviving postsynaptic receptors with direct muscarinic agonists to compensate for the loss of cholinergic input to the cortex and hippocampus. The previous two categories (precursor and anti-AChE) of drugs require intact, functioning presynaptic terminals. They would be less effective in improving cholinergic tone or restoring the balance of the CNS than the drug that interact with the postsynaptic cholinergic receptors. The muscarinic agonist arecoline has been tested in aged monkey. It showed a significant improvement in a delayed recall task.[106] Dr. Tam will review their study on compound that lower the threshold of AChE release which is currently in the stage of pre-clinical trial for AD.

Approaches of using somatostatin, serotonin and angiotensin II to modulate ACh function have been considered. Somatostatin proves not to be a promising therapeutic approach since there is no correlation between the decrease in somatostatin binding sites in brain and CAT activity. With serotonin, evidence suggests that serotonin receptors are located on cholinergic projections. Behavioral study suggests serotonin modulates memory function. There is indication that angiotensin converting enzyme inhibitors can facilitate ACh release and also enhances cognitive activity. This postulates that piracetam, an angiotensin covering enzyme inhibitor, may prevent age-related decrease in ACh receptor density.[107] Restoration of memory following septo-hippocampal grafts in rats has been reported. It has been proposed that grafting of embryonic septo-hippocampus may be a viable technique for treating AD.[108] Yet numerous problems in the grafting procedure still exist.

Other than cholinergic hypothesis of memory dysfunction, various neurotransmitters and degeneration of noncholinergic neurons have also been reported in brains of AD patients. There are changes in noradrenergic[109–111], dopaminergic,[109,112] serotonergic[110,113], and somatostatin-sensitive neurons.[114–117] Tests for evaluating the dopamine receptor agonist apomorphine, GABA receptor agonist muscimol, and the alpha-adrenergic receptor agonist clonidine, have failed to produce any significant effects in aged monkey.[118]

Gangliosides have been reported to play an important role in neurotransmitter functions and in central nervous system regeneration. Dr. N.M.K. Ng Ying Kin will discuss their findings

Gangliosides have been reported to play an important role in neurotransmitter functions and in central nervous system regeneration. Dr. N.M.K. Ng Ying Kin will discuss their findings on the abnormalities in the contents of gangliosides and their biosynthetic precursors in the brain of AD patient. This might suggest an alternative therapeutic strategy.

The last few years, emphasis for AD research has been in understanding the role of beta-amyloid protein, a major component of the abnormal structures, plaques, in the AD brains. An intriguing question is whether release of beta-amyloid precursor protein (APP) may contribute to Alzheimer's pathology.[119–122] The APP is embedded in cell membrane.[123,124] The discovery that one end of the beta-amyloid formation and plaque deposition are secondary events. The plaque deposit occurs after the cells are dead and membranes degenerated, and released APP.[125] The onset of AD before age 60 is infrequent and possibly caused by either a mutation in the APP gene located on chromosome 21 or, more commonly, by an unidentified gene on chromosome 14.[126–129] There is evidence of the involvement of chromosome 19 in late onset of AD.[130] The apolipoprotein E (APOE) locus is found to genetically associate with AD.[131,132] The APOE4, the codes for one of the three major versions of the APOE found in human, is the first genetic risk factor found for late-onset AD.[133] These findings may provide a clue for genetic involvement in AD which hopefully will lead to promising therapy.

OTHER NEUROLOGICAL DISORDERS (MENTAL PROBLEMS, ETC.)

Of all the neurological disorders perhaps depression and schizophrenia are the most common problems of all society in the world. There were no effective treatment and the etiologies are unclear. In the beginning of this century, electric shock treatment and lobectomy had been tried with some success but with many serious damages and controversy. Since the discovery of neurotransmitters in the brain and more and more studies on neurochemicals, these therapeutic strategies have been stopped and treatments have been emphasized on supplying various chemicals, such as phenothiazine, tricyclics and lithium. Serotonin has been found to be involved with depression and suicidal tendency. Dr. David T. Wong presents their findings on the effect of serotonin (5HT) uptake inhibitor as antidepressant drug. Also Dr. Eric H.F. Wong discusses their study on the heterogeneity of $5HT_3$ receptor binding sites to assist in developing effective antidepressants. Dr. Luo Hechun from Beijing, China reports their experience in treating manic depression and schizophrenia by computer controlled electro-acupuncture. Stimulation of certain acupoints can affect release of neurotransmitters and hormones in the brain of rodent.

Neuronal cell death after cerebral ischemia, epilepsy, metal poisoning and various neurodegenerative diseases is another problem that kept the clinician's hands tide. Scientists have been investigating the connection of neurochemicals such as opiates, glutamate, metals and neurotoxins to cell death and neurological disorders. Dr. Honore reviews the findings of excitatory amino acid receptors, the gated ion channel and the G-protein that coupled to these receptors. He presents evidence of excessive glutamate release be the cause of delayed nerve cell death after cerebral ischemia. Other scientists also present their studies in treating neurological disorders with various Chinese traditional medicine in this symposium.

Myasthenia Gravis, an autoimmune neuromuscular disorder, exhibits effects on the acetylcholine presynaptic membrane as well as the neuromuscular junction. Receptor binding antibody is found in 87% of patients. It present similarity with curare (nicotinic muscrinic cholinergic antagonist) poisoning and responds to antincholinesterase drugs. There are reports that acetylcholine release in skeletal muscle from patients with Myasthenia Gravis is increased compared to that in the muscle from control patients without muscular disease. Botulium inhibits quantal acetylcholine, newly synthesized acetylcholine and depolarization.[134,135]

1. Neurotransmitter-gated channel	Nicotinic acetylcholine, glutamate, gaba, glycine Na^+, Ca^{++}, K^+
2. Voltage-gated channels	Peptides, + charge residues directly involved with action Na^+, Ca^{++}, K^+
3. Cyclic-nucleotide-gated	cGMP gated channel
4. Intracellular membrane channel (calcium release channel)	IP-gated
5. G-protein-coupled receptor	Rhodopsin, ACh-M, Adrenergic, dopaminergic, serotonergic, tachykinin, angiotensine, LH, TSH Na^+-Ca^{++}, Ca^{++}-K^+

Figure 2. Different families of receptors and ion channels.

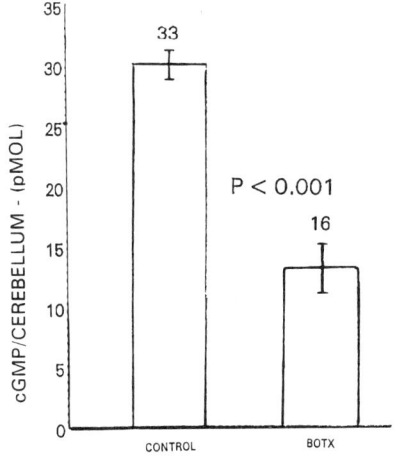

Figure 3. cGMP in the cerebellum of mice treated with botulinum toxin or saline.

Botulinum A has been used to treat Myasthenia Gravis, strabismus, blepharospasm, facial, cervical and extraperimidal dystonia with satisfaction.[136–142]

ION CHANNELS, ION FLUXES, SECOND MESSENGER AND PROJECTION OF FUTURE DEVELOPMENT OF THERAPEUTIC STRATEGIES FOR NEUROLOGICAL DISORDERS

Neurotransmitters bind to the receptors and open or close the ion channels coupled to the receptor. Cation enter the cell and stimulates or inhibits the cyclase activity. Dopamine stimulates the adenylate cyclase and increases the cyclic AMP whereas, cholingergic agents stimulate the guanylate cyclase to produce cyclic GMP. We had demonstrated that muscarinic cholinergic agonists bind to the receptor on the cell membrane, promoting excess calcium into the cell and enhanced the production of cGMP.[143–145] Dr. Numa had classified the sodium channels into five types. (Fig 2) Nicotinic cholinergic agonists act on the sodium channel. Neurotoxins had been reported to be excellent tools for studying the functions of sodium channels in the brain.[146] Our preliminary data showed that thioridazine acts on the anesthetic site of the action potential sodium channel. It opens the sodium channel and allowing sodium to enter the synaptosome prepared from various regions of the mouse brain. Our results indicate that thioridazine can protect the mice from botulinum toxicity. We tested our theory that botulinum toxin inhibits acetylcholine release may decrease the influx of sodium which in turn affects production of cGMP in the brain causing death. We measured the cGMP concentration of the cerebellum of the mice 48 hours after being treated with toxic dose of botulinum toxin and control mice by ether extraction and radioimmunoassay. (Fig. 3) The cGMP levels in the toxin infected mice were significantly reduced. Further experiment is warranted to understand the mechanism.

The basic mechanism of drug receptor interaction and signal transduction are of great interest and importance. All the neurotransmitter receptors are membrane bond. There is no existing receptor crystal available for determining x-ray crystalography for atomic coordinates. The development of 3-D high performance graphics computer and software provide promising tools for understanding theoretically and experimentally the actions of neurotransmitters on their receptors. Molecular receptor modelling is to model molecule interaction at molecular level in terms of steric interaction and electro-interaction. This provides models with well defined properties and boundaries. Basically molecular modeling is to use 3-D graphics computer as a tool applying physicochemical principles incorporate with biological actions of chemical studies to understand the chemical action on the biological substrate in the molecular level. This technique can provide a rational, systematic and efficient system for drug development. In the 2000 year era, this new state of art and new biotechnologies will further the success of our future drug design in obtaining more efficient and economical therapeutic agents.

REFERENCE

1. Blascho,H: (1959) Pharmacol. Rev. 11:307–316.
2. Schuman,HJ: (1960) In: adrenergic Mechanisms. Ed. by J.R. Vane,R.E.W. Wolstenholme and M. O'corner. J. and A. churchill Lit. London.
3. Montagu,KA: (1957) Nature, London 180:244–245.
4. Weil-Malherbk,H, Bone,A: Nature, London 180:1050–1051.
5. Carlsson,A, Lindquist,M, Magnusson,T, Waldeck,B: (1958) Science 127:471.
6. Bertler,A, Rosengren,E: (1959) Experimentia 15:10–11.
7. Carlsson,A: (1959) The occurrence, distribution and physiological role of catecholamines in the nervous system. Pharmacol. rev. 11:490–493.
8. Ehringer,H, Hornykiewicz,O: (1960) Klin. Wochenschr. 38:1236–1239.
9. Poirier,LJ, sourkes,TL: (1965) Influence of the substantia nigra on the catecholamine content of the striatum. Brain 88:181–192.
10. Sano,I, Gamo,T, Kakimoto,Y, Taniguchi,K, Takesada,M and Nishinuma,K: biochem. biophys. Acta 32:586–587.
11. Hornykiewicz,O: (1966) Dopamine (3-hydroxytyramine) and brain function. Pharmacol. Rev. 18:925–964.
12. Delay,J, Dinker,P: (1958) In Handbook of clinical Neurology, ed. P.J. Vinkers, G.W. Bryun, 6:248–266. Amsterdam & North Holland.
13. Axelrod,J, Kopin,IJ, and Mann,JD: (1959) 3-methoxy-4 hudroxyphenylglycol sulfate, a new metabolite of epinephrine and norepinephrine. biochim. biophys. Acta 36: 576–577.

16. Tang, LC: (1984) A presonal and scientific biography of Dr. Cotzias C. Cotaizs. Neurotoxicology 5:5–12.
17. Cotzias, GC, Tang,LC, Ginos, jz, Nicholson,AR,Jr. and Papavasiliou, PS.: (1971) Block of cerebral actions of L-dopa with methyl receptor substances. Nature 231:533–535.
18. Cotzias, GC, Papavasiliou, PS and Gellene, R: (1969) Modification of parkinsonism - Chronic treatment with L-dopa. N. Engl. J. Med. 280:337–345.
19. Papavasliou, PS, Cotzias, GC, Duby, SE, Steck, AJ, Fehling, C. and Bell, MA: (1972) Levodopa in parkinsonism: Potentiation of central effects with a peripheral inhibitor. N. Engl. J. Med. 286:8–14.209.
20. Hutton, JT, Morris, JL, Bush, DF, Smith, ME, Liss, CL, Reines, S: (1989) Multicenter controlled study of sinemet CR vs. Sinemet (25/100) in advanced Parkinson's disease patients. Neurology 39(Suppl,2):67–72.
21. Jankovic, J, Schwartz, K, Vander Linden, C: (1989) Comparison of sinemet CR4 and standard Sinemet: double blind and long-term open trial in parkinsoian patients with flutuations. Mov. Disord. 4:303–309.
22. Olanow, CW, Nakano, K, Nausieda, P, Tetrud, JA, Manyam, B, et al: (1991) An open multicenter trial of Sinemet CR in levodopa-naive Parkinson's disease patients. Clin. Neuropharmacol. 14:235–240.
23. Sage, JI, mark, MH: (1992) The rationale fror continuous dopaminergic stimulation in patients with Parkinson's disease. Neurology 42(Suppl. 1):23–28.
24. Chia, LG and Liu, LH: (1992) Parkinson's disease in Taiwan: an analysis of 215 patients. Neuroepidemiology 11:113–120.
25. Barbeau, A: (1973) Treatment of Parkinson's disease with L-dopa and Ro 4–4602: Review and present status. Adv. in Neurology 2:173–198.
26. Nutt, JG: (1989) Levodopa-induced duskinesia: review, observations, and speculations. Neurology 40:340–345.
27. Caraceni, T, Scigliano,G, Musicco,M: (1991) the coccurrence of motor fluctuations in Parkinsonian patients treated longterm with levodopa. Neurology 41:380–384.
28. Cedarbaum,JM, gandy, SE, McDowell, FH: (1991) "early" initiation of levodopa treatment does not promote the development of motor response fluctuations, dyskinesias, or dementia in Parkinson's disease. Neurology 41:622–629.
29. Jankovic, J: (1985) Long-term use of dopamine agonists in Parkinson's disease. Clin. Neuropharmacol. 8:131–140.
30. Nutt, JG, Fellman, JH: (1984) Pharmacokinetics of levodopa. Clin. Neuropharmacol. 7:35–49.
31. Laihinen, A, Rinne, UK, Suchy, I: (1992) Comparison of Lisuride and bromocriptine in the treatment of advanced Parkinson's disease. Acta Neurol Scand (Denmark) 86:593–595.
32. Duby, SE, cotzias, GC, Papavasiliou, PS, and Lawrence, WH: (1972) Injected apomorphine and oral levodopa in parkinsonism. Achives of Neurology 27:474–480.
33. Cotzias, GC, Tang, LC, and Ginos, JZ: (1974) Proc. Natl Acad. Sci (USA) 71:2715–2719.
34. Parkinson Study Group: (1989) Effect of deprenyl on the progression of disability in early Parkinson's disease. N. Engl. J. Med. 321:1364–1371.
35. The Parkinson Group: (1993) Effects of tocopherol and deprenyl on the progression of disability in early Parkinson's disease. N. Engl. J. Med. 328:176–183.
36. Olanow, CW: (1993) MAO-B unhibitors in Parkinson's disease. Adv. Neuro. 60:666–71.
37. Limousin, P, Pollak, P, Gervason-Tournier, CL, Hommel, M, Perret, JE: (1993) Ro 40–7592, a COMT inhibitor, plus levodopa in Parkinson's disease [letter]. Lancet 341 (8860):1605.
38. Cooper, JA, Sagar, HJ, Doherty, SM, Jordan, N, Tidsawell, P, sullivan, EV: (1992) Different effects of dopaminergic and anticholinergic therapies on cognitive and motor function in Parkinson's disease. Afollow-up study of untreated patients. Brain (England) 115:1701–1725.
39. Sandyk, R: (1992) L-trytophan in neuropsychiatric disorders: a review. Int. J. Neurosci. (England) 67:127–144.
40. Van Spaendonck, KP, Berger, HJ, Horstink, MW, buytenhyijs, EL, cools, WR: (1993) Imparied cognitive hifting in parkinsonian patients on anticholinergic therapy. Neuropsychologia 31:407–411.
41. Factor,SA, Brown,D: 1992) Clozapine prevents recrurrence of psychosis in Parkinson's disease. Mov. disord. 7:125–131.
42. Hesselink, JM: (1993) Serotonin and Parkinson's disease [letter] Am J Psychiatry 150:843–844.

40. Van Spaendonck, KP, Berger, HJ, Horstink, MW, buytenhyijs, EL, cools, WR: (1993) Imparied cognitive hifting in parkinsonian patients on anticholinergic therapy. Neuropsychologia 31:407–411.
41. Factor,SA, Brown,D: 1992) Clozapine prevents recrurrence of psychosis in Parkinson's disease. Mov. disord. 7:125–131.
42. Hesselink, JM: (1993) Serotonin and Parkinson's disease [letter] Am J Psychiatry 150:843–844.
43. FritzGerald,PM, Jankovic,J : (1990) Nondopaminergic therapy in Parkinson's disease. In therapy of Parkinson's disease, ed. W.C. Koller, G. Paulson. New York/Basel: Marcel Dekker pp.583.
44. Greenamyre, JT: (1993) Glutamate-dopamine interactions in the basal ganglia: relationship to Parkinson's disease. J. Neural. Transm. Gen. Sect. 91:255–269.
45. Zipp, F, Baas, H, Fisher, PA: (1993) Lamotrigine--antiparkinsonian activity by blockade of glutamate release? J. Neural. Transm. Park. Dis. Dement. Sect. 5:67–75.
46. Kornhuber, J, Riederer, P: (1993) N-methyl-D-aspartate (NMDA) antagonists in Parkinson's disease. J. Neurol. Neurosurg. Psychiatry 56:427.
47. Langston, JW, Ballard, PA, Tetrud, JW, Irwin, I: (1983) Chronic parkinsonism in humans due to a product of meperidine analogue synthesis. Science 219:979–980.
48. Burns,RS, Chiueh,CC, Markey,SP, Ebert,MH, Jacobwitz,DM and Kopin,IJ: (1983) A primate model of parkinsonism: Selective destruction of dopaminergic neurons in the pars compacta of the substantia nigra by n-methyl-4-phenyl-1,2,3,6-tetrahydopyridine. Proc. Natl. Acad. Sci. USA 80:4546–4550.
49. Tang, LC and Cotzias, GC: (1976) Modification of the actions of some neuroactive drugs by growth hormone. Archives of Neurology 33:131–134.
50. Cotzias, GC, Tang, Lc and Mena,I: (1973) Effects of inhibitors and stimualtion of protein synthesis on the cerebral actions of levodopa. In: Neuroscience Research vol 5, ed. by I.J. Kopin and S. Ehrenpreis. Academic Press, New York. pp. 97–105.
51. Hirata, H, Asanuma, M, Kondo, Y, Ogawa, N: (1992) Influence of protein-restricted diet on motor response fluctuations in Parkinson's disease. Rinsho shinkeigaku 32:973–978.
52. Weiss, R: (1993) Promising protein for Parkinson's [news; comment] Science 260:1072–1073.
53. Parkinson Study Group. (1989) Effect of deprenyl on the progression of disability in early Parkinson's disease, N. Engl J Med 321:1364–71.
54. Laine, E, Blond, S, Caparros-Lefebvre, D: (1992) Novel possibilities for treatment of parkinsonian tremor and other abnormal movements by stimulation of the nucleus ventralis intermedius thalami. Bull Acad Natl Med (France) 176:1147–1156.
55. Nishimura, H, Hirai, T: (1993) A new operational method of functional neurosurgery combining micro-recording and MRI stereotaxy for the treatment of Parkinson's disease. No To Shinkei (Japan) 45:144–155.
56. Emerich,DF, Winn,SR, christenson,L., Palmatier,MA, Gentile,FT, Sanberg,PR: (1992) A novel approach to neural transplantation in Parkinson's disease: use of polymer-encapsulated cell therapy. Neurosci Biobehav Rev 16:437–447.
57. Diedrich,N, Goetz,CG, Stebbins,GT et al: (1992) Blinded evaluation confirm lont-term asymmetric effect of unilateral thalamotomy or subthalamotomy on tremor in Parkinson's disease. Neurology 42:1311–1314.
58. Bergman,H, Wichman,T, Delong,MR: (1990) Reversal of experimental parkinsonism by lesion of the subthalamic nucleus. science 249:1436–1438.
59. Laitinen,LV, Bergenheim,AT, Hariz,MI: (1992) Leksell's posteroventral pallidotomy in the treatment of Parkinson's disease. J. Neurosurg. 76:53–61.
60. Goetz,CG, stebbins,GT, Klawans,HL, Koller,WC, Grossman,TG et al: (1991) United Parkinson Foundation Neurotransplantation Registry on adrenal medullary treansplants: presurgical and 1-and 2-year follow up. Neurology 41:1719–1722.
61. Zhang,WC: (1992) Long-term effects of intracerebral implantation of adrenal medulla in the treatment of Parkinsonism. Chung Hua Wai Ko Tsa Chih (China) 30:355–357.
62. Widner,H, Rehncrona,S: (1993) Transplantation and surgical treatment of parkinsonian syndromes. Curr. Opin. Neurol. Neurosurg. 6:344–349.
63. Anglade,P, Hirsch,EC, Brandel,JP, Javoy-Agid,F, Duyckaerts,C, Hauw,JJ, Agrid,Y: (1993) Adrenal transplant, dopaminergic neurons, and Parkinson's disease [letter] Ann. Neurol. 33:662–663.

66. Freed,CR, Breeze,RE, Rosenberg,NL, et al.: (1992) Survival of implanted fetal dopamine cells and neurological improvement 12 to 46 months after transplantation for Parkinson's disease. N. Engl. J. Med. 327:1549-1555.
67. Gage,FH: (1993) Parkinson's disease. Fetal implants put to the test [new]. Nature:361:405-406.
68. Freed,CR, Breeze,RE, Rosenberg,NL, Schneck,SA: (1993) Embryonic dopamine cell implants as a treatment for the second phase of Parkinson's disease. Replacing failed nerve terminals. Adv. Neurol. 60:721-728.
69. Walters,AM, Clarke,DJ, Bradford, HF, Stern,GM: (1992) The properties of culture fetal human and rat brain tissue and its use as grafts for the relief of the Parkinsonian syndrome. Neurochem. Res. 17:893-900.
70. Iacono,RP, Tang,ZS, Mazziotta,JC, Grafton,S, Doehn,M: (1992) Bilateral fetal grafts for Parkinson's disease: 22 months' results. Stereotact. funct. Neurosurgery 58:84-87.
71. Landau,WM: (1993) Clinical neuromythology X. Faithful fashion: survival status of the brain transplant cure for parkinsonism. Neurology 43:644-649.
72. Hitchcock,ER: (1993) Fetal transplant update [letter]. Science 259:442-443.
73. Pezzoli,G, Zecchinelli,A, Ricciardi,S, Burke,RE, Fahn,S, et al.: (1991). Intraventricular infusion of epidermal growth factor restores dopaminergic pathway in hemiparkinsonian rats. Mov. Disord. 6:281-287.
74. Scheider,JS, Pope,A. Simpson,K, Taggert,J, smith,MC, DeStephano,L: (1992) Recovery from experimental parkinsonism in primates with GM1 ganglioside treatment. Science 256:843-846.
75. Jiao,S, Gurevich,V, Wolff,JA: (1993) Lont-term correlation of rat model of Parkinson's disease by gene therapy. Nature 362:450-453.
76. Mouton, PR, Meyer,EM & Arendash,GW: (1989) Intracortical AF64A: memory impairments and recovery from cholinergic hypofunction. Pharmacol Biochem. Behav. 32:841-848.
77. Bowen,DM, Smith,CB, White,P & Davison,AN: (1976) Beurotransmitter-related enzyme and indices of hypoxia in senile dementia and other abiotrophies. Brain:459-496.
78. Perry,EK, Perry,RH, Blessed,G, Tomlinson,SE: (1977) Necropsy evidence of central cholinergic deficits in senile dementia [letter]. Lancet 1:189.
79. Coyle,JT, Price,DL, Delong,MR: (1983) Alzheimer's disease: a disorder of cortical cholinergic innervation. Science 219:1184-1190.
80. Johnston,MV, Grzanna,R, Coyle,JT: (1979) Methylazoxymethanol treatment of fetal rats results in abnormally dense noradrenergic innervation of neocortex. Science 203:369-371.
81. Sims,NR, Bowen,DM et al: (1983) Presynaptic cholinergic dysfunction in patients with dementia. J. Neurochem. 40:503-539.
82. Sims,NR, Bowen,DM, Smith,CC et al: (1980)Glucose metabolism and acetylcholine synthesis in relation to neuronal activity in Alzheimer's disease. Lancet: 1:333-335.
83. Lippa,AS, Pelham,RW et al: (1980) Brain cholinergic dysfunction and memory in aged rats. Neurobiol. Aging 1:13-19.
84. Gibson,GE, Peterson,C, Jenden,DJ: (1981) Brain acetylcholine synthesis declines with senescence. Science 213:674-676.
85. Mesulam,MM, Mufson,EJ, Levet,AI, Wainer,BH: (1983) Cholinergic innervation of cortex by the basal forebrain: cytochemistry and cortical connections of the septal area, diagonal band nuclei, nucleus basalis (substantia innominata), and hypothalamus in the rhesus monkey. J Comp. Neurol. 214:170-197.
86. Mastropaolo,J, Nadi,NS et al: Proc. Natl Acad Sci USA 85:9841-9845.
87. Beal,MF, Clevens,RA, et al: (1988) J Neurochem. 51:1935-1941.
88. Bartus,RT, Dean,RL, Pontecorvo,MJ, Flicker,C: (1985) The cholinergic hypothesis: a historical overview, current perspective, and future direction. Ann N.Y. Acad. Sci. 444:332-358.
89. Cohen,EL. & Wurtman,RJ: (1975) Brain acetylcholine: increase after systemic choline administration. Life Sci 16:1095-1102.
90. Haubrich,DR, Wang,PF, Clody,DE, Wedeking,PW: (1975) Increase in rat brain acetylcholine induced by choline or deanol. Life Sci 17:975-980.
91. Wurtman,RJ, Fernstrom,JD: (1976) control of brain neurotransmitter synthesis by precursor availability and nutritional state. Biochem Pharmacol 25:1691-1696.

92. Hirsch,MJ, Wurtman,RJ: (1978) Lecithin consumption increases acetylcholine concentrations in rat brain and adrenal gland. Science 202:223–235.
93. Bartus, RJ, Dean,RL et al: (1982) Science 217:408–417.
94. Bartus,RJ, dean,RL & Beer,B: In Nutrition in Gerontology. J.M.Ordy, D.Harman & R.Alfin-Slater,Eds. Raven Press, N.Y. 1984.
95. Drachman,DA: In Psychopharmacology: A Generation of Progress. M.A.Lipton, A.DiMascio & K.F. Killan, Eds. Raven Press, N.Y., 1978.
96. Drachman,DA, Sahakian,BJ: (1980) Memory and cognitive function in the elderly. A preliminary trial of physostigmine. Arch Neurol 37:674–675.
97. Smith,CM, Swash,M: (1979) Physostigmine in Alzheimer's disease [letter] Lancet 1:42.
98. Goodnick,P, Gershon,S: (1984) Chemotherapy of cognitive disorders in geriatric subjects. J Clin Psychiatry 45:196–209.
99. Bartus,RT: (1979) Four stimulants of the central nervous system: effects on short-term memory in young versus aged monkeys. J Am Geriatr Soc 27:289–297.
100. Davis,KL, Mohs,RC: (1982) Enhancement of memory processes in Alzheimer's disease with multiple-dose intravenous physostigmine. Am J Psychiatry 139:1421–1424.
101. Davis,KL, Mohs,RC, Tinklenberg,JR: (1978) Physostigmine: improvement of long-term memory processes in normal humans. Science 201:272–274.
102. Christie,JE: In Alzheimer's Disease: A Report of Progress in Research. S.Corkin,K.L.Davis,J.H.Growdon,E.Usdin & R.J.Wurtman,Eds. Raven Press. N.Y., 1982.
103. Thal,LJ, Fuld,PA, Masur,DM, Sharpless,NS: (1983) Oral physostigmine and lecithin improve memory in Alzheimer's disease. Ann Neurol 13:491–496.
104. Mandel,RJ, Chen,AD, Connor,DJ, Thal,LJ: (1989) Continuous physostigmine infusion in rats with excitotoxic lessions of the nucleus basalis magnocellularis: effects on performance in the water maze task and cortical cholinergic markers. J pharmacol Exp ther 251:612–619.
105. Bartus,RT, Dean,RL, Goas,JA, Lippa,AS: (1980) Age-related changes in passive avoidance retention: modulation with dietary choline. Science 209:301–303.
106. Costall,B, Barnes,JM et al: (1990) Pharmacopsychiatry 23 suppl 2:85–88.
107. Bond,NW, Walton,J, Pruss,J: (1989) Restoration of memory following septo-hippocampal grafts: a possible treatment for Alzheimer's disease. Biol Psychol 28:67078.
108. Adolffson,R, Gottfries,CG et al: In Alzheimer's Disease: Senile Dementia and Related Disorders. R.Katzman,R.D.Terry & K.L.Bick, Eds. Raven Press, N.Y. 1978.
109. Arai,H, Kosaka,K, Iizuka,R: (1984) Changes of biogenic amines and their metabolites in postmortem brains from patients with Alzheimer-type dementia. J. Neurochem 43:388–393.
110. Cross,AJ, Crow,TJ, Johnson,JA, Joseph,MH et al: (1983) Monoamine metabolism in senile dementia of alzheimer type. J. Neurol Sci 60:383–392.
111. Owen,F, Baker,HF et al: (1981) Effect of cheronic amphetamine administration on central dopaminergic mechanisms in the vervet. Psychopharmacology 74:213–216.
112. Soininen,HE, MacDonald,E et al: (1981) Acta Neurol Scandinav. 64:101–107.
113. Sims,NR,Bowen,DM et al: (1983) Metabolic processes in Alzheimer's disease: adenine nucleotide content and production of $^{14}CO_2$ from [u-14]glucose in vitro in human neocortex. J. Neurochem. 41:329–334.
114. Davies,P, Katzman,R, Terry,RD: (1980) Reduced somatostatin-like immunoreactivity in cerebral cortex from cases of Alzheimer's disease and Alzheimer senile dementia. Nature 288:279–280.
115. Davies,P, Terry,RD: (1981) Cortical somatostatin-like immunoreactivity in cases of Alzheimer's disease and senile dementia of the alzheimer-type. Neurobiol Aging 2:9–14.
116. Rossor,MN, Emson,PC, et al: (1980) Reduced amounts of immunoreactive somatostatin in the temporal cortex in senile dementia of alzheimer type. Neurosci Lett 20:373–377.
117. Bartus,RT, Dean,RL, Beer,B: (1983) An evaluation of drugs for improving memory in aged monkeys: implications for clinical trials in humans. Psychopharmacol Bull 19:168–184.
118. Catteruccia,N, Willingale-Theune,J et al: (1990) Ultrastructure localization of the putative precursors of the A4 amyloid protein associated with Alzheimer's disease. Am J Pathol 137:19–26.

117. Bartus,RT, Dean,RL, Beer,B: (1983) An evaluation of drugs for improving memory in aged monkeys: implications for clinical trials in humans. Psychopharmacol Bull 19:168–184.
118. Catteruccia,N, Willingale-Theune,J et al: (1990) Ultrastructure localization of the putative precursors of the A4 amyloid protein associated with Alzheimer's disease. Am J Pathol 137:19–26.
119. Unterbeck,A, Bayney,RM et al: Review of Biological Research in Aging 4:139–162. M.Rothstein, Eds. Wilet-Liss N.Y., 1990.
120. Sisodia,SS, Koo,EH et al: (1990) Evidence that beta-amyloid protein in Alzheimer's disease is not derived by normal processing. Science 248:492–495.
121. Muller-Hill,B, Beyreuther,K: (1989) Molecular biology of Alzheimer's disease. Annu Rev Biochem 58:287–307.
122. Goldgaber,D, Lerman,MI et al: (1987) characterization and chromosomal localization of a cDNA encoding brain amyloid of Alzheimer's disease. Science 235:877–880.
123. de Sauvage,F, Octave,JN: (1989) A novel mRNA of the A4 amyloid precursor gene coding for a possibly secreted protein. Science 245:651–657.
124. Abraham,CR, Van Nostrand,W et al: Second International Conference on alzheimer's disease and Related Disorders: Neurobiol Aging 11:303, 1990.
125. Price,DL: (1986) New perspective on alzheimer's disease. Annu Rev Neurosci 9:489–512.
126. St George-Hyslop,PH et al: (1987) Science 235:885-
127. Schellenberg,GD etal: (1992) Science 258:668-
128. St George-Huslop et al: (1992) Nature Genet 2:330-
129. Goate,A et al: (1991) Nature 349:704-
130. Pericak-Vance,MA et al: (1991) Am J Hum Genet 48:1034-
131. Strittmatter,WJ et al: (1993) Proc Natl Acad Sci USA 90:1977-
132. Penicak-Vance,MA et al: (1988) Exp Neurol 102:271-
133. Corder,EH, Saunders,AM, Strittmatter,WJ et al: (1993) Gene dose of apolipoprotein E type 4 allele and the risk of alzheimer's disease in late onset families. Science 261:921–923. 134. Bigalke,H et al: (1981) Tetanus toxin amd botulinum A toxin inhibit acetylcholine release from but not calcium uptake into brain tissue. Naun Schmied Arch Pharmacol 315:143–8.
135. Bandyopadhay,S et al: (1987) Role of heavy and light chains of botulinum neurotoxin in neuromuscular paralysis. J Biol Chem 62:2660–3.
136. Lande,S et al: (1989) Effects of botulinum neurotoxin and Lambert-eaton myasthenic syndrome IgG at mouse nerve terminals. J Neural Transm 1:229–42.
137. Jankovic,J & Brin,MF: (1991) Therapeutic uses of botulinum toxin. N Engl Med 324:1186–94.
138. Cohen,DA et al: (1986) Botulinum injection therapy for blepharospasm: a review and report of 75 patients. Clin Neuropharmacol 9:415–29.
139. Elston,J & Lee,J: (1988) Clinical use of botulinum toxin. Lancet 2:1139.
140. Fahn,S et al: (1985) Double-blind controlled study of botulinum toxin for blepharospasm. Neurology 35:271–2.
141. Newman,NJ & Lambert,SR: (1992) Botulinum toxin treatment of supernuclear ocular mobility disorders. Neurology 42:1391–3.
142. Brin,MF et al.: (1988) Localized injections of botulinum toxin for treatment of focal dystonia and hemifacial spasm. In: Fahn,S et al, ed. Advances in Neurology Bol 50 Dystonia 2 New York,: Raven Press; p 599–608.
143. Tang,LC, Schoomaker,EB $ Weissman,W: (1984) Cholinergic agonists stimulate calcium uptake and cGMP formation in human RBC. Biochem Biophys Acta 772:235–8.
144. Tang,LC: (1986) Identification and characterization of human erythrocyte muscarinic receptors. General Pharmacology 17:281–5.
145. Tang,LC: (1991) Human erythrocyte as a model for investigating muscarinic agonists and antagonists. General Pharmacology 22:485–90.
146. Le vine,H & Cuatrecasas,P: (1981) An overview of toxin-receptor interactions. Pharmacol Ther 12:167–207.

SELEGILINE INDUCES "TROPHIC-LIKE" RESCUE OF DYING NEURONS WITHOUT MAO INHIBITION

William G. Tatton[1], K. Ansari[1], W. Ju[1], P. T. Salo[1] and Peter H. Yu[2]

[1] Center for Research in Neurodegenerative Diseases
University of Toronto
Toronto, Canada
[2] Neuropsychiatry Research Unit
Department of Psychiatry
University of Saskatchewan
Saskatoon, Saskatchewan, Canada

Selegiline has been claimed to slow the progression of motor deficits in Parkinson's Disease (PD)[1] and cognitive decline in Alzheimer's Disease (AD).[2] It is controversial whether the slowing represents a symptomatic action due to improved dopaminergic neurotransmission or neuroprotection due to a decrease in the production of toxic oxidative radicals consequent on the inhibition of monoamine oxidase B (MAO-B). We propose a third action for selegiline—a "trophic-like" rescue of dying neurons.

We first reported that selegiline reduces the death of MPTP-damaged, murine dopaminergic nigrostriatal neurons (dNSns) by 60–70 % when selegiline treatment is delayed until after the conversion of MTPT to MPP$^+$ by MAO-B is completed.[3] We now report that delayed selegiline doses of 0.01 mg/Kg/2days produce the same reductions in the death of MPTP-damaged dNSns but are insufficient to inhibit MAO-A or MAO-B. In parallel research, we found that murine dNSNS die gradually after MPTP-exposure with highest rates of death occuring between 5 and 15 days after the exposure. The gradual death is in keeping with reports that MPTP damage can induce apoptotic neuronal death[4] which has been related to a loss of adequate trophic support to neurons.[5]

Accordingly, we examined the death of immature rat facial motoneurons (FMns) after axotomy of their peripheral axons.[6] Most of the trophic support to the immature FMns is derived from their target muscles. We found that an average 27% of FMns axotomized at postnatal day 14 (PNd14) survive to PNd35 in animals treated with intraperitoneal saline every second day beginning on the day of axotomy. In animals treated with 10 or 0.01 mg/Kg of selegiline on the same schedule, averages of 52–56% the axotomized FMns survived (p < 0.001) showing that selegiline therapy can compensate, in part, for a loss of target-derived trophic support. Doses of 0.001 mg/Kg of selegiline did not increase the survival of the axotomized FMns.

We measured the inhibition of MAO-A and MAO-B in the brainstem of FNd14 rats for doses of both (−)-deprenyl (selegiline) and (+)-deprenyl ranging from 0.001 to 50 mg/Kg. The ED_{50} for MAO-B inhibition caused by the (−)-enantiomer was 0.09 mg/Kg and 8 mg/Kg for the (+)-enantiomer, less than 100 fold difference. Doses of (+)-deprenyl of 10 mg/Kg did not increase the survival of the FMns axotomized at PNd14 but caused 70% MAO-B inhibition. In comparison, doses of 0.01 mg/Kg of (−)-deprenyl increased FMn survival but only caused 10% MAO-B inhibition. Hence the induction of the compensation for loss of target-derived trophic support by selegiline is sterospecific at greater than 1000 fold and independent of MAO-B inhibition.

Measurements of the process growth of dopaminergic neurons in ventral mesencephalic striatal organotypic cultures have shown the selegiline concentrations as low as 10^{-9} M increase process growth in an average neuron by 37% over the initial 12 days in culture. Therefore, these newly discovered, "trophic-like actions" of selegiline may involve a high affinity, sterospecific interaction with an unknown protein. The reactive hyperplasia of astrocytes appears to contribute to the survival of axotomized motoneurons and other damaged neurons.[7] Work in our laboratory indicates that selegiline increases the hyperplasia of astrocytes in the immediate vicinity of axotomized FMns, suggesting the possibility that reactive astrocytes are an intermediary in the selegiline-induced rescue of dying neurons.

Neuronal rescue by selegiline could contribute to the reported slowing of the progression of AD and PD and should be possible at markedly lower doses than those required to provide neuroprotection through MAO-B inhibition.

REFERENCES

1. The Parkinson Study Group: (1989) Effect of deprenyl on the progression of disability in early Parkinson's disease. New Engl. J. Med. 321:1364–1371.
2. Mangoni, A, Grassi, MP, Frattola, L, Piolti, R, Brassi, S, Motta, A., Marcone, A and Smirne, S: (1991) Effects of a MAO-B inhibitor in the treatment of Alzheimer's disease. Eur. Neurol. 31:100–107.
3. Tatton, WG and Greenwood, CE: (1991) Rescue of dying neurons: a new action for deprenyl in MPTP Parkinsonism. J. Neurosci. Res. 30:666–672.
4. Dipasquale, B, Marini, AM and Youle, RJ: (1991) Apoptosis and DNA degradation induced by 1-methl-4-phenypyridinium in neurons. Biochem Biophys Res. Comm. 181:1442–1448.
5. Raff, MC: (1992) Social controls on cell survival and cell death. Nature 356:397–400.
6. Salo, PT and Tatton, WG: Deprenyl reduces the death of motoneurons caused by axtomy. J. Neurosci Res. 31:394–400.
7. Graeber, MB and Kreutzberg, GW: (1986) Astrocytes increase in glial fibrillary acidic protein during retrograde changes of ficial motor neurons. J. Neurocytol 15:363–374.

ALIPHATIC PROPARGYLAMINES, A NEW SERIES OF POTENT SELECTIVE, IRREVERSIBLE NON-AMPHETAMINE-LIKE MAO-B INHIBITORS

Their Structures, Function and Pharmacological Implications

Peter H. Yu, Bruce A. Davis, and Alan A. Boulton

Neuropsychiatric Research Unit
Department of Psychiatry
University of Saskatchewan, Saskatoon
Saskatchewan, Canada

Abstract

l-Deprenyl, a selective irreversible MAO-B inhibitor, has been shown to prolong the onset of disability in Parkinson's patients and to improve cognitive behavior in Alzheimer's disease. It has been claimed that l-deprenyl exhibits neuroprotective and neurorescue effects in several animal models. The precise mechanism of these effects is unknown. It is yet to be established whether or not the effects are unique to l-deprenyl; a drug which possesses, in addition to inhibition of MAO-B activity, an amphetamine moiety. Based on the fact that several N-methylpropargylamine derivatives have been shown to be MAO inhibitors and that aliphatic amines are typical MAO-B substrates with a high affinity for the enzyme, we have synthesized a series of aliphatic propargylamines which have turned out to be highly potent, selective and irreversible MAO-B inhibitors, structurally unrelated to amphetamine. The potency of these inhibitors is related to their chain length and the substitution of a hydrogen on the terminal carbon of the aliphatic chain. MAO-I activity, as assessed *in vitro,* increased as the aliphatic carbon chain length increased; substitution of the hydrogen at the aliphatic chain terminal by hydroxyl, carboxyl or carboethoxyl groups or replacement of the methyl group on the nitrogen atom by an ethyl group considerably reduced their inhibitory activity. Stereospecific effects were observed with the R-(-)-enantiomer being 20-fold more active than the S-(+)-enantiomer. Inhibitors with relatively short carbon chain lengths (i.e. four to six carbons) were found to be more potent at inhibiting brain MAO-B activity *in vivo* especially after oral administration. M-2-PP [N-methyl-N-(2-pentyl)-propargylamine] and 2-HxMP [N-(2-hexyl)-N-methyl-propargylamine], for example, are approximately 5 fold more potent and selective inhibitors of mouse brain MAO-B activity than l-deprenyl after oral administration. Like l-deprenyl, chronic low dose administration of the aliphatic propargylamines caused a slight cumulative inhibition of MAO-A activity in the mouse brain. These new inhibitors selectively inhibited

MAO-B activity *in vivo*, i.e. they increased 2-phenylethylamine levels substantially, but did not affect the levels of dopamine, DOPAC, HVA, 5-HT and 5-HIAA. Both 2-HxMP and M-2-PP have been shown to be capable of protecting against MPTP-induced nigrostriatal dopamine depletion and against DSP-4 [N-(2-chloroethyl)-N-ethyl-2-bromobenzylamine] induced noradrenaline depletion in the hippocampus of the mouse. These new aliphatic MAO-B inhibitors seem to be nontoxic and may be useful in the treatment of certain neuropsychiatric disorders.

INTRODUCTION

l-Deprenyl (selegiline, Eldepryl), a typical MAO-B inhibitor, has been shown to be useful in the treatment of Parkinson's and Alzheimer's diseases (PD and AD respectively). It was first used as an adjunct drug to potentiate l-DOPA in the treatment of PD (i.e. to reduce the oxidation of dopamine) (1). l-Deprenyl reduces the requirement for l-DOPA and thus reduces side effects related to the use of l-DOPA. The problem with l-DOPA chemotherapy, however, is that its efficacy lasts only about 3–5 years and its side effects can gradually become intolerable. l-Deprenyl has been shown to be capable of prolonging the efficacy of l-DOPA in the more advanced stages of treatment (i.e. it eases the "on-off" side effect) (2). The treatment of PD with l-DOPA, however, seems to be mostly symptomatic; it does not appear to cure or prevent the progression of the illness. Dopamine neurons continue to die.

Recently it has been reported that l-deprenyl can, by itself, significantly delay the onset of disability associated with early, otherwise untreated, cases of PD (3,4). These findings have been confirmed in several centres around the world (5,6). l-Deprenyl has also been claimed to improve the clinical condition of some Alzheimer's patients (7,8) and in depression (9). In addition l-deprenyl has been shown to prolong the life span and improve the sexual activity of rodents (10) and perhaps humans (11). Unlike MAO-A inhibitors, MAO-B inhibitors do not usually cause hypertensive crises, and thus they possess the potential to become very useful neuropsychiatric and geriatric drugs. l-Deprenyl probably possesses other functions in addition to its ability to potentiate dopamine function. It has been shown that l-deprenyl can protect against neurotoxins, such as MPTP (12) and DSP-4 (13). Even chemical and physical assaults on nerve cells can be rescued by l-deprenyl after the insult (14,15). The mechanism of this neurorescue effect is unclear.

MPTP is a neurotoxin which can damage the nigra-striatal nerve pathway and induce PD symptoms in humans (16) and primates (17,18). These neuronal damage is responsive to l-DOPA treatment (19). l-Deprenyl, as well as other MAO inhibitors, has been shown to be capable of preventing MPTP-induced Parkinson-like neurotoxicity in animals (12). MPTP itself is not directly involved in the toxic action, but in the brain it is converted to the distal toxin MPP^+ by MAO-B in glial cells (20). It has been suggested that PD might be caused by MPTP-like substances existing in the environment which might be ingested or absorbed by PD patients (21); alternatively, it has been suggested that such neurotoxins might be formed endogenously. Several compounds, such as N-methylated tetrahydroisoquinoline and tetrahydro-carboline, have been proposed as such candidates (22). Blocking MAO-B activity therefore prevents such neurotoxic action. l-Deprenyl also exhibits neuroprotective effects against another neurotoxin, namely, DSP-4, which causes the depletion of noradrenaline in the nerve terminals of the hippocampus (13). Unlike MPTP, DSP-4 is not metabolized by MAO-B. Exactly how l-deprenyl works to protect against this neurotoxic damage is not understood.

It has been proposed that MAO-catalyzed deamination reactions can enhance oxidative stress and that the resultant free radicals may cause damage to nerve cells (23). Such oxidations lead to the production of hydrogen peroxide, and this in turn can be converted into the hazardous hydroxyl free radical in the presence of ferrous ions. Inhibition of MAO activity will reduce this oxidative stress and thus slow down any associated damage. It is also possible that certain neurotrophic factors may be stimulated by l-deprenyl. We have recently observed that l-deprenyl may be involved in the regulation of the cell cycle where it could possibly prevent the senescence of astroglial cells, which presumably act by providing trophic factor support to the nerve cells (24). In addition it has been shown that MAO-B inhibitors stimulate the gene expression of the amine synthesis enzyme L-aromatic amino acid decarboxylase (25) and inhibit the expression of glial fibrillary acidic protein (26).

Although l-deprenyl does rescue "stressed" nerve cells, it only partially restores some of them. Questions therefore that arise are: Is l-deprenyl a unique compound of this kind? Is a special form of MAO-B present and involved in cell death or in the regulation of neural regeneration? Is it possible that l-deprenyl-like compounds will exhibit similar or more potent effects than does l-deprenyl itself? Since we recently discovered that aliphatic amines are readily deaminated by MAO-B, we have pursued this avenue of research and synthesized and assessed a series of structurally related aliphatic N-propargylamine compounds. Some of these compounds have been found to be not only highly potent MAO-B inhibitors, e.g. N-(2-hexyl)-N-methyl-propargylamine (2-HxMP) and N-methyl-N-(2-pentyl)-propargylamine (M-2-PP), but more selective than l-deprenyl *in vivo*. These aliphatic propargyl MAO-B inhibitors are also very effective in protecting nerve cells against MPTP and DSP-4 induced damage. Preliminary results also indicate that 2-HxMP possesses neurorescue effects in mouse facial axotomy paradigm. These investigations are most encouraging and point to the possible discovery of even more effective neuroprotective and neurorescue agents.

DEAMINATION OF ALIPHATIC AMINES BY MAO-B

MAO-B has been shown to be involved in the conversion of the antiepileptic prodrug, 2-n-pentyl-aminoacetamide (Milacemide) (27). Milacemide can cross the blood brain barrier, where it is then oxidized by MAO to form glycinamide, which is subsequently cleaved to glycine. The delivery of glycine is presumably related to the anticonvulsant activity of Milacemide (28) and the consequent apparent improvement in learning and memory (29). We have recently shown that 2-propyl-pentylamine (2-propyl-1-aminopentane) and 2-propyl-pentylglycinamide can similarly be deaminated by rat liver MAO-B followed by conversion to 2-propylpentaldehyde and then valproic acid (VPA) (30,31) both *in vitro* and *in vivo*. Although these compounds can serve as prodrug delivering valproic acid to the brain, they appear unfortunately to be toxic.

Aliphatic monoamines are known to be metabolized by MAO-B (32). The km values with respect to straight chain aliphatic amines with carbon numbers between 5 and 10 are very low, indicating that these amines possess a high affinity for the active site of MAO-B. This property has been exploited by us in the design of potential MAO-B inhibitors.

ALIPHATIC PROPARGYLAMINES AS MAO-B INHIBITORS

l-Deprenyl, a structural analogue of amphetamine, is catabolized to produce desmethyldeprenyl, methamphetamine and amphetamine (33); this has caused some concern since its amphetamine-like properties may be associated with its clinical efficacy, although it is a fact that the l-form of deprenyl as well as its l-form metabolites are behaviorally much less active than the d-forms. Different MAO-B inhibitors, not possessing any amphetamine moiety or amphetamine-like properties, should therefore be assessed (34). The high affinity of some aliphatic amines for the active site of MAO-B has now been exploited in the design of some specific MAO-B inhibitors. For this reason we have synthesized a series of aliphatic propargylamine derivatives (35). Some of them, such as 2-HxMP and M-2-PP, are highly potent, selective and irreversible MAO-B inhibitors. MAO inhibitory activity appears to be correlated with the lipophilicity of these compounds. The length of the carbon chain of the N-alkyl group is not only related to inhibitory potency, but it also affects the relative selectivity towards MAO-A and MAO-B. When the terminal carbon is substituted with a hydroxyl group, MAO-B inhibitory activity is markedly reduced. Substitution with a carboxy or carbethoxy group at the terminal carbon also causes a considerable reduction in the MAO inhibitory activity. The inhibitory activity of these compounds is stereospecific; for example, the (R) stereoisomer of N-(2-butyl)-N-methylpropargylamine is about 20 fold more potent than the (S) enantiomer in the inhibition of MAO-B activity; this is quite similar to the stereospecific effect exhibited by l-deprenyl (36).

Aliphatic propargylamines with shorter carbon chain lengths appear to be more potent than their longer chain analogs in the inhibition of brain MAO-B activity (i.e. as assessed from their ED_{50} values) following intraperitoneal administration. These shorter chain length molecules are more easily absorbed (into lipids, membranes, etc.) and/or more readily transported into the brain. In comparison to compounds with longer carbon chain lengths, they are less lipophilic and therefore less likely to bind to, or associate with, peripheral lipophilic components. 2-HxMP and M-2-PP were found to be even more potent at blocking MAO-B activity in the brain following oral administration than was l-deprenyl. It is also interesting to note that the aliphatic propargylamines, such as M-2-PP, are considerably more stable in aqueous solution than is l-deprenyl.

A series of substituted N-alkylpropargylamines with N-2-butyl-N-methylpropargylamine (2-BuMP) as the base compound have been synthesized and compared for the effects of structural modification on the effectiveness and selectivity of the inhibition of MAO activities. When the N-methyl group was replaced by a hydrogen atom, an ethyl group or a propargyl group, MAO inhibitory activity is abolished. The modification of the propargyl group, e.g. to 3-butynyl, N-cyanomethyl or to allyl, also destroys the inhibitory activity. The potency of the inhibitors is related to the carbon chain length of the alkyl group as well as to the substitution of the alpha or the terminal carbon atoms. Substitution of hydroxyl, carboxyl or carboethoxyl groups on the terminal carbon of the alkyl chain reduces considerably the inhibitory activity. An increase in MAO inhibitory activity was observed for molecules posssessing a single methyl group substitution on the alpha carbon in comparison to those substituted with two hydrogen or two methyl groups. Other branched alkyl N-methylpropargylamines, e.g. N-methyl-N-(3-pentyl)propargylamine, appear to be slightly less selective in MAO-B inhibitory activity (37).

Chronic effect of aliphatic propargylamine MAO inhibitors (via intraperitoneal injection or oral administration) on mouse brain MAO activity levels has been assesssed. Inhibition of MAO-A and MAO-B was dependent both on the inhibitor and the dose applied. Both 2-HxMP and M-2-PP at a low dose (0.25 mg/Kg) were without effect on either MAO-A or MAO-B 24 hours after a single i. p. injection. This dose was effective on MAO-B, however, after 13 days of treatment. At higher acute doses (2 mg/Kg) M-2-PP selectively inhibited MAO-B activity. After 10 and 21 daily treatments a greater inhibition of MAO-B was observed, but by these times MAO-A had also become slightly inhibited. At lower oral doses (i.e. 1 and 10 µg/mL in drinking water), selective inhibition of MAO-B activity was achieved, while at higher doses (i.e. 100 µg/mL), MAO-A also became inhibited following three weeks of treatment. A prolonged inhibition of MAO-B activity following a single higher acute dose (2 mg/Kg, i. p.) of several aliphatic propargylamines (e.g. M-2-PP) was observed, confirming that the inhibition of MAO-B by aliphatic propargylamines is, like l-deprenyl, also irreversible *in vivo* and that the synthesis of new brain mitochondrial MAO appears to be an extremely slow process.

NEUROCHEMICAL EFFECTS

2-HxMP does not affect caudate levels of DA, NE or 5-HT after an acute dose of up to 50 mg/Kg. At the relatively higher doses (i.e. doses above 10 mg/Kg), however, some amine metabolites, such as DOPAC and HVA in the caudate, are significantly reduced. There is little effect on 5-HIAA levels. Similar results were obtained also with the other aliphatic N-propargylamines possessing shorter carbon side chains, e.g. M-2-PP and N-(2-butyl)-N-methylpropargylamine.

Trace amines in caudate tissues have been assessed following i. p. administration of different doses of M-2-PP, 2-HxMP and l-deprenyl. PE (a typical MAO-B substrate) levels increased substantially even at doses as low as 0.5 mg/Kg. Both aliphatic N-methylpropargylamines seem to be more potent than l-deprenyl in causing an increase of PE levels. The levels of p-TA, which is a mixed-type substrate for MAO, however, were only slightly increased and even then higher doses (i.e. 20 mg/Kg) were requried. Similar results were observed with respect to the effects on m-TA.

Mice have been chronically orally treated with M-2-PP and l-deprenyl. Both inhibitors were made up fresh on a daily basis and included in the drinking water (10 µg/mL). After three weeks MAO activities in the lateral cortex and the amines and their metabolites in the striatum were analyzed. This chronic treatment causes a selective inhibition of mouse brain MAO-B activity; MAO-A activity was not affected. Both M-2-PP and l-deprenyl caused a significant increase in DA and 3-MT levels in the mouse striatum. It is important to note here that l-deprenyl has been shown to lose about 30% of its activity in the drinking water each 24 hour. M-2PP is quite stable in water for at least one week.

NEUROPROTECTIVE EFFECTS

Several inhibitors of this series have been found to be capable of protecting against MPTP-induced depletion of striatal dopamine neurons as well as against DSP-4-induced depletion of noradrenaline in the hippocampus. The effect of M-2-PP and M-1-PP (N-methyl-N-(1-pentyl)-propargylamine) on MPTP-induced depletion of DA levels and the number of DA uptake sites in the neostriatum has been investigated (38). MPTP reduces striatal DA levels to 37 % of its control value within 3 days following treatment. This reduction is prevented by M-2-PP at relatively low doses, i.e. 0.5 and 2.5 mg/Kg. A structural analogue of M-2-PP, namely M-1-PP is less efficient at inhibiting MAO-B activity; it only protected striatal DA depletion at the higher dose of 10 mg/Kg. The MPTP lesion causes a 30 % reduction in the number of DA uptake sites, and this was also prevented by both doses of M-2-PP but only the higher dose of M-1-PP. The result of the uptake study is consistent with the results of restoring of striatal DA levels in MPTP treated mice.

DSP-4 induces a considerable depletion in hippocampal NE in CD1 Swiss white mice one week after drug administration. When the MAO inhibitors, M-2-PP and 2-HxMP, are injected two hours before DSP-4 treatment, they, like l-deprenyl, exhibit a protective effect against the DSP-4 induced reduction of noradrenaline in the nerve terminals of the mouse hippocampus. These two aliphatic N-propargyl inhibitors do not affect the transport of noradrenaline into the hippocampus and so the protective effect of l-deprenyl on DSP-4 induced neurotoxicity seems not to be due to the inhibition of DSP-4 uptake.

CONCLUSION

Aliphatic N-propargylamines, such as M-2-PP and 2-HxMP, are highly potent, irreversible and selective MAO-B inhibitors both *in vitro* and *in vivo*. Whilst being similar to l-deprenyl they do not possess any amphetamine moiety within their structure. They are able to protect nigrostriatal dopamine neurons and hippocampal noradrenaline nerve terminals against MPTP and DSP-4 induced depletions. They may prove to be useful in studies on the mechanism of action in neuroprotection and neurorescue and in the treatment of neurodegenerative disorders, such as Parkinson's and Alzheimer's diseases.

REFERENCES

1. Birkmayer W, Knoll J, Riederer P and Youdim MBH (1983) (-)-Deprenyl leads to prolongation of l-Dopa efficacy in Parkinson's disease. Mod. Pbl. Pharmacopsychiatr. 19: 170–176.
2. Lieberman A. and Fazzini E. (1991): Experience with selegiline and levodopa in advanced Parkinson's disease. Acta Neurol. Scand. 84 (Suppl.) 136: 66–69.
3. The Parkinson Study Group. (1989) Effect of deprenyl on the progression of disability in early Parkinson's disease. New Engl. J. Med. 321: 1364–1371.
4. he Parkinson Study Group (1993) Effects of tocopherol and deprenyl on the progression of disability in early Parkinson's disease. New Engl. J. Med. 328: 176–183.
5. Tetrud J. W. and Langston J. W. (1989) The effect of deprenyl (selegiline) on the natural history of Parkinson's disease. Science 245: 519–522.
6. Allain H., Cougnard J., Neukirch H. C., the FMST members. (1991) Selegiline in de novo parkinsonian patients: The French selegiline multicenter trial (FSMT). Acta Neurol. Scand. 84 (Suppl.) 136:73–78.
7. Mangoni A., Grassi M. P., Frattola L., Piolti R., Brassi S., Motta A., Marcone A. and Smirne S. (1991) Effects of a MAO-B inhibitor in the treatment of Alzheimer disease. Eur. Neurol. 31: 100–107.
8. Tariot P. N., Cohen R. M., Sunderland T., Newhouse P. A., Yount D., Mellow A. M., Weingartner H., Mueller E. A. and Murphy D. L. (1987) L-Deprenyl in Alzheimer's disease-preliminary evidence for behavioral change with monoamine oxidase B inhibition. Arch. Gen. Psychiatry 44: 427–433.
9. Quitkin FM, Liebowitz MR, Stewart JW, McGrath P. J., Harrison W., Rabkin J. G., Markowitz J. and Davies S. O. (1984) l-Deprenyl in atypical depressives. Arch. Gen. Psychiatry 41:777–781.
10. Knoll J., Dallo J. and Yen T. T. (1989) Striatal dopamine, sexual activity and lifespan longevity of rats treated with (-)-deprenyl. Life Sciences 45: 525–531.
11. Birkmayer W., Knoll J., Riederer P., Hars V. and Marton J. (1985) Increased life expectancy resulting from addition of l-deprenyl to Madopar treatment in Parkinson's disease: A long term study. J. Neural Transm. 64: 113–127.
12. Heikkila R. E., Hess A. and Duvoisin R. C. (1984) Dopaminergic neurotoxicity of 1-methyl-4-phenyl-1,2,3,6-tetrahydropyridine in mice. Science 224: 1451–1453.
13. Finnegan K. T., Skratt J. S., Irwin I., DeLanney L. E. and Langston J. W. (1990) Protection against DSP-4-induced neurotoxicity by deprenyl is not related to its inhibition of MAO-B. Eur. J. Pharmacol. 184: 119–126.
14. Salo P. T. and Tatton W. G. (1991) Deprenyl reduces the death of motoneurons caused by axotomy. J. Neurosci. Res. 31: 394–400.
15. Tatton W. G. and Greenwood C. E. (1991) Rescue of dying neurons: a new action for deprenyl in MPTP Parkinsonism. J. Neurosc. Res. 30: 666–672.
16. Langston J. W., Ballard P. A., Tetrud J. W. and Irwin I. (1983) Chronic parkinsonism in human due to product of meperidine-analog synthesis. Science 219: 979–980.
17. Burn R. S., Chiueh C. C., Markey S. P., Ebert M. H., Jacobowitz D. M. and Kopin I. J. (1983) A primate model of Parkinsonism: selective destruction of dopaminergic neurons in the pars compacta of the substantia nigra by N-methyl-4-phenyl-1,2,3,6-tetrahydropyridine. Proc. Natl. Acad. Sci. USA 80: 4546–4550.

18. Langston J. W., Forno L. S., Rebert C. S. and Irwin I (1984) Selective nigral toxicity after systemic administration of 1-methyl-4-phenyl-1,2,3,6-tetrahydropyridine in the squirrel monkey. Brain Res. 292: 390–394.

19. Davis G. C., Williams A. C., Markey S. P., Evert M. H., Caine E. D., Reichert C. M. and Kopin I. J. (1979) Chronic Parkinsonism secondary to intravenos injection of meperidine analogues. Psychiat. Res. 1: 249–254.

20. Chiba K., Trevor A. and Catagnoli N. Jr. (1984) Metabolism of the neurotoxic tertiary amine, MPTP, by brain monoamine oxidase. Biochem. Biophys. Res. Comm. 120: 574–578.

21. Snyder S. H. and D'Amato R. J. (1986) MPTP: a neurotoxin relevant to the pathophysiology of Parkinson's diasease. Neurology 36: 250–258.

22. Ohta S., Kohno M., Makino Y., Tachikawa O. and Hirobe, M. (1987) Tetrahydroisoquinoline and 1-methyltetrahydroisoquinoline are present in the human brain: relation to parkinson's disease. Biomed. Res. 8: 453–456.

23. ohn G, and Spina M. B. (1989) Deprenyl suppresses the oxidant stress associated with increased dopamine turnover. Amer. Neurol. Assoc. 26: 689–690.

24. Skibo G., Ahmed I., Yu P. H., Boulton A. A. and Fedoroff S. (1992) l-Deprenyl, a monoamine oxidase-B (MAO-B) inhibitor, acts on the astroglia cell cycle at the G1/G0 boundary. Am. Soc. Cell Biology.

25. Li X. M., Juorio A. V., Paterson I. A., Zhu M. Y. and Boulton A. A. (1992) Specific irreversible monoamine oxidase B inhibitors stimulate gene expression of aromatic l-amino acid decarboxylase in PC12 cells. J. Neurochem. 59: 2324–2327.

26. Li X. M., Qi J., Juorio A. V. and Boulton A. A. Reduction in GFAP mRNA abundance induced by (-)-deprenyl and other MAO-B inhibitors in C6 glioma cells. J. Neurochem. (in press).

27. De Varebeke P. J., Cavalier R., David-Remacle M. and Youdim M. B. H. (1988) Formation of the neurotransmitter glycine from the anticonvulsant milacemide is mediated by brain monoamine oxidase-B. J. Neurochem. 50: 1011–1016.

28. van Dorsser W., Barris D., Cordi A. and Roba J. (1983) Anticonvulsant activity of milacemide. Arch. Int. Pharmacodyn. 266, 239–249.

29. Handelmann G. E., Nevins M. E., Mueller L. L., Arnolde S. M. and Cordi A. A. (1989) Milacemide, a glycine prodrug, enhances performance of learning tasks in normal and amnestic rodents. Pharmacol. Biochem. Behav. 34, 823–828.

30. Yu, P. H. and Davis, B. A. 2-Propyl-1-aminopentane, its deamination by monoamine oxidase and semicarbazide-sensitive amine oxidase, conversion to valproic acid and behavioral effects. *Neuropharmacology 30* (1991) 507–515.

31. Yu, P. H. and Davis, B. A. Simultaneous delivery of valproic acid and glycine to the brain; deamination of 2-propylpentylglycinamide by monoamine oxidase B. Mol. Chem. Neuropathol. 15 (1991) 37–49.

32. Yu, P. H. (1989) Deamination of aliphatic amines of different chain lengths by rat liver monoamine oxidase A and B. J. Pharm. Pharmacol. 41, 205–208.

33. Heinonen E. H., Myllyla V., Sotaniemi, K., Lammintausta, R., Salonen, J. S., Anttila, M., Savijarvi, M. and Rinne, U. K. (1989) Pharmacokinetics and metabolism of selegiline. Acta Neurol. Scand., 126: 93–99.

34. Langston J. W. (1990) Selegiline as neuroprotective therapy in Parkinson's disease: concepts and controversies. Neurology 40(Suppl.): 61–66.

35. Yu P. H., Davis B. A. and Boulton A. A. (1992) Aliphatic propargylamines: potent selective irreversible monoamine oxidase B inhibitors. J. Med. Chem. 35 : 3705–3713.

36. Robinson B. J. (1985) Stereoselectivity and isozyme selectivity of monoamine oxidase inhibitors: enantiomers of amphetamine, N-methylamphetamine and deprenyl. Biochem. Pharmacol. 34: 4105–4108.

37. Yu P. H., Davis B. A and Boulton A. A. Effect of structural modification of alkyl N-propargylamines on the selective inhibition of monoamine oxidase B activity. Biochem. Pharmacol. (in press)

38. Yu P. H., Davis B. A. and Boulton A. A. Neurochemical and neuroprotective effects of some aliphatic propargylamine MAO-B inhibitors. J. Neurochem. *(in press)*

MEDICAL TREATMENT OF PARKINSON'S DISEASE

D. L. Xu

Shanghai Second Medical University
Department of Neurology
Rui Jin Hospital
Shanghai, China

Parkinson's disease (PD) is a progressive neurodegenerative disease. It's main feature is the nigrostriatal neuronal cell loss resulting in striatal dopamine (DA) deficiency. L-dopa + peripheral amino acid decarboxylase inhibitor (DDI:Sinemet or Madopar) is still the most effective agent for the symptomatic treatment of PD, but the dopa therapy is far from perfect. For advanced cases especially for patients with wearing off of the drug effect, fluctuation of response and "on-off," the conventional preparation such as Sinemet or Madopar due to their short T1/2 is no longer suitable for clinical use. Thus controlled release preparation e.g. Sinemet CR or Madopar HBS is preferable.

Following agents may be used as adjunct to L-dopa:

1. DA agonists: bromocriptine, pergolide, lisuride, apomorphine, etc.

2. MAO-B inhibitors (MAOBI): L-deprenyl–selegiline, aminoethyl-chloro-benzamide Ro-16–6491, aminoethylchloropyridine carboxamide Ro-19–6327;

3. Amantadine and anticholnergics: budipine, dexetimide, trihexyphenidyl;

4. Others: COMT inhibitor (Ro-40–7592), threo DOPS, tetrabiopterine, nicotin-amide adenine dinucleotid (NADH) and PLG.

PLG (MIF$_1$), L-prolyl-L-leucyl-glycinamide is an antagonist of MSH (melanocytes stimulating hormone). PLG potentiates the action of L-dopa in 6-OHDA animals and partially protects MPTP neurotoxicity in Rhesus monkey. It is used clinically for advanced PD.

For psychiatric complications, one may use clozapine and molindone. For insomnia, anxiety and depression, short acting benzodiazepine or soporific antidepressants may be used.

Meanwhile the reasonable treatment of PD would be DA agonist + MAOBI + CR dopa.

NEUROPHARMACOLOGICAL EFFECTS OF (−)-STEPHOLIDINE AND ITS ANALOGUES ON BRAIN DOPAMINERGIC SYSTEM

Guo-Zhang Jin and Bao-Cun Sun

Shanghai Institute of Materia Medica
Chinese Academy of sciences
Shanghai 200031, China

Tetrahydroprotoberberines (THPBs), a series of natural products isolated from the chinese herbs Corydalis and Stephania, have been proven to be novel dopamine (DA) receptor antagonists. among them, (−)-stepholidine (SPD) is a leading compound due to its strong affinity to DA receptors and its peculiar pharmacological characteristics.

(−)-SPD has high affinity to both DA D_1 and D_2 receptors with preference to D_1 receptors. It antagonizes the DA receptor agonist-induced behavioral responses. Biochemical experiments show that (−)-SPD significantly increases L-dopa accumulation and facilitates the release of DA from nerve terminals and it decreases the content of store acetylcholine (ACh) in the striatum. Electrophysiological results demonstrate that (−)-SPD reverses and/or antagoniszes the DA receptor agonist-induced inhibition of DA neurons in substantia nigra pars compacta (SNC) and ventral tegmental area (VTA). All these indicate that (−)-SPD is a DA D_2 receptor antagonist. In addition, (−)-SPD inhibits DA-stimulated cAMP formation, which reflects its D_1 receptor antagonistic effect.

However, on unilateral 6-hydroxyldopamine (6-OHDA)-lesioned rotational behavioral, (−)-SPD induces contralateral rotation similar to DA receptor agonists. (−)-SPD challenged rotation is preferentially blocked by D_1 receptor antagonist SCH 23390, but not D_2 receptor antagonist and it has same slow development process as D_1 agonist SKF 38393. These results indicate that (−)-SPD possesses D_1 receptor agonistic property in the unilateral 6-OHDA-lessioned supersensitive model. On the other hand, when in presence of D_1 antagonist to occupy the D_1 receptors, (−)-SPD antagonizes the D_2 receptor agonist N-0437-induced rotation. This suggests that the D_2 antagonistic effects of (−)-SPD remain in 6-OHDA-lesioned rats, but its previous D_1 antagonistic effects converses into D_1 agonistic action under the same conditions.

Reserpinized rat (reserpine 1 mg/Kg X 6 dose) is another supersensitive model of dopaminergic system. It has been found that D_1 receptor agonist, although having little effects on SNC DA cell firing in normal rats, can express its inhibitory effects on DA cells in the reserpinized rats. In this case, (−)-SPD antagonizes and/or reverses mixed DA agonist apomorphine (APO)-induced inhibition of SNC DA cell firing rate. Furthermore, (−)-SPD reverses both D_1 and D_2 receptor agonist-induced inhibition. These results indicate that (−)-SPD shows both D_1 and D_2 receptor antagonistic effects in reserpinized rats. These results imply that the supersensitivities induced by 6-OHDA lesions and reserpine have different mechanisms.

VTA is another important nuclei in the brain containing DA neurons, and it plays an important role in the controling and modulating of psychomotility. (−)-SPD sensitively reverses apomorphine (APO)-induced inhibition of VTA DA cell firing similar to its effects on SNC DA cells. Yet, large doses of (−)-SPD induced depolarization inactivation (DI) of VTA DA cells, while (−)-SPD never induces DI of SNC DA cells, even at very large doses. Chronic experiments also show that (−)-SPD only induced DI of VTA DA cells. This discrepancy is considered as a common characteristic of atypical neuroleptics. These results, thus, indicate that (−)-SPD may serve as a leading compound for searching new kinds of atypical neuroleptics.

NEUROTOXICITY OF MPTP AND UPTAKE OF MPPT INTO DOPAMINE AND NOREPINEPHRINE NEURONS IN MICE

E. H. Y. Lee* and K. T. Lu

Institute of Biomedical Sciences
Academia Sinica
Taipei (11529), Taiwan
The Republic of China

We have earlier demonstrated that 1-methyl-4-phenyl-1,2,3,6-tetrahydropyridine (MPTP) produces a direct toxicity on DA neurons in the substantia nigra (SN) and norepinephrine (NE) neurons in the locus coeruleus (LC) other than in the terminals. These results suggest that DA neurons in the SN and NE neurons in the LC are possibly able to uptake MPP^+ through the DA and NE uptake systems, respectively, presumably via DA and NE dendrites. The present study examined this hypothesis. Adult male BALB/c mice were used. In the first part, animals received various combinations of nomifensine (NOM, a specific DA uptake blocker) and MPTP treatment. NOM was infused to the SN or striatum at 1.5 µg and 3.0 µg per day for a total of seven days. The infusion volume was 0.25 and 0.5 µl, respectively. MPTP was given systemically (IP) at 30 mg/kg per day for a total of seven days. On each day, NOM was given 30 min prior to MPTP injection. In the second part, animals received the same combinations of desimipramine (DMI, a specific NE uptake blocker) and MPTP treatments. DMI was infused to the LC and hippocampus at 1.25 µg and 2.5 µg per day 30 min before MPTP injection for a total of 7 days. The volumes of infusion were 0.25 µl in the LC and 0.5 µl in the hippocampus. Animals were subject to locomotor activity test 7 days after the last MPTP, MPP^+ (saline) injections. They were then sacrificed and brain tissues of the striatum and hippocampus were subject to DA and NE analyses with HPLC fluorescence detection. Results indicated that MPTP consistently and markedly decreased DA level in the striatum and NE level in the hippocampus. It also impaired locomotor activity and produced long-lasting tremor in mice. NOM pretreatment in the striatum completely prevented MPTP's toxicity, while NOM pretreatment in the SN only partially, but significantly, prevented MPTP's toxicity on both DA and motor activity. Similar results were obtained with DMI and MPTP studies.

---

* Send correspondence to: Dr. Eminy H. Y. Lee, Institute of Biomedical Sciences, Academia Sinica, Taipei (11529), Taiwan, The Republic of China. Tel: 886–2–7899125. Fax: 886–2–7822835.

DMI pretreatment in the hippocampus completely prevented MPTP's toxicity, while DMI pretreatment in the LC partially protected against MPTP's toxicity on NE and motor function. In vitro uptake studies have indicated that MPP$^+$ and DA competed for the same uptake site in both the SN and striatum; and MPP$^+$ and NE competed for the same uptake site in both the LC and the hippocampus. These results together suggest that DA and NE dendrites (and possibly short axons) in the SN and LC can also uptake DA and NE, respectively, and MPP$^+$ possibly shares the same uptake systems with DA and NE. These results provide the first in vitro and in vivo evidence of dendritic uptake of DA and NE in the brain.

INTRODUCTION

1-methyl-4-phenyl-1,2,3,6-tetrahydropyridine (MPTP) has been demonstrated to produce selective dopamine (DA) neuron degeneration along the nigrostriatal pathway (2, 14, 21). Chronic or subchronic MPTP treatments decrease DA and DA metabolite concentrations (44), decrease tyrosine hydroxylase (TH) activity (27) and TH immunoreactivity in the striatum (13). MPTP also produces Parkinson-like symptoms in humans (41), primates (2, 19) and certain species of rodents (6, 14, 17). Therefore, it has been commonly used as a pharmacological agent for animal studies of parkinsonism. MPTP is believed to produce its neurotoxicity through conversion to its oxidative metabolite 1-methyl-4-phenylpyrinidium (MPP$^+$) (5, 18) by monoamine oxidase B (MAOB) (15), primarily in glia cells (31) as well as in the mitochondria. MPP$^+$ was then suggested to be selectively taken up by DA nerve terminals in the striatum (5) with retrograde transport to DA cells in the substantia nigra (SN) (3). It is then suggested to inhibit Complex I enzyme of the respiratory chain within the mitochondria, to produce free radicals and finally causes cell death (For review, see 9, 37). Other than its neurotoxicity in the striatum, we and others have found that MPTP also produces a direct toxicity on DA neurons in the SN (16, 40). Other reports have shown that MPTP is also toxic to DA neurons in the ventral tegmental area (28, 40). Furthermore, we (16) and others (8, 28) have reported that other than DA neurons, MPTP also exerts a toxicity on norepinephrine (NE) neurons in the locus coeruleus (LC). Moreover, these effects seem to be mediated through astrocytes in the SN and LC locally since direct application of the selective gliotoxin α-aminoadipic acid prevents MPTP's toxicity in these two areas (4). These results together suggest that other than in the striatum, MPP$^+$ may also be directly taken up into DA neurons in the SN, presumably through DA dendrites. The same implication applies to the noradrenergic system. It is likely that NE terminals in the hippocampus and dendrites (as well as short axons) of NE neurons in the LC are also able to uptake MPP$^+$, possibly through the NE uptake system. Numerous studies investigating MPTP's toxicity and MPP$^+$ uptake using in vitro methods have been done, while relatively fewer in vivo experiments have been conducted. In addition, more biochemical and histochemical studies have been carried out, while fewer behavioral observations have been made with MPTP and MPP$^+$. The present study was aimed to use the pharmacological, biochemical and behavioral approaches, adopting both in vivo and in vitro methods to examine the neurotoxicity of MPTP and the mechanism of MPP$^+$ uptake in catecholamine neurons.

MATERIALS AND METHODS

Subjects

Adult male BALB/c mice (2–3 months old, 20–24 g) bred in the Institute of Biomedical Sciences, (Academia Sinica, Taiwan, The Republic of China) were housed five per cage. After surgery, they were then housed singly. They were maintained on a 12/12 hr light/dark cycle (with light on at 6:30 am) in a temperature-regulated(23 ± 2 °C) animal room with food and water continuously available. Drug injections were given at different times of the day whereas behavioral measures were always conducted during the light phase of the diurnal cycle.

Drugs and Reagents

MPTP, MPP$^+$, nomifensine (NOM) and desimipramine (DMI) were purchased from Research Biochemical Inc (Wayland, MA, USA). [^{3}H]DA and [^{3}H]NE were purchased from Amershan Company, UK. All other chemical reagents were purchased from Merck Company of the highest grade.

Surgery

Mice were subject to stereotaxic surgery under sodium pentobarbital anesthesia (40 mg/kg, ip). Twenty-three gauge stainless steel thin-wall cannulae (12 or 15 mm long) were implanted bilaterally into the SN or the LC. The cannula tip was aimed at the dorsal surface of SN, LC, striatum or hippocampus. The coordinates for SN are: AP, -2.2 mm from bregma; ML ±1.6 mm from midline; and DV, -2.9 mm below the skull surface. The coordinates for LC are: AP, -4.0 mm from bregma; ML, ±0.6 mm from midline; and DV, -2.9 mm below the skull surface. The coordinates for the striatum are: AP, +0.5 mm from bregma; ML, ±2.5 mm from midline; and DV, -2.5 mm below the skull surface. The coordinates for the hippocampus are: AP, -1.8 mm from bregma; ML, ±1.0 mm from midline; and DV, -1.0 mm below the skull surface according to the atlas by Lehmann (23) of mice. The tooth bar was at -1.0 mm. Two small stainless-steel screws serving as anchors were implanted over the right frontal and left posterior cortices. The cannulae were affixed on the skull with dental cement. A stylet was inserted into each cannula to maintain patency.

Drug Infusions

Drug infusions started seven to ten days after recovery from the surgery. In the first part of the study, mice received bilateral infusions of saline, NOM or DMI into the striatum, the hippocampus, the SN or the LC when animals were awake and held gently by the experimenter. All infusions were given for 7 days continuously at one infusion per day. Intraperitoneal MPTP injections were always given 30 min following saline or drug infusions each day for a total of 7 days. In the second part of the study, animals received either saline or MPP$^+$ infusions to the SN or the LC also for 7 days continuously. The infusion was administered through a 30-gauge injection needle connected to a 1 µl Hamilton syringe by 0.5 m polyethylene tubing (PE-20). The injection needle was bent at a length such that, when inserted into the cannula, the needle tip would protrude 1.2 mm beyond the tip of the cannula. Drug solutions were delivered into the above areas manually at a rate of 0.25 µl/min. MPTP, MPP$^+$, NOM and DMI were dissolved in saline. A volume of 0.25 µl was injected into each SN or LC and a volume of 0.5 µl was injected into each striatum or hippocampus in these experiments. Behavioral measures were conducted 7 days after the last MPTP (saline) injections or the last MPP$^+$ (saline) infusions throughout all experiments.

High-performance Liquid Chromatography with Fluorescence Detection of DA, NE and 5HT

The chromatographic system used was a 5-µm Ultrasphere ODS reverse phase column (4.6 mm by 15 cm, Beckman Instruments, Fullerton, CA) with an Altex pump (Beckman

Instruments) and Shimadzu RF-530 spectromonitor (Shimadzu, Tokyo, Japan). The excitation and emission wavelengths were set at 290 and 330 nm, respectively. The flow was maintained at 1.2 ml/min. The sensitivity for these amines was in the nanogram range. The mobile phase consisted of 0.02 M potassium phosphate monobasic containing1 g/l of 1-heptanesulfonic acid sodium salt (pH 3.3) and a mixture of methanol-H_2O (3:2). Output was recorded with a Shimadzu C-R3A Integrator (Shimadzu). DA and NE were estimated according to the method of Peat and Gibb (30) with some modifications. Tissue was weighed while still frozen and homogenized in 5 volumes of 0.1 M perchloric acid containing 4 mM sodium metabisulfite. The homogenate was then centrifuged at 6000g for 15 min using a refrigerated centrifuge, and the clear supernatant (20 μl) was injected directly into the chromatographic system. NE and DA were dissolved at 3.3 and 4.5 min, respectively.

Activity Monitor

The behavioral apparatus used was described in detail elsewhere (22). Briefly, there were two digiscan activity monitors (Coulbourn Instruments, PA) approximately 16 inches square with 16 x 16 horizontal by vertical infrared sensors. These sensors were used to localize the animal's floor position as well as many behavioral categories. In the present study, only horizontal activity was recorded. Horizontal activity is measured by total number of beam breaks in an X-Y plane recorded every 10 msec. Only one animal was placed in each activity chamber per measurement period.

Procedure

Seven days after the last MPTP infusions, animals were placed in the activity monitor for a 30-min behavoral test. At the end of the behavioral testing, animals were sacrificed by decapitation and their brains were taken for biochemical assays of DA, NE and 5HT. Brain regions of SN, LC, striatum and hippocampus were dissected out by a free-hand procedure according to the method of Segal and Kuczenski et al. (35). Tissue of SN was further dissected out by a punch (2 mm in diameter) and LC was cut with a slicer (1 mm in width). Tissues were frozen quickly by dry ice and kept at -80°C until biochemical assays could be performed.

Tremor Monitor

The tremor monitor was purchased from the Columbus Instrument Company (Tremor Monitor II) consisting of the sensor platform, the control unit and the PC-800 printer. Normal animal's activity consists of a vertical motion component. When this information is viewed on the frequency and amplitude domains, a spectral profile is generated. Tremor has a unique spectral profile that is it lacks the normal low frequency component. During behavioral testing, animal's vertical activities were recorded. These data were later compared with the installed parameters with system gain set at 75% (to amplify the signal); trigger rate at 35% (to set up the optimal amplitude level); dead band at 10% (to reduce noise) and corner frequency at 74% (to set the criteria of 18.5 cps as tremor frequency. Normal animals exhibit less than 10 cps). Animals were subject to a 30 min activity test and the number of tremor sessions was printed out every 10 min automatically.

Histology

Separate animals were used for histological verification of cannula and needle placement in the SN, LC, striatum and hippocampus. Methylene blue dye (3 mg/ml, 0.25 μl) was injected into each SN and LC. The volume for injection into each striatum and hippocampus was 0.5

μl. The brain were frozen-sectioned in a cryostat and 20-μm thick sections were taken at 40-μm intervals through these regions. They were then stained with Evan's blue dye and checked with the atlas by Lehmann (23) for histological identification. We have previously demonstrated that at a volume of 0.5 μl, the infusion was confined to the SN and LC areas without diffusion (16). Those animals whose needle placements were not in the correct area were deleted from the present study. The histological diagrams have been shown in detail elsewhere (16).

Synaptosome Preparation

Synaptosomes were prepared from thawed tissues of the striatum, the hippocampus, the SN and the LC of mice in 0.32 M sucrose solution. The subcellular fractionation procedure used was according to that of Gray and Whittaker (10) with some modifications. Because of small sizes of the SN and LC, tissues from these two areas were pooled from four animals together for synaptosome preparation and uptake studies. The supernatant from the first centrifugation was layered onto 1.2 M sucrose and centrifuged at 160,000 g for 15 min. The interface material, containing synaptosomes, myelin and mitochondria, were removed and diluted threefold with 0.32 M sucrose before being layered onto 0.8 M sucrose and centrifuged at 160,000 g for another 15 min. The pellet was resuspended in 0.32 M sucrose to give a 200–300 mg of protein/ml solution.

Uptake Procedures

The buffer used for uptake studies was the Krebs Ringer phosphate medium adjusted to pH 7.4 with HCl, containing 140 mM Na^+, 22mM HPO_4^{-2}, 5.9 mM K^+, 125 mM Cl^-, 1.2 mM $H_2PO_4^-$, 1.2mM Mg^{+2}, 1.2 mM SO_4^{-2}, and 0.75 mM Ca^{+2}. Experiments were conducted to examine the effects of unlabelled MPP^+ on [^{3}H]DA and [^{3}H]NE uptake in SN, striatum, LC and hippocampus. To determine the specific competition between DA and MPP^+ uptake (as well as NE and MPP^+ uptake), effects of unlabelled MPP^+ on [^{3}H]5HT uptake in the striatum and hippocampus were also performed. Incubations were performed at 37°C with constant shaking and were initiated by addition of 100 μl of synaptosomal suspension to 900 μl of incubation medium. Termination of uptake was achieved by rapid filtration onto Whatman GF/B filters using a Brandel cell harvester 5 min later. The filters were washed three times, each with 1 ml volumes, of isotonic NaCl and then transferred to vials for radioactivity measurements using liquid scintillation counting. Protein content determinations were made using the method of Lowry et al. (24).

Statistics

All the biochemical data and tremor frequency were analyzed with one-way analysis of variance (ANOVA). Locomotor activity data were analyzed using two-way ANOVA with "group" as the between-subject variable and "time" as the repeated measure. Specific comparisons between each experimental group and the control group at different time points were made with two-tailed Dunnett's method. Specific comparisons between experimental groups were made with the Tukey's method (36). The uptake results were analyzed with ANOVA and linear regression.

Repeated NOM Infusions to the SN and Striatum and Repeated MPTP Injections

Animals were randomly divided into eight groups to receive various treatments. Group I (n = 4) received saline infusions to the SN and saline injections (ip); Group II (n = 4) received saline infusions to the striatum and saline injections (ip); Group III (n = 4) received saline infusions to the SN and MPTP injections (ip); Group IV (n = 4) received saline infusions to the striatum and MPTP injections (ip). Groups I and II were later combined as the control group; Groups III and IV were later combined as the MPTP group. Group V (n = 6) received NOM infusions to the SN and saline injections (ip); Group VI (n = 7) received NOM infusions to the SN and MPTP injections (ip); Group VII (n = 6) received NOM infusions to the striatum and saline injections (ip) and Group VIII (n = 7) received NOM infusions to the striatum and MPTP injections (ip). The concentrations of NOM infused to the SN and the striatum were 1.5 µg and 3.0 µg, respectively. The concentration of MPTP was 30 mg/kg. Infusions were given 30 min prior to systemic injections every day for a continuation of 7 days in all.

Repeated DMI Infusions to the LC and Hippocampus and Repeated MPTP Injections

Similar to the design of the previous experiment, animals were randomly divided into eight groups. Group I (n = 4) received saline infusions to the LC and saline injections (ip); Group II (n = 4) received saline infusions to the hippocampus and saline injections; Group III (n = 4) received saline infusions to the LC and MPTP injections; Group IV (n = 4) received saline infusions to the hippocampus and MPTP injections. Similarly, Groups I and II were later combined as the control group; Groups III and IV were later combined as the MPTP group. Group V (n = 6) received DMI infusions to the LC and saline injections; Group VI (n = 7) received DMI infusions to the LC and MPTP injections; Group VII (n = 6) received DMI infusions to the hippocampus and saline injections and Group VIII (n = 7) received DMI infusions to the hippocampus and MPTP injections. The concentrations of DMI delivered to the LC and the hippocampus were 1.25 µg and 2.5 µg, respectively. The concentration of MPTP was 30 mg/kg. Infusions were also given 30 min prior to systemic injections for 7 days continuously.

Repeated MPP^+ Infusions to the SN

To examine whether MPP^+ exerts a direct and irreversible toxicity on DA neurons in the SN, animals were randomly divided into two groups. Group I (n = 10) received bilateral saline infusions to the SN; Group II (n = 10) received 1.2 µg MPP^+ infusions to the SN bilaterally. The infusion was given once per day at 0.25 µl each side for a total of 7 days.

Repeated MPP^+ Infusions to the LC

To examine whether MPP^+ also produces a toxicity on NE neurons in the LC, animals were randomly divided into two groups. Group I (n = 10) received bilateral saline infusions to the LC; Group II (n = 10) received 1.2 µg MPP^+ infusions to the LC bilaterally. The infusion volume and experimental designs were the same as that in the above experiment.

Repeated MPTP Injections on Tremor Frequency

To examine whether MPTP produces consistent tremor in experimental animals resembles that in PD patients, animals were randomly distributed to four groups. Group I (n = 9) received daily saline injections (1 ml/kg) for 7 days and were subject to tremor measures at 1 day, 3 days or 7 days after the last injections (n = 3 each); Groups II (n = 6), III (n = 6) and IV (n = 6) all received daily MPTP injections (30 mg/kg, ip, one injection each day) for 7 days continuously and tremor was measured at 1 day, 3 days and 7 days after the last MPTP injections, respectively.

[³H]DA and MPP⁺ Uptake Studies

Experiments in this part were designed to examine the effects of unlabelled MPP⁺ on [³H]DA uptake in synaptosomes from the striatum and the SN. Tissue homogenates from all animals were pooled to get homogeneous synaptosomal fractions. The concentrations of [³H]DA used were 12.5, 25 and 50 nM. The concentrations of MPP⁺ added were 10, 100 and 1000 nM. To determine the specific competition between [³H]DA and MPP⁺ uptake, effects of unlabelled MPP⁺ on [³H]5HT (50 nM) uptake were also performed in the striatum.

[³H]NE and MPP⁺ Uptake Studies

Experiments in this section were designed to study the effects of unlabelled MPP⁺ on [³H]NE uptake in synaptosomes from the hippocampus and the LC. Similarly, tissue homogenates were pooled. The concentrations of [³H]NE used were also 12.5, 25 and 50 nM. The concentrations of MPP⁺ added were 1, 10 and 100 nM. The possible nonspecific uptake of MPP⁺ by 5HT terminals were also examined in synaptosomes from the hippocampus at the same concentrations of MPP⁺ used above. The concentration of [³H]5HT added was also 50 nM.

RESULTS

Effects of Repeated NOM and MPTP Administrations on Monoamines and Behavior

Biochemical results of this experiment are shown in Table 1. There was an overall significant effect of MPTP on striatal DA ($F(5,36) = 3.01$, $P < 0.05$). Further analyses revealed that repeated MPTP injections consistently and significantly decreased DA concentration in the striatum ($t = 2.87$, $P < 0.01$). NOM alone in the striatum did not significantly alter DA level ($t = 0.59$, $P > 0.05$), while it partially prevented MPTP's toxicity on DA in this area ($t = 2.01$, $P < 0.05$ when comparing the NOM + MPTP group with the MPTP group and $t = 2.18$, $P < 0.05$ when comparing the NOM + MPTP group with the control group). On the other hand, repeated NOM infusions alone in the striatum also did not markedly alter DA concentration ($t = 0.47$, $P > 0.05$), however, it completely antagonized the neurotoxicity of MPTP on DA neurons ($t = 2.74$, $P < 0.05$ when comparing the NOM + MPTP group with the MPTP group and $t = 0.76$, $P > 0.05$ when comparing the NOM + MPTP group with the control group). Behavioral results were similar to that of the biochemical results. There was an overall significant effect of MPTP on locomotion in mice ($F(5,36) = 2.96$, $P < 0.05$). Further analyses indicated that repeated MPTP injections markedly impaired locomotor activity ($t = 3.18$, $P < 0.01$). NOM alone in the SN did not markedly alter motor activity ($t = 0.93$, $P > 0.05$); however, it partially prevented the motor-impairing effect of MPTP ($t = 1.99$, $P < 0.05$ when comparing the NOM + MPTP group with the MPTP group and $t = 2.13$, $P < 0.05$ when comparing the NOM + MPTP group with the control group). Similarly, NOM alone in the striatum did not markedly affect motor activity ($t = 0.81$, $P > 0.05$), while it completely prevented MPTP-induced motor impairment ($t = 3.26$, $P < 0.01$ when comparing the NOM + MPTP group with the MPTP group and $t = 0.64$, $P > 0.05$ when comparing the NOM + MPTP group with the control group).

Effects of Repeated DMI and MPTP Administrations on Monoamines and Behavior

Biochemical results of this experiment are shown in Table 2. There was also a significant overall effect of MPTP on hippocampal NE ($F(5,36) = 2.83$, $P < 0.05$). Further, repeated MPTP

Table 1. Interactive Effects of Repeated NOM Infusions (1.5 µg/0.25 µl in the Substantia Nigra and 3.0 µg/0.5 µl in the Striatum

Treatment	n	DA	% of control
Sal + Sal	8	9856 ± 713	
Sal + MPTP	8	6869 ± 594**a	70
NOM + Sal (SN)	6	10097 ± 806	102
NOM + MPTP (SN)	7	8070 ± 679*a,b	82
NOM + Sal (ST)	6	9844 ± 882	100
NOM + MPTP (ST)	7	9421 ± 903**b	96

* p<0.05 and ** P<0.01. a: when compared with the control group, b: when compared with the MPTP group

Statistical significance was evaluated by ANOVA, Dunnett's method and Tukey's method. Data are expressed as nanograms per gram tissue and are not corrected for recovery. Values are means ±S.E.M.

injections significantly decreased NE concentration in this area (t = 3.09, P < 0.01). Repeated DMI infusions alone in the LC did not markedly alter NE level (t = 0.43, P > 0.05), while it partially prevented the neurotoxicity of MPTP on NE in the hippocampus (t = 1.98, P < 0.05 when comparing the DMI+ MPTP group with the MPTP group and t = 2.24, P < 0.05 when comparing the DMI + MPTP group with the control group). Repeated DMI treatment alone in

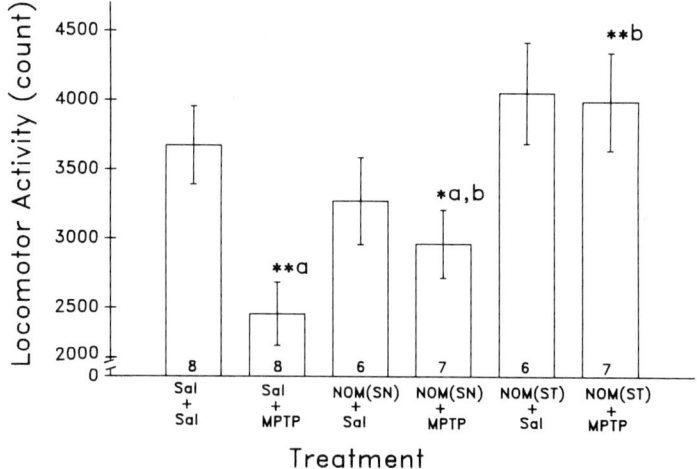

Figure 1. Interactive effects between chronic NOM infusions to the SN and striatum and chronic MPTP injections on locomotor activity in mice (n = 6–8 in each group). Data are means ± S.E.M. * P < 0.05 and ** P < 0.01. a: when compared with the control group, b: when compared with the MPTP group.

Table 2. Interactive Effects of Repeated DMI Infusions (1.25 µg/0.25 µl in Locus Coeruleus and 2.5 µg/0.5 µl in Hippocampus) and MPTP Injections (30 mg/kg, ip) on NE Concentration in the Hippocampus

Treatment	n	NE	% of control
Sal + Sal	8	909 ± 78	
Sal + MPTP	8	600 ± 51**a	66
DMI + Sal (LC)	6	875 ± 92	96
DMI + MPTP (LC)	7	727 ± 64*a,b	80
DMI + Sal (hippo)	6	931 ± 90	102
DMI + MPTP (hippo)	7	904 ± 74**,b	99

* $p<0.05$ and ** $P<0.01$. a: when compared with the control group, b: when compared with the MPTP group

Statistical significance was evaluated by ANOVA, Dunnett's method and Tukey's method. Data are expressed as nanograms per gram tissue and are not corrected for recovery. Values are means ±S.E.M.

the hippocampus did not significantly alter NE level ($t = 0.29$, $P > 0.05$), while it completely blocked the neurotoxicity of MPTP on NE in this area ($t = 2.93$, $P < 0.01$ when comparing the DMI + MPTP group with the MPTP group and $t = 0.26$, $P > 0.05$ when comparing the DMI + MPTP group with the control group). Behavioral results yielded a similar pattern and are shown in Figure 2. There was a significant overall effect of MPTP on locomotor activity in mice ($F(5,36) = 2.91$, $P < 0.05$). Moreover, repeated MPTP injections markedly impaired motor activity ($t = 2.71$, $P < 0.01$). DMI treatment in LC alone did not affect motor activity ($t = 1.01$, $P > 0.05$), while it partially prevented the motor-impairing effect of MPTP ($t = 2.11$, $P < 0.05$ when comparing the DMI + MPTP group with the MPTP group and $t = 1.96$, $P < 0.05$ when comparing the DMI + MPTP group with the control group). Similarly, DMI infusions in the hippocampus alone did not markedly affect locomotion ($t = 0.73$, $P > 0.05$), while it completely prevented the motor-impairing effect of MPTP in mice ($t = 2.35$, $P < 0.01$ when comparing the DMI + MPTP group with the MPTP group and $t = 0.68$, $P > 0.05$ when comparing the DMI + MPTP group with the control group).

Effects of Repeated MPP$^+$ Infusions to the SN on Monoamines and Behavior

Biochemical results of this experiment are summarized in Table 3. One-way ANOVA has revealed that repeated MPP$^+$ infusions to the SN significantly decreased DA concentrations in both the SN ($F(1,17) = 4.71$, $P<0.05$) and its corresponding projection site, the striatum ($F(1,17) = 4.58$, $P<0.05$). Behavioral measures yielded a similar profile. Figure 3 illustrated that repeated MPP$^+$ infusions to the SN markedly impaired locomotor activity in mice ($F(1,17) = 4.61$, $P<0.05$).

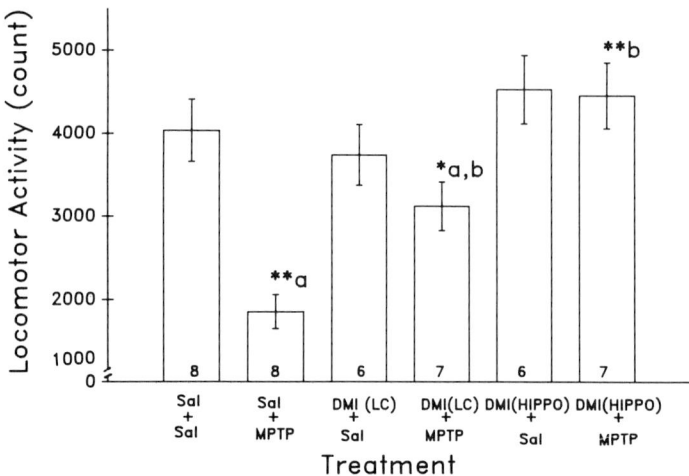

Figure 2. Interactive effects between chronic DMI infusions to the LC and hippocampus and chronic MPTP injections on locomotor activity in mice (n = 6–8 in each group). Data are means ± S.E.M. * P < 0.05 and ** P < 0.01. a: when compared with the control group, b: when compared with the MPTP group.

Effects of Repeated MPP$^+$ Infusions to the LC on Monoamines and Behavior

Biochemical results of this experiment are shown in Table 4. One-way ANOVA indicated that repeated MPP$^+$ infusions to the LC significantly decreased NE concentrations in both the LC ($F(1,17) = 4.45$, $P < 0.05$) and the hippocampus ($F(1,17) = 4.52$, $P < 0.05$). As revealed from Figure 4, MPP$^+$ infusions to the LC also markedly impaired locomotor activity in mice ($F(1,17) = 4.46$, $P < 0.05$).

Effects of Repeated MPTP on Tremor Frequency

Effects of repeated MPTP treatment on tremor sessions measured at 1 day, 3 days and 7 days after withdrawal of the last MPTP (saline) injections are shown in Figure 5. As indicated in this figure, there was a significant overall effect of MPTP on the number of tremor sessions ($F(3,23) = 3.21$, $P < 0.01$). Further analyses revealed that chronic MPTP injections significantly

Table 3. Effects of Repeated MPP$^+$ Infusions to the Substantia Nigra (SN) (1.2 µg/0.25 µl) on DA Concentration in the SN and the Striatum

Treatment	n	[DA] in SN	[DA] in striatum
Sal	10	2364 ± 195	9057 ± 812
MPP$^+$	9	1537 ± 148*	5616 ± 487*

* $p<0.05$.

Statistical significance was evaluated by ANOVA. Data are expressed as nanograms per gram tissue and are not corrected for recovery. Values are means ±S.E.M.

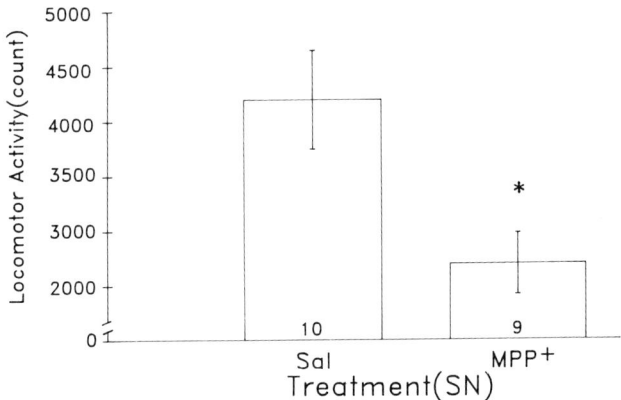

Figure 3. Effects of chronic MPP$^+$ infusions to the SN on locomotor activity in mice (n = 9–10 in each group). Data are means ± S.E.M. * P < 0.05 when compared with the control group.

increased tremor sessions at each of these time intervals examined (t = 3.68, P < 0.01; t = 3.03, P < 0.01 and t = 2.91, P < 0.01 at 1 day, 3 days and 7 days after withdrawal of MPTP, respectively).

[^{3}H]DA and MPP$^+$ Uptake Studies

Effects of MPP$^+$ on [^{3}H]DA uptake in the striatum and the SN are illustrated in Figures 6(a) and 6(b). As shown in these figures, the amount of [^{3}H]DA uptake increased as the concentration of [^{3}H]DA increased (F(2,45) = 3.31, P < 0.05 for the striatum and F(2,45) = 3.89, P < 0.05 for the SN). Meanwhile, linear regression analyses revealed that there was a significant competitive inhibition between MPP$^+$ and [^{3}H]DA uptake in both the striatum and the SN. When the concentration of MPP$^+$ added was 1000 nM, the amount of [^{3}H]DA uptake decreased to approximately 58% of the control on the average in the striatum and 45% of the control in the SN. Further, there was not a competitive inhibition between MPP$^+$ and [^{3}H]5HT uptake up to the concentration of MPP$^+$ reaching 1000 nM in the striatum. When MPP$^+$ was at the concentration of 1000 nM, [^{3}H]5HT uptake was significantly decreased (Figure 6(c), t = 2.74, P < 0.05).

Table 4. Effects of Repeated MPP$^+$ Infusions to the Locus Coeruleus (LC) (1.2 µg/0.25 µl) on NE Concentration in the LC and the Hippocampus

Treatment	n	[NE] in LC	[NE] in hippocampus
Sal	10	688 ± 56	867 ± 67
MPP$^+$	9	495 ± 51*	572 ± 48*

* p<0.05.
Statistical significance was evaluated by ANOVA. Data are expressed as nanograms per gram tissue and are not corrected for recovery. Values are means ±S.E.M.

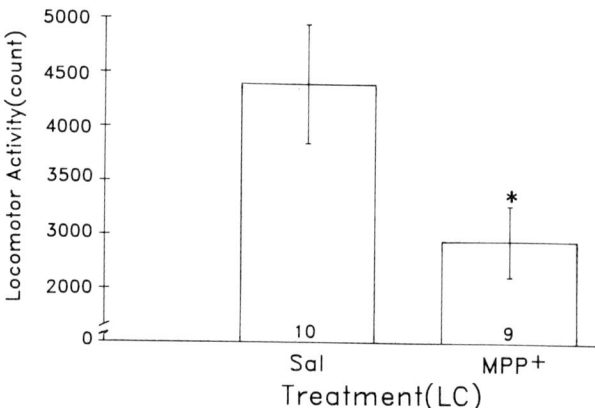

Figure 4. Effects of chronic MPP+ infusions to the LC on locomotor activity in mice (n = 9–10 in each group). Data are means ± S.E.M. * $P < 0.05$ when compared with the control group.

[^{3}H]NE and MPP+ Uptake Studies

Effects of MPP+ on [^{3}H]NE uptake in the hippocampus and the LC are shown in Figures 7(a) and 7(b). Similarly, the amount of [^{3}H]NE uptake increased as the concentration of [^{3}H]NE increased ($F(2,45) = 4.43$, $P < 0.05$ for the hippocampus and $F(2,45) = 5.22$, $P < 0.01$ for the LC). Linear regression analyses also indicated that there was a competitive inhibition between MPP+ and [^{3}H]NE uptake in both the hippocampus and the LC. As the concentration of MPP+ reached 100 nM, the amount of [^{3}H]NE uptake decreased to approximately 45% of the control on the average in the hippocampus and 33% of the control in the LC. Moreover, there was not a competitive inhibition between MPP+ and [^{3}H]5HT uptake up to the concentration of MPP+ reaching 100 nM in the hippocampus. When MPP+ was at the concentration of 100 nM, [^{3}H]5HT uptake was markedly decreased (Figure 7(c), $t = 2.67$, $P < 0.05$).

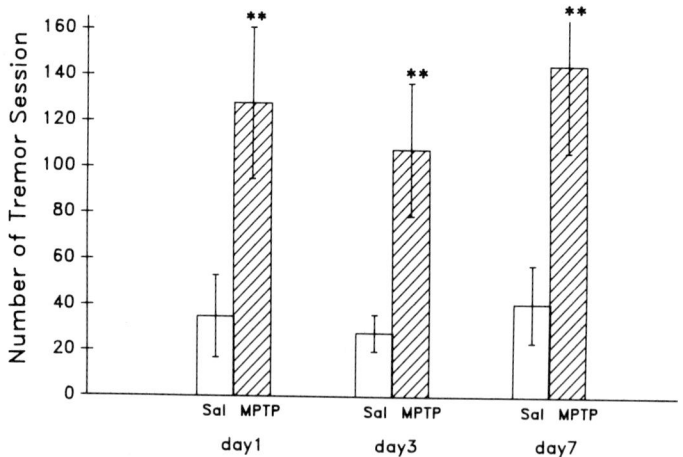

Figure 5. Effects of chronic MPTP injections on tremor frequency at 1 day, 3 days and 7 days after the last MPTP injection in mice (n = 6–9 in each group). Data are means ± S.E.M. ** $P < 0.01$ when compared with the corresponding control group.

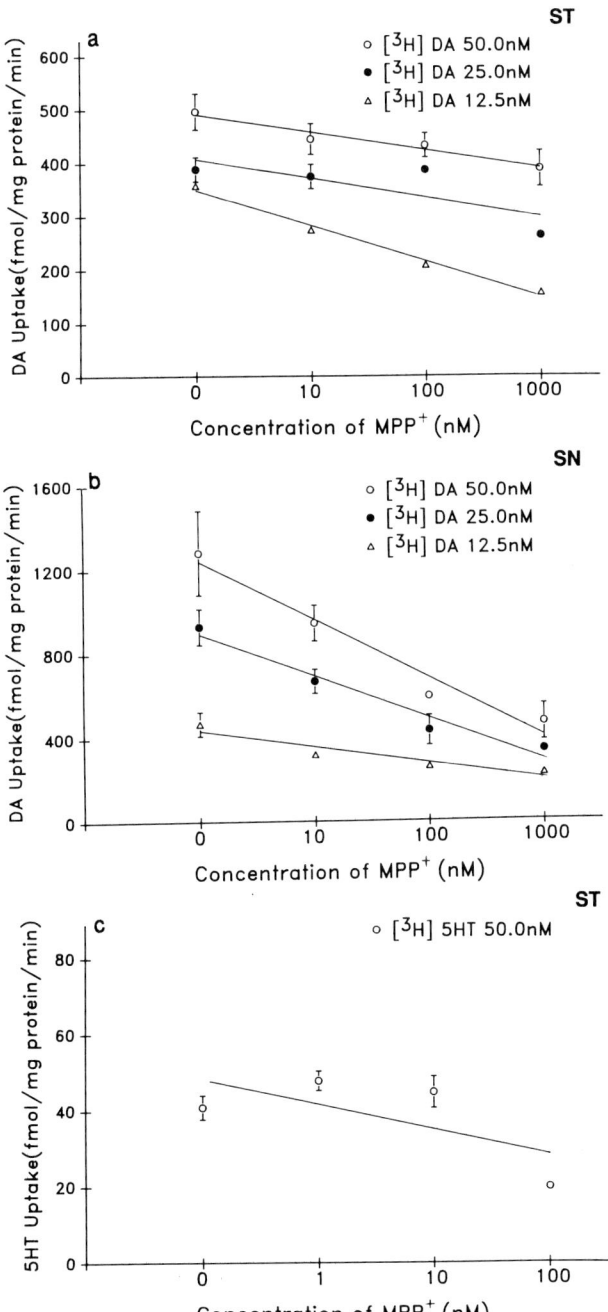

Figure 6. Effects of MPP$^+$ on (a) [^{3}H]DA uptake in the striatum, (b) [^{3}H]DA uptake in the SN and (c) [^{3}H]5HT uptake in the striatum in mice. n = 4 (repeat number) at each concentration of MPP$^+$. Data are means ± S.E.M.

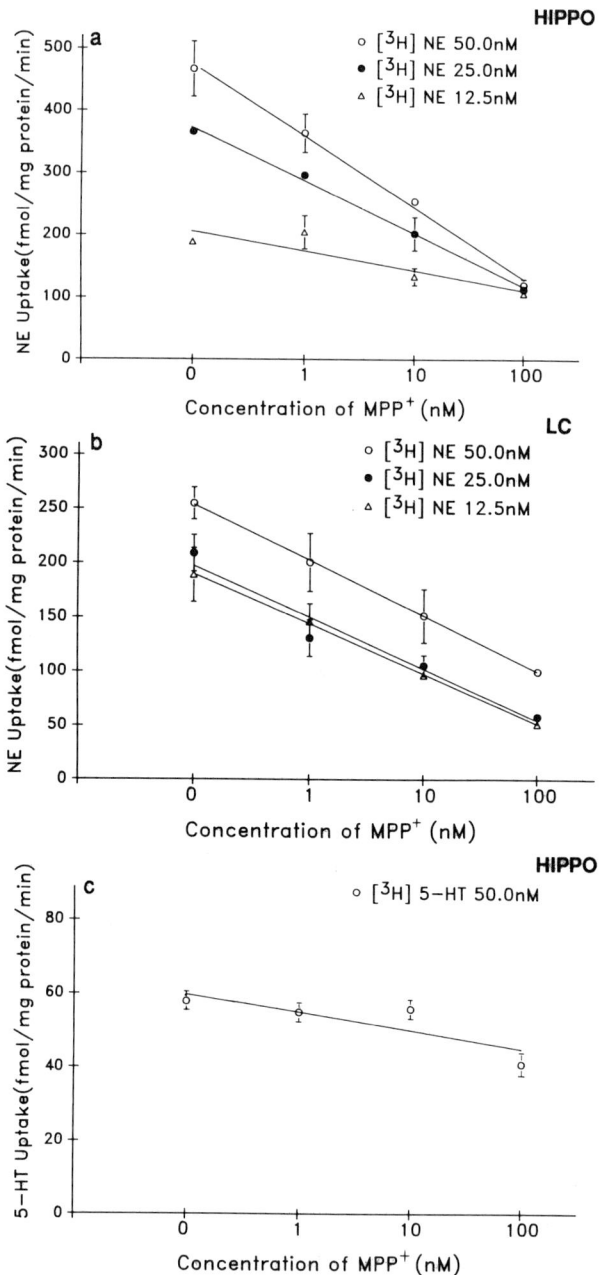

Figure 7. Effects of MPP+ on (a) [³H]NE uptake in the hippocampus, (b) [³H]NE uptake in the LC and (c) [³H]5HT uptake in the hippocampus in mice. n = 4 (repeat number) at each concentration of MPP+. Data are means ± S.E.M.

DISCUSSION

The present results confirm our previous findings that MPTP exerts a direct toxicity on both DA neurons in the SN and NE neurons in the LC (16). Consistent with the biochemical results, repeated MPTP infusions to the SN and LC also produced long-lasting and irreversible behavioral impairments in BALB/c mice. These results are congruent with other reports showing that MPTP causes a long-term motor deficit in C57BL/6 mice (1, 38). We have currently also found that repeated MPTP injections produced significant and long-lasting tremors in mice; although there are also reports showing a dismatch between the biochemical and behavioral effects of MPTP (7). This discrepancy can be explained by the possible involvement of other neurotransmitter systems in MPTP's toxicity (20). Furthermore, when MPP^+ was directly infused into the SN and LC, it also produced a significant decrease of both DA concentration in the striatum and NE concentration in the hippocampus. It also markedly impaired locomotor activity. Although the extent of DA depletion alone caused by MPTP and MPP^+ was approximately 35% only which may not be sufficient to explain the significant motor deficit observed, while there was also an average of 35% depletion of NE produced by MPTP and MPP^+, and the combined depletions of both catecholaminergic systems may yield the behavioral results. The present results also suggest that it is MPP^+, instead of MPTP, exerts the toxicity on catecholamine neurons, and MPTP, other than in the striatum, can be locally converted to MPP^+ in both the SN and LC. This notion is supported by recent findings that local glial lesions in the SN and LC abolishes the toxicity of MPTP in these two areas (4). It further suggests that MAO (probably both the A and B forms) in the SN and LC play an important role in mediating MPTP's toxicity.

MPP^+ is suggested to exert its toxicity through the DA uptake system in the striatum (33, 43). This suggestion extends to the current in vivo findings that NOM infusion to the striatum prevented MPTP's toxicity. Meanwhile, we have also found that NOM pretreatment in the SN also partially but significantly abolished MPTP's toxicity on DA neurons, suggesting that dendrites and short axons of DA cells in the SN may also uptake DA. Further, DA and MPP^+ may also share the same uptake system in the SN. This suggestion is further supported by the present in vitro uptake studies. MPP^+ produced a dose-dependent competitive inhibition of $[^3H]DA$ uptake in both the SN and striatum (Figures 6(a) and 6(b)). Dendritic release of DA in the SN has been reported (29, 34), while we have provided the first evidence of dendritic (and short axonal) uptake of DA in DA neurons. Similar results were observed in the hippocampus and the LC. DMI pretreatment in the hippocampus completely antagonized, while DMI pretreatment in the LC partially antagonized the toxicity of MPTP on NE neurons. These results also suggest that, other than in the terminal areas, dendrites (and possibly short axons) located on NE cells in the LC are also able to uptake NE, and NE and MPP^+ shares the same uptake system. Moreover, these uptake systems are quite specific because MPP^+ uptake was not observed in 5HT neurons except at high concentrations. These results are inconsistent with the reports of Takada et al. (40) and Gupta et al. (12) showing that MPTP is also toxic to 5HT neurons in raphe nuclei. However, they are congruent with most of the literatures that MPTP affects the serotonergic system to a much lesser extent. Although these results together strongly suggest that catecholaminergic cells are also able to uptake locally released DA and NE in addition to their terminal sites, however, the possible uptake of DA and NE by glia cells and axons can not be excluded. Moreover, the involvement of both DA and NE neurons (at least) in MPTP's toxicity further supports the notion from clinical observations that Parkinson's disease may be a heterogeneous disease also involves central noradrenergic system (25, 26, 32, 42).

In the present study, high doses of NOM (3.0 μg) and DMI (2.5 μg) infused to the striatum and hippocampus tended to increase DA and NE concentrations in these areas, respectively, and inclined to enhance locomotor acitivity. This is probably due to chronic inhibition of DA and NE re-uptake from the terminals by NOM and DMI and, therefore, elevates synaptic concentrations of DA and NE as well as the behavioral functions they mediate. In a preliminary study, we have found that at higher doses (4.5 μg for NOM and 3.5 μg for DMI), both NOM and DMI significantly increased locomotor activity (unpublished observations). On the other hand, if the same doses of NOM (3.0 μg) and DMI (2.5 μg) were given to the SN and LC, DA level in the striatum and NE level in the hippocampus tended to be decreased, and locomotor activity tended to be impaired. It is likely that NOM and DMI also prevents the re-uptake of DA and NE in cell bodies, consequently, they increased extracellular concentrations of DA and NE in these areas. Extracellular DA and NE were expected to decrease DA and NE neuronal firing through autoreceptor inhibition (11, 39). Decreased neuronal firing leads to decreased neurotransmitter release and, consequently, reduces the motor functions they govern.

Many studies have reported that MPTP does cause a consistent lesion of certain neurotransmitter systems, while tremor and regidity were less frequently and consistently observed. We have earlier reported and presently also found that MPTP produced phasic tremor and regidity, but tremor was more limited to a certain duration after MPTP administrations by visual observations (16). We therefore adopted the tremor monitoring system which detects low amplitude but high frequency movements. These results demonstrated a significant and long-lasting increase of tremor activity by repeated MPTP (Figure 5). Further experiments are ongoing by using the electrophysiological method to study the effects of MPTP on muscle tension and reflex function. Finally, consistent and irreversible biochemical depletions and behavioral impairments, including tremor, verifies the usefulness and reliability of using MPTP as a pharmacological agent for parkinsonism research in experimental animals.

In conclusion, we have presently found that repeated MPTP injections and MPP^+ infusions to the SN and LC both produced an irreversible toxicity on DA and NE neurons, respectively. MPTP also caused long-lasting behavioral impairments including decrease of locomotor activity and increase of tremor frequency. Pretreatment with NOM and DMI (chronic and local infusions) protected against MPTP's toxicity on both catecholaminergic cell bodies and terminal regions as well as on motor functions. These results suggest the role of dendritic uptake of DA and NE by SN and LC neurons, respectively, in vivo, and it is confirmed by results of in vitro uptake studies. Serotonin neurons were affected by MPTP to a much lesser extent. These results also suggest that MPP^+ shares the same DA uptake system in both the SN and striatum; and it shares the same NE uptake system in both the LC and the hippocampus.

ACKNOWLEDGMENT

This work was supported by a Grant from the National Science Council of Taiwan, the Republic of China (NSC81–0412-B- 001–07).

REFERENCES

1. ARAI, N., MISUGI, K., GOSHIMA, Y. AND MISU, Y.: Evaluation of a 1-methyl-4-phenyl-1,2,3,6-tetrahydropyridine (MPTP)-treated C57 black mouse model for parkinsonism. Brain Res. 515:57–63, 1990.
2. BURNS, R. S., CHIUEH, C. C., MARKEY, S. P., EBERT, M. H., JACOBWITZ, D. M. AND KOPIN, I. J.: A primate model of parkinsonism: Selective destruction of dopaminergic neurons in the pars compacta of the substantia nigra by N-methyl-4-phenyl-1,2,3,6-tetrahydropyridine. Proc. Natl. Acad. Sci. U. S. A. 80: 4546–4550, 1983.
3. CAMPBELL, K. J., TAKADA, M. AND HATTORI, T.: Evidence for retrograde axonal transport of MPP^+ in the rat. Neurosci. Lett. 118: 151–154, 1990.

4. CHANG, F. W., WANG, S. D., LU, K. T. AND LEE, E. H. Y.: Differential interactive effects of gliotoxin and MPTP in the substantia nigra and the locus coeruleus in BALB/c mice. Brain Res. Bull. 31: 253–266, 1993.
5. CHIBA, K., TREVOR, A. J. AND CASTAGNOLI, N., JR.: Activie uptake of MPP^+, a metabolite of MPTP, by brain synaptosomes. Biochem. Biophys. Res. Commun. 128: 1228–1232, 1985.
6. CHIUEH, C. C., JOHANNESSEN, J. N., CHESSELET, M. F. AND MARKEY, S. P.: Neurotoxic mechanism of 1-methyl-4-phenyl-1,2,3,6-tetrahydropyridine (MPTP) and its oxidative metabolites in the nigrostriatal system of C57BL6 mice. Fed. Proc. 44: 893, 1985.
7. COLOTLA, V. A., FLORES, E., OSCOS, A., MENESES, A. AND TAPIA, R.: Effects of MPTP on locomotor activity in mice. Neurotoxicol. Teratol. 12: 405–407, 1990.
8. FORNO, L. S., LANGSTON, J. W., DELANNEY, L. E., IRWIN, I. AND RICAURTE, G. A.: Locus ceruleus lesions and eosinophilicinclusions in MPTP-treated mondeys. Ann. Neurol. 20: 449–455, 1986.
9. GERLACH, M., RIEDERER, P., PRZUNTEK, H. and YOUDIM, B. H.: MPTP mechanisms of neurotoxicity and their implications for parkinson's disease. Eur. J. Pharmacol. 208: 273–286, 1991.
10. GRAY, E.G. AND WHITTAKER, V. P.: The isolation of nerve endings from brain: an electron microscopic study of cell fragments divided by homogenization and centrifugation. J. Anat. 96: 79–88, 1962.
11. GROVES, P. M.. WILSON, C. J., YOUNG, S. J. AND REBEC, G. V.: Self-inhibition by dopaminergic neurons. Science, 190: 522–529, 1975.
12. GUPTA, M., FELTEN, D. L. AND GASH, D. M.: MPTP alters central catecholamine neurons in addition to the nigrostriatal system. Brain Res. Bull. 13: 737–742, 1984.
13. HALLMAN, H., LANGE, J., OLSON, L., STROMBERG, I. AND JONSSON, G.: Neurochemical and histochemical characterization of neurotoxic effects of 1-methyl-4-phenyl-1,2,3,6-tetrahydropyridine on brain catecholamine neurons in the mouse. J. Neurochem. 44: 117–127, 1985.
14. HEIKKILA, R. E., HESS, A. AND DUVOISIN. R. C.: Dopaminergic neurotoxicity of 1-methyl-4-phenyl-1,2,3,6-tetrahydropyridine in mice. Science (Wash. DC) 224: 1451–1453, 1984a.
15. HEIKKILA, R. E., MANZINO, L., CABBAT, F. S. AND DUVOISIN, R. C.: Protection against the dopaminergic neurotoxicity of 1-methyl-4-phenyl-1,2,3,6-tetrahydropyridine by monoamine oxidase inhibitors. Nature (Lond.) 311: 467–469, 1984b.
16. HU, S. C., CHANG, F. W., SUNG, Y. J., HSU, W. M. AND LEE, E. H. Y.: Neurotoxic effects of 1-methyl-4-phenyl-1,2,3,6-tetrahydropyridine in the substantia nigra and the locus coeruleus in BALB/c mice. J. Pharmacol. Exp. Ther. 259: 1379–1387, 1991.
17. JARVIS, M. F. AND WAGNER, G. C.: 1-Methyl-4-phenyl-1,2,3,6-tetrahydropyridine-induced neurotoxicity in the rat; characterization and age-dependent effects. SYNAPSE, 5: 104–112, 1990.
18. JAVITCH. J. A., D'AMATO, R. J., STRITTMATTER, S. M. AND SNYDER, S. H.: Parkinsonism-inducing neurotoxin, N-methyl-4-phenyl-1,2,3,6-tetrahydropyridine: Uptake of the metabolite N-methyl-4-phenylpyridine by dopamine neurons explains selective toxicity. Proc. Natl. Acad. Sci. U. S. A. 82: 2173–2177, 1985.
19. JENNER, P., RUPNIAK, N. M. J., ROSE, S., KELLY, E., KILPATRICK, G., LESS, A. AND MARSDEN, C. D.: 1-Methyl-4-phenyl-1,2,3,6-tetrahydropyridine-induced parkinsonism in the common marmoset. Neurosci. Lett. 50: 85–90, 1984.
20. KURIYAMA, T., TAGUCHI, J. I. AND KURIYAMA, K.: Functional alterations in striatal cholinergic and striatonigral gabaergic neurons following 1-methyl-4-phenyl-1,2,3,6-tetrahydropyridine (MPTP) administration. Neurochem. Int. 16: 319–329, 1990.
21. LANGSTON, J. W., FORNO, L. S., REBERT, C. S. AND IRWIN, I. 1-methyl-1,2,5,6-tetrahydropyridine causes selective damage to the zona compacta of the substantia nigra in the squirrel monkey. Brain Res. 292: 390–394, 1984.
22. LEE, E. H. Y., LIN, Y. P. AND YIN, T. H.: Effects of lateral and medial septal lesions on various activity and reactivity measures in rats. Physiol. Behave. 42: 97–102, 1987.
23. LEHMANN, A.: Atlas stereotaxique du cerveau de la souris. 1974.
24. LOWRY, O. H., ROSEBROUGH, N. J., FARR, A. L. AND RANDALL, R. J.: Protein measurement with the folin phenol reagent. J. Biol. Chem. 193: 265–275, 1951.
25. MANN, D. M. A. AND YATES, P. O.: Pathological basis for neurotransmitter changes in Parkinson's disease. Neuropathol. Appl. Neurobiol. 9: 3–19, 1983.

26. MANN, D. M. A. YATES, P. O. AND HAWKES, J.: The pathology of thehuman locus ceruleus. Clin. Neuropathol. 2: 1–7, 1983.

27. MATSUDA, L. A., SCHMIDT, C. J., HANSON, G. R. AND GIBB, J. W.: Effect of 1-methyl-4-phenyl-1,2,3,6-tetrahydropyridine (MPTP) on striatal tyrosine hydroxylase and tryptophan hydroxylase in rat. Neuropharmacology, 25: 249–255, 1986.

28. MITCHELL, I. J., CROSS, A. J., SAMBROOK, M. A. AND CROSSMAN, A. R.: Sites of the neurotoxic action of 1-methyl-4-phenyl-1,2,3,6-tetrahydropyridine in the macaque monkey include the ventral tegmental area and the locus coeruleus. Neurosci. Lett. 61: 195–200, 1985.

29. NISSBRANDT, H., SUNDSTROM, E., JONSSON, G., HJORTH, S. and CARLSSON, A.: Synthesis and release of dopamine in rat brain: comparison between substantia nigra pars compacta, pars reticulata, and striatum. J. Neurochem. 52: 1170–1182, 1989.

30. PEAT, M. A. AND GIBB, J. W.: High performance liquid chromatographic determination of indoleamines, dopamine and norepinephrine in rat brain with fluorimetric detection. Anal. Biochem. 128: 275–280, 1983.

31. RANSOM, B. R., KUNIS, D. M., IRWIN, I. AND LANGSTON, J. W.: Astrocytes convert the parkinsonism inducing neurotoxin, MPTP, to its active metabolite, MPP$^+$. Neurosci. Lett. 75: 323–328, 1987.

32. REID, W. G. J., BROE, G. A., HELY, M. A., MORRIS, J. G. L., WILLIAMSON, P. M., O'SULLIVAN, D. J., RAIL, D., GENGE, S. AND MOSS, N. G.: The neuropsychology of de novo patients with age on onset. Int. J. Neurosci. 48: 205–217, 1989.

33. RICAURTE, G. A., LANGSTON, J. W., DELANNEY, L. E., IRWIN, I. AND BROOKS, J. D.: Dopamine uptake blockers protect against the dopamine depleting effect of 1-methyl-4-phenyl-1,2,3,6-tetrahydropyridine (MPTP) in the mouse striatum. Neurosci. Lett. 59: 259–264, 1985.

34. ROBERTSON, G. S., DAMSMA, G. and FIBIGER, H. C.: Characterization of dopamine release in the substantia nigra by in vivo microdialysis in freely moving rats. J. Neurosci. 11: 2209–2216, 1991.

35. SEGAL, D. S. AND KUCZENSKI, R.: Tyrosine hydroxylase activity: Regional and subcellular distribution in brain. Brain Res. 68: 261–266, 1974.

36. SIEGEL, M.: Nonparametric Statistics for the Behavioral Science, McGraw-Hill Company, New York, 1965.

37. SINGER, T. P. AND RAMSAY, R. R.: Mechanism of the neurotoxicityof MPTP. FEBS Lett. 274: 1–8, 1990.

38. SUNDSTROM, E., FREDRIKSSON, A. AND ARCHER, T.: Chronic neurochemical and behavioral changes in MPTP-lesioned C57BL/6 mice: A model for Parkinson's disease. Brain Res. 528: 181–188, 1990.

39. SVENSSON, T. H., BUNNEY, B. S. AND AGHAJANIAN, G. K.: Inhibition of both noradrenergic and serotonergic neurons in brain by the α-adrenergic agonist clonidine. Brain Res. 92: 291–306, 1975.

40. TAKADA, M., LI, Z. K. AND HATTORI, T.: Intracerebral MPTP injections in the rat cause cell loss in the substantia nigra, ventral tegmental area and dorsal raphe. Neurosci. Lett. 78: 145–150, 1987.

41. TETRUD J. W. AND LANGSTON, J. W.: Tremor in MPTP-induced parkinsonism. Neurology, 42: 407–410, 1992.

42. VAN DOGNEN, V. P. A. M.: The human locus coeruleus in neurology and psychiatry (Parkinson's, Lewy body, Hallervorden-Spatz, Alzheimer's and Korsakoff's disease, (pre) senile dementia, schizophrenia, affective disorders, psychosis). Prog. Neurobiol. 17: 97–139, 1981.

43. WILLOUGHBY, J., COWBURN, R. F., HARDY, J. A., GLOVER, V. AND SANDLER, M.: 1-Methyl-4-phenylpyridine uptake by human and rat striatal synaptosomes. J. Neurochem. 52: 627–631, 1989.

44. WILSON, J. A., DOYLE, T. J. AND LAU, Y. S.: MPTP, MPDP$^+$ and MPP$^+$ cause decreases in dopamine content in mouse brain slices. Neurosci. Lett. 108: 213–218, 1990.

LINOPIRDINE

A Depolarization-Activated Releaser of Transmitters for Treatment of Dementia

S. William Tam and Robert Zaczek

Central Nervous System Diseases Research
The DuPont Merck Pharmaceutical Company
Wilmington, Delaware

INTRODUCTION

Advances in medicine are giving rise to extended life expectancy and thus diseases which usually affect the elderly population will increase in prevalence. Alzheimer's disease (AD), the leading cause of dementia in the aged population, is devastating to the patients as well as their family members. This is a costly disease to the individuals and to society. Although an enormous effort is being made to understand the cause of this disease and to develop therapies for it, AD remains one of the foremost challenges to medical research today.

The dementia occuring in AD is associated with dysfunction and death of neurons in a variety of cell populations, including cholinergic, monoaminergic, and peptidergic systems[1–3]. A substantial amount of data support a central role of acetylcholine (ACh) in the cognitive dysfunction seen in AD and aging. Cerebral cortical ACh synthesis declines as a function of age in animals[4,5]. One of the earliest observed neurochemical changes in AD is the profound loss of neocortical cholinergic innervation.[6–9] This loss has been found to be correlated with the degree of dementia found in the disease. Lesioning of cholinergic cell bodies in the nucleus basilis magnocellularis projecting to the neocortex induces marked deficits in cognitive performance in experimental animals[3,10,11] and these cognitive deficits can be attenuated by cholinergic agonists, ACh releasing agents, and acetylcholinesterase inhibitors.[2,12] These observations have led to what has been called the *cholinergic hypothesis* of AD which suggests that the cholinergic losses observed in AD lead directly to the cognitive and memonic deficits observed in the disease. However, with the wide range of neurochemical alterations now documented in AD[13–15] the *cholinergic hypothesis* appears to be an oversimplification.

Strategies to enhance cholinergic function in brain (e.g. precursor loading, acetylcholinesterase inhibitors, and cholinergic receptor agonists) represent the predominant approaches that are currently being clinically evaluated to reverse some of the cognitive deficits seen in AD (for a review see Davis et al.[16]). Precursor loading with choline or phosphatidylcholine failed to have a significant effect on AD symptoms.[17,18] Treatment with acetylcholinesterase inhibitors such as physostigmine[19,20] and tetrahydroaminoacridine [21,22] to retard the degradation of ACh has had some limited benefit, but side effects have also been noted. Initial attempts to treat AD with direct cholinergic agonists were limited by low efficacy and side effect issues.[23,24] However, the identification of multiple subtypes of muscarinic receptors has stimulated interests in developing M_1 receptor agonists to treat AD. Finally, an alternative approach to treat AD is to stimulate cholinergic function by enhancing the release of ACh.

A number of compounds, such as the aminopyridines, increase both basal (release in the absence of a stimulus) and stimulus-evoked release of ACh. This type of non-specific activation may lead to untoward events such as neurotransmitter depletion, overload toxicity, and desensitization. Therefore a compound which specifically enhances evoked release and not basal release should enhance normal synaptic activity and increase the signal-to-noise ratio during neuronal transmittion. During the evaluation of a number of compounds that had the potential to reduce cholinergic system dysfunction and enhance cognitive function by increasing ACh levels in the brain, linopirdine (DuP 996; AVIVA®) 3,3-bis(4-pyrindinylmethyl)-1-phenylindolin-2-one) was identified. Linopirdine represents a novel class of compounds which are depolarization-activated releasers of neurotransmitters.[25] This chapter describes the preclinical results of linopirdine.

SYNTHESIS

Linopirdine was synthesized according to the method[26] shown in Figure 1. Diphenylamine was added to a solution of oxalyl chloride and distilled and refluxed to form indoline-2, 2-dione. 4-Picoline in acetic acid was then added, followed by acetic anhydride, and then water and isopropanol. The mixture was cooled and filtered to yield 3-(4-pyridinylmethylidene)-1-phenylindolin-2-one (I). Sodium borohydride pellets were added to a slurry of I in methanol and then 10 N sodium hydroxide was added. A solution of 4-picolychloride HCl was added to this mixture and followed by 10 N sodium hydroxide. The solids were collected by filtration, washed with water, and dried in a vacuum. The resultant crude product was recrystallized in isopropanol and water to yield the product.

NEUROTRANSMITTER RELEASE ENHANCEMENT

Linopirdine enhancement of K^+-stimulated [^{3}H]ACh release was demonstrated using a two pulse K^+-depolarization paradigm in superfused rat brain slices preloaded with [^{3}H]choline.[27,28] The effect of 10^{-5} M linopirdine on the release of ACh *in vitro* is shown in Figure 2. Addition of linopirdine to brain slices has no effect on the basal release of ACh. The ACh release enhancing effect of linopirdine is dose dependent (Figure 3). A significant effect on enhancement of ACh release can be observed at 10^{-6} M and reaches maximum at 10^{-5} M. Similar effects of linopirdine are observed using slices of rat cerebral cortex, hippocampus, and striatum. The release enhancing effects of linopirdine is not limited to cholinergic processes alone. The release of tritium from brain slices preloaded with [^{3}H]dopamine (DA), [^{3}H]serotonin (5-HT), [^{3}H]glutamate (Glu) and [^{3}H]d-aspartate, as well as , [^{3}H]choline, is enhanced by linopirdine.[28,29] Linopirdine enhances the release of the newly synthesized pool of radiolabeled neurotransmitters as well as enhances the release of both endogenous ACh from neocortical slices and DA from striatal slices.[30] Since the neuronal deterioration which is involved in AD involves axons and terminals emanating from multiple neuronal cell types,[13–16] the ability of linopirdine to enhance the release of multiple neurotransmitters offers an advantage over current AD therapies which are aimed at stimulating the cholinergic system alone.

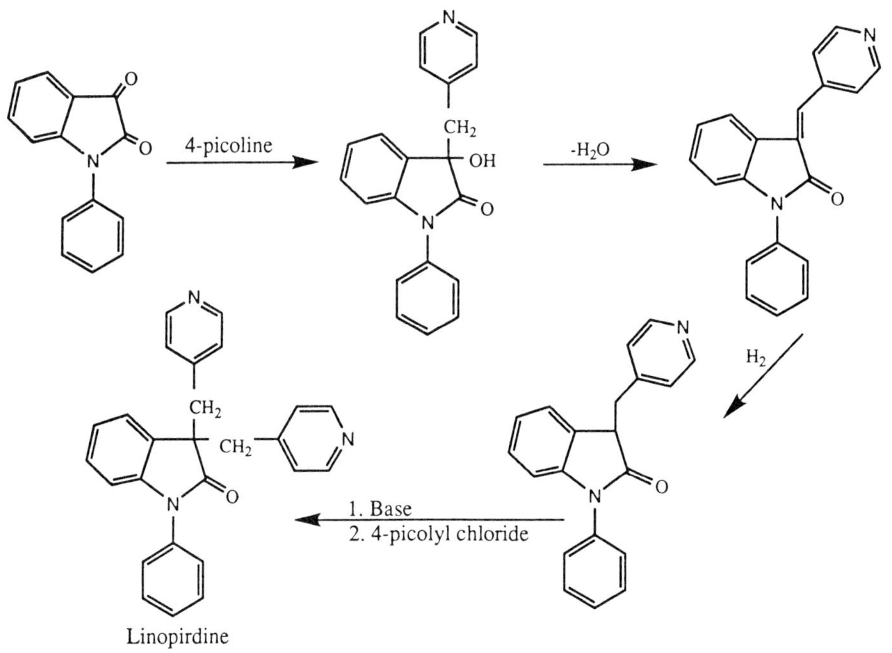

Linopirdine

Figure 1. The scheme of synthesis of linopirdine.

There is some degree of neurotransmitter specificity for linopirdine release enhancement. While linopirdine enhances the release ACh, DA, 5-HT and Glu, the K^+-evoked release of norepinephrine (NE) from neocortical and hippocampal slices is not altered in the presence of the drug.[28,31] A summary of effects of linopirdine on enhancement of depolarization-evoked release of different neurotransmitters from various regions of the brain is shown in Figure 4. The drug apparently decreases the depolarization threshold of neurotransmitter release in response to depolarization by increased concentrations of potassium. We have found that [^{3}H]NE release from neocortical and hippocampal slices is much more sensitive to small

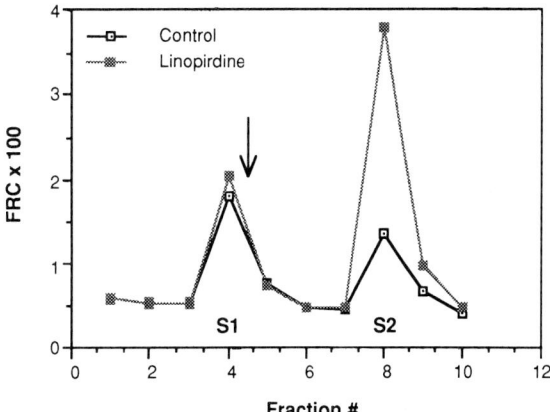

Figure 2. Effects of 10 mM linopirdine on 25 mM K+-evoked ACh release from rat cerebral cortex *in vitro*. First stimulation (S1) in the absence of linopirdine or vehicle (added at arrow). Second stimulation (S2) in the presence of linopirdine or vehicle. FRC is equal to the dpm of the fraction divided by the total dpm in the slices at the beginning of the fraction. Each point represents the mean of 4 observations. Note no increase in basal release is induced in the presence of linopirdine before the K^+ stimulus at S2.

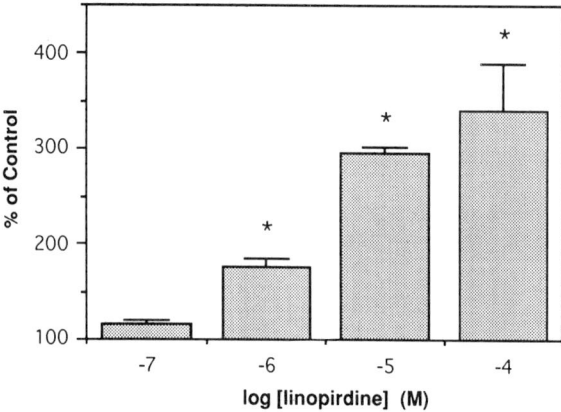

Figure 3. Effects of various concentrations of linopirdine on 25 mM K+-evoked ACh release from rat cerebral cortex *in vitro*. Results are S2/S1 ratios, expressed as percentage of concomitantly-run controls. Data represent means of 4 observations ± S.E.M. *Significantly different from control, p, one-way ANOVA followed by Duncan's Multiple Range test.

changes in potassium concentrations than the release of [^{3}H]ACh [^{3}H]DA and [^{3}H]5-HT and this may explain the lack of effect of linopirdine on [^{3}H]NE release.[31]

Linopirdine has been shown to elevate extracellular ACh levels in *in vivo* experiments using both epidural cup[28] and *in vivo* microdialysis[32,33] procedures. In awake, freely moving rats, linopirdine at 5 and 10 mg/kg i.p. was shown to increase extracellular ACh levels by 42% and 59% over baseline, respectively.[32] A reasonable correlation exists between effective drug levels *in vitro* (EC50 ≈ 5 µM) and *in vivo* since the concentration of linopirdine in brain reaches 1 µM after the administration of the drug at 10 mg/kg.[32]

BEHAVIORAL EFFECTS OF LINOPIRDINE

Linopirdine has been tested in a number of behavioral models used to measure learning and memory (see Table 1). In an active avoidance paradigm in which mice are given a 50-trial session where they are required to press a lever to avoid receiving a footshock, linopirdine at 0.085 to 2.5 mg/kg, s.c., significantly increases the number of avoidance responses.[34] In the mouse one-trial passive avoidance procedure using a light/dark box, linopirdine at a dose of 0.01 mg/kg significantly enhances step-through latencies.[35] In rats, linopirdine partially reverses the hypoxia-induced passive avoidance deficits with potency similar to the acetylcholinesterase inhibitor physostigmine and much more potently than tetrahydroaminoacridine (THA).[34,36] Linopirdine also partially reverses CO_2-induced amnesia in rats in a passive avoidance test.[34] The performance of rats in a pole-jump one-way active avoidance test is improved by linopirdine.[34] Linopirdine affects neither the threshold at which rats react to noxious footshock nor on spontaneous locomotor activity.[34] Linopirdine improves the performance of rats in lever press acquisition for food whereas THA and physostigmine are not active in this test.[34] In the Morris swim maze test, linopirdine improves the performance of aged rats[37] and rats which have received electrolytic lesions of the medial septum.[35]

The data in several rodent behavioral models of learning and memory are consistent with the cognition enhancing effects of linopirdine. However, the effective doses of linopirdine in some of these models is substantially lower than the approximately 1 µM in the brain and hippocampus that consistently lead to increases in extracellular ACh *in vivo*.[32,33] The reasons for the apparent discrepancy between the effective dose range of linopirdine in rodent behavioral models and that required to enhance ACh release are unclear and could be the result of combined effect of small increases of multiple neurotransmitters *in vivo*.

OTHER EFFECTS OF LINOPIRDINE

Linopirdine administration in naive rats at 0.01 to 1.0 mg/kg, s.c., does not significantly affect cerebral glucose metabolism in rats.[38] Since linopirdine protects against hypoxia-induced passive avoidance deficits in rats, the effect of hypoxia with and without linopirdine on cerebral glucose metabolism was studied. Hypoxia causes a small increase of glucose metabolism in some brain regions and linopirdine administered after hypoxia significantly decreases glucose metabolism in the hippocampus, limbic cortex, ventral hippocampal commissure, medial septum, striatum, subthalamic nucleus, zona incerta, lateral habenula, cerebral cortex, cerebellar vermis, and a few thalamic nuclei compared to the hypoxia controls.[38] Linopirdine after hypoxia brings the level of cerebral glucose metabolism close to that of non-hypoxia controls. These linopirdine effects are observed in limbic areas associated with learning and memory. The relationship of linopirdine's effects on cerebral glucose metabolism and neurotransmitter release enhancement is unclear, since doses which lead to significantly increased ACh release in vivo (10 mg/kg) were not tested.

Linopirdine has been shown to have effects on c-fos immediate early gene expression in rat brains.[39] Linopirdine (10 mg/kg i.p.) increases c-fos expression primarily in cerebral cortical regions of old rats (24 months). The c-fos induction is an indication of cellular activation. A similar pattern of increased c-fos expression has also been observed with administration of acetylcholinesterase inhibitors which are a class of compounds under consideration as cognitive enhancers. The implications for these effects of linopirdine on c-fos induction and changes in cerebral glucose metabolism remain unclear.

Linopirdine produces changes in the cerebral cortical electroencephalographic (EEG) recordings in both rats and humans. At doses of 0.3 to 3 mg/kg s.c., linopirdine significantly increases the faster frequency bands (α nd β) and decreases slower frequency bands δ and τ in

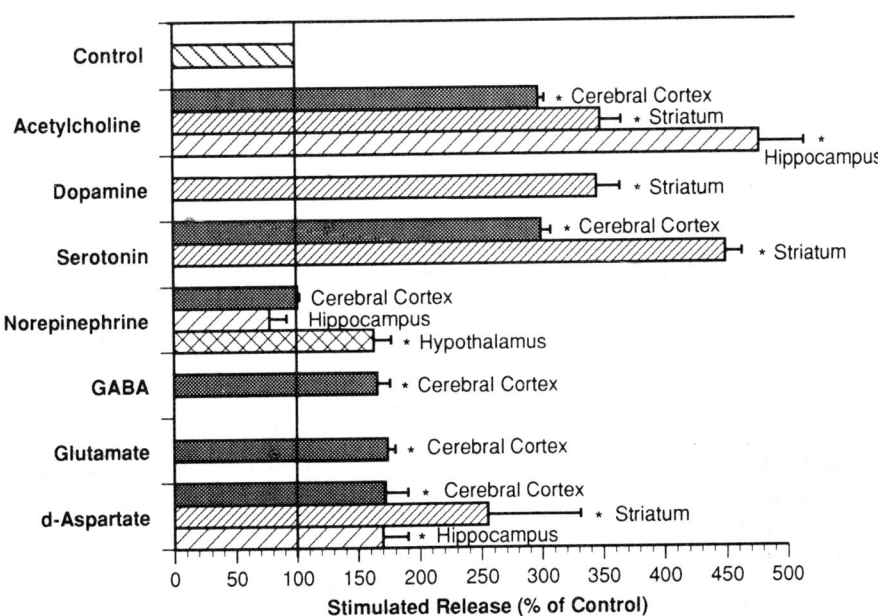

Figure 4. Summary of effects of linopirdine (10 mM) on K$^+$-evoked release of various neurotransmitters from rat brain slices in vitro. Data represent means ± S.E.M. *Significantly different from control, p<0.05 one-way ANOVA followed by Duncan's Multiple Range test.

rats.[34] The maximum effect is reached at 1 mg/kg. The increase in faster frequency bands and decrease in slower frequency bands are consistent with a vigilance enhancing effect. The EEG effects of linopirdine in humans are similar to those found in rats. In healthy male volunteers (18–40 years of age) 30 mg of linopirdine or 20 mg b.i.d. of linopirdine increases the total power as well as the power of the α and α-adjacent β bands in the EEG suggest vigilance-improving effects.[40,41]

MECHANISM OF ACTION FOR LINOPIRDINE

While much has been done to study linopirdine's effects at cellular, tissue, and animal levels, the precise mechanism by which linopirdine enhances depolarization activated release of transmitters remains unknown. Since linopirdine has no effect on basal release of neurotransmitters, direct effects on second messengers systems are not likely responsible for the action of this drug. The neurotransmitter release enhancement of linopirdine requires membrane depolarization. Linopirdine is effective in enhancing the release of ACh evoked by increased potassium concentration,[42] by the sodium channel agonist veratridine and by the potassium channel blocker 3,4-diaminopyridine.[42] However, the drug does not affect the release of ACh evoked by the excitatory amino acid N-methyl-d-aspartic acid (NMDA).[42] This suggests that linopirdine acts at processes which precede the elevation of cytosolic free calcium since NMDA receptor activation leads to opening of the coupled calcium channel. Somewhat more puzzling is the fact that linopirdine does not enhance ACh release evoked by electrical stimulation.[42,43] Understanding the reasons for the differential effects of linopirdine on transmitter release stimulated by elevated K^+ concentrations and veratridine on the one hand and direct pulsatory current application on the other will be important in unraveling the mechanism of action of the drug. This *in vitro* release enhancing activity of linopirdine is not an artifact of the *in vitro* assay condition since the drug also enhances ACh release *in vivo*.

Linopirdine appears to block potassium channels in neurons. It attenuates potassium currents in hippocampal slices[44] and in dispersed neocortical cells in culture.[45] In addition, linopirdine lowers the threshold of ACh release induced by increases in potassium concentrations but does not affect maximal release at very high concentrations of potassium.[46] These data are consistent with the possibility that linopirdine blocks potassium channels. However, m channel antagonist as well as other K^+ channel antagonists, 4-aminopyridine and tetraethylammonium, neither mimic nor block the action of linopirdine suggesting that the K^+ channels blocked by these standard blockers are not the site of action for linopirdine.[47] Ca^{++} channel antagonists (L-, N-, and P-types) have no effect on the ACh release enhancement by linopirdine suggesting that these channels are not the primary site of action for linopirdine.[47,50]

Table 1. Effects of Linopirdine in Animal Models of Learning and Memory

Behavioral Program	Effective Doses (mg/kg)	Reference
Lever-press active avoidance (mouse)	0.085 to 2.5 s.c	34
One-trial passive avoidance (mouse)	0.01 i.p	35
Pole-jump active avoidance (rat)	0.085 to 0.85 s.c.	34
Reversal of hypoxia-induced passive avoidance deficit (rat)	0.01 to 0.1 s.c.	34,36
Reversal of CO_2-induced passive avoidance deficit (rat)	0.085 and 0.25 s.c	34
Lever-press acquisition for food (rat)	0.03 to 0.1 p.o.	34
Water maze (aged rats)	0.25 and 2.5 p.o	37
Water maze (septal lesioned rats)	0.01 and 03 i.p .	35

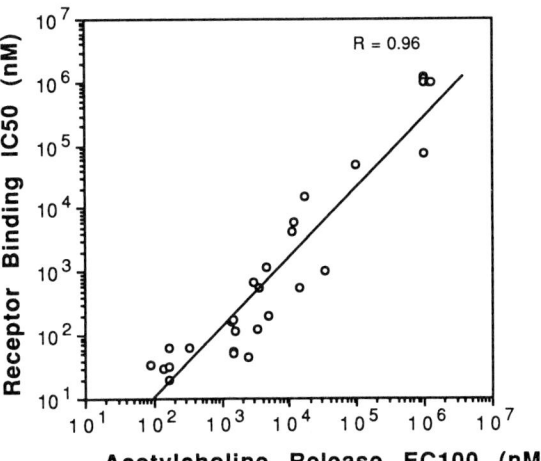

Figure 5. Correlation of [3H]linopirdine receptor binding affinity of linopirdine and its 30 structural analogs with the potency of these compounds to enhance K+-stimulated ACh release by 100% (EC100). This figure is reproduced from Tam et al.48 with permission from The American Society for Pharmacology and Experimental Therapeutics.

Linopirdine in dose ranges that are effective in enhancing neurotransmitter release does not inhibit acetylcholinesterase activity or bind to muscarinic or nicotinic cholinergic receptors.[28] Choline uptake in synaptosomes is also not affected by linopirdine.[29]

It has been suggested that the pharmacologic activity of linopirdine is related to a novel specific binding site labeled with [3H]linopirdine.[48] [3H]Linopirdine binds with high affinity to rat brain membranes (K_D = 19 nM; B_{max} = 102 fmol/mg protein). Binding to this site is specific, saturable, reversible, and time-, temperature-, and pH-dependent. This binding site appears to be novel since a wide variety of pharmacological agents and neuropeptides are unable to displace the [3H]linopirdine binding. Importantly, analogues of the linopirdine structural series demonstrate an excellent correlation between binding affinity to this site and the potency to enhance ACh release *in vitro*[48] (Figure 5). Autoradiographic studies with [3H]linopirdine indicates the highest densities of binding sites are localized in the hippocampus (CA1 to CA3 pyramidal cell layers and the granule cells of the dentate gyrus), the cerebral cortex (lamina IV), the dorsal raphe nucleus and the interpeduncular nucleus.[49] The localization of [3H]linopirdine binding sites in the brain areas implicated in cognitive processes and affected in Alzheimer's disease suggest that this drug may have therapeutic potential for the treatment of cognitive deficits seen in dementia.

SUMMARY

Linopirdine (DuP 996, AVIVA®), currently in Phase III clinical trial for the treatment of Alzheimer's disease, is a representative of a class of novel molecules which enhances the stimulus-evoked but not basal release of several neurotransmitters including ACh, DA, 5-HT and Glu. Linopirdine has been shown to enhance ACh release in the hippocampus *in vivo*. In addition, linopirdine produces a number of effects including EEG patterns of enhanced vigilance, induction of c-fos expression in cerebral cortex, reduction of the increase of cerebral glucose utilization induced by hypoxia, and improved performance in animal models of learning and memory. The specific action of linopirdine on depolarized neurons but not on basal release suggests that compounds of this class will enhance normal brain activity and not lead to a non-specific activation. Furthermore, the effect of linopirdine on multiple neurotransmitter systems that are deficient in Alzheimer's disease suggests that this class of agents may be more efficacious in the treatment of dementia than therapies aimed at individual neurotransmitters systems.

REFERENCES

1. C.G. Gottfries, Alzheimer's disease and senile dementia: biochemical characteristics and aspects of treatment, *Psychopharmacology* 86:245–252 (1985).
2. V. Haroutunian, P. Kanof and K.L. Davis, Pharmacological alleviation of cholinergic lesion induced memory deficits in rats, *Life Sci* 37:945–952 (1985).
3. D.J. Hepler, G.L. Wenk, B.L. Cribbs, D.S. Olton and J.T. Coyle, Memory impairments following basal forebrain lesions, *Brain Res* 346:8–14 (1985).
4. F. Pedata, J. Slavikova, A. Kotas and G. Pepeu, Acetylcholine release from rat cortical slices during postnatal development and aging. *Neurobiol Aging* 4:31–35 (1983).
5. N.R. Sims, K.L. Marek, D.M. Bowen and A.N. Davison, Production of [^{14}C]acetylcholine and [^{14}C]carbon dioxide from [U-^{14}C]glucose in tissue prisms from aging rat brain, *J Neurochem* 38:488–492 (1982).
6. P. Davis and A.J.F. Maloney, Selective loss of central cholinergic neurons in Alzheimer's disease, *Lancet* 11:1403 (1976).
7. E.K. Perry, B.E. Tomlinson, G. Blessed, K. Bergman, P.H. Gibson and R.H. Perry, Correlation of cholinergic abnormalities with senile plaques and mental test scores in senile dementia, *Br Med J* 2:1427–1429 (1978).
8. P. Whitehouse, D. Price, R. Struble, A. Clark, J.T. Coyle and M. DeLong, Alzheimer's disease and senile dementia: loss of neurons in the basal forebrain, *Science* 215:237–239 (1982).
9. J.T. Coyle, D. Price and M. Delong, Alzheimer's Disease: a disorder of cholinergic innervation, *Science* 219:1184–1190 (1983).
10. B. Lerer, J. Warner, E. Friedman, G. Vincent and E. Gamzu, Cortical cholinergic impairment and behavioral deficits produced by kainic acid lesions of rat magnocellular basal forebrain, *Behav Neurosci* 99:661–677 (1985).
11. M. Watson, T.W. Vickroy, H.C. Fibiger, W.R. Roeski and H.I. Yamamura, Effects of bilateral ibotenate-induced lesions of the nucleus basalis magnocellularis upon selective cholinergic biochemical markers in the rat anterior cerebral cortex, *Brain Res* 346:387–391 (1985).
12. R.M. Ridley, T.K. Murray, J.A. Johnson and H.F. Baker, Learning impairment following lesion of the basal nucleus of Neynert in the marmoset: modification by cholinergic drugs, *Brain Res* 376:108–116 (1986).
13. D.L. Price, New perspectives on Alzheimer's disease, *Ann Rev Neurosci* 9:489–512 (1986).
14. R.J. D'Amato, R.M. Zweig, P.J. Whitehouse, G.L. Wenk, H.S. Singer, R. Mayeux, D.L. Price and S.H. Snyder, , Aminergic systems in Alzheimer's disease and Parkinson's disease, *Ann Neurol* 22:229–236 (1987).
15. R.G. Struble, R.E. Powers, M.F. Casanova, C.A. Kitt, E.C. Brown and D.L. Price, Neuropeptidergic systems in plaques of Alzheimer's Disease, *J Neuropathol Exp Neurol* 46:567–584 (1987).
16. R.E. Davis, M.E. Emmerling, J.C. Jaen, W.H. Moos and K. Spiegel, Therapeutic intervention in dementia, *Crit Rev Neurobiol* 7:41–83 (1993).
17. A.R. Little, P. Levy, P. Chaqui-Kidd and D. Hand, A double-blind placebo controlled trial of high dose lecithin in Alzheimer's disease, *J Neurol Neurosurg Psychiatry* 48:736–742 (1985).
18. A. Heyman, D. Schmechmel, W. Wilkinson, H. Rogers, R. Krishnan, D. Holloway, K. Schultz, L. Gwyther and R. Peoples, Failure of long term high-dose lecithin to retard progression of early-onset Alzheimer's disease, *J Neural Transm Suppl* 24:279–283 (1987).
19. K.L. Davis and R.C. Mohs, Enhancement of memory processes in Alzheimer's Disease with multiple-dose intravenous physostigmine, *Am J Psychiatry* 139:1421–1424, 1982.
20. L.J. Thal, P.A. Fuld, D.M. Masur and N.S. Sharpless, Oral physostigmine and lecithin improve memory in Alzheimer's disease, *Ann Neurol* 13:491–496 (1983).
21. N. Sitaram, Cholinergic hypothesis of human memory: Review of basic and clinical studies, *Drug Develop Res* 4:481–488 (1984).

22. W.K. Summers, L.V. Majovski, G.M. Marsh, K. Tachiki and A. Kling, Oral tetrahydrozminoacridine in long-term treatment of senile dementia, Alzheimer type, *N Eng J Med* 315:1241–1245 (1986).
23. M.M. Mouradian, E. Mohr, J.A. Williams and T.N. Chase, No response to high dose muscarinic agonist therapy in Alzheimer's disease, *Neurol* 38:606–608 (1988).
24. K.L. Davis, E. Hollander, M. Davidson, B.M. Davis, R.C. Mohs and T.B. Horvath, Induction of depression with oxotremorine in patients with Alzheimer's disease, *Am J Psychiatry* 144:468–472 (1987).
25. R.A. Earl, M.J. Myers, A.L. Johnson, R.M. Scribner, M.A. Wuonola, G.A. Boswell, W.W. Wilkerson, V.J. Nickolson, S.W. Tam, D.R. Britelli, R.J., Chorvat, R. Zaczek, L. Cook, C. Wang, X. Zhang, R. Lan, B. Mi and H. Wenting., Acetylcholine-releasing agents as cognition enhancers, structure-activity relationships of pyridinyl pendant groups on selected core structures, *Bioorganic and Medicinal Chemistry Letters* 2:851–854 (1992).
26. W.M. Bryant, III, and G.F. Huhn, Process for preparing 3,3-disubstituted indolines, U.S. Patent 4,806,651 (1989).
27. V.J. Nickolson, S.W. Tam, M.J. Myers, and L. Cook, DuP 996 enhances the K^+-stimulated release of acetylcholine in rat brain *in vitro* and *in vivo*. *FASEB J* 3:A931 (1989).
28. V.J. Nickolson, S.W. Tam, M.J. Meyers and L. Cook, DuP 996 (3,3(4-pyrindylmethyl)1-phenylindolin-2-one) enhances the stimulus evoked release of acetylcholine from rat brain *in vitro* and *in vivo*, *Drug Dev Res* 19:285–300 (1990).
29. R. Zaczek, W.J. Tinker, A.R. Logue, G.A. Cain, C.A. Teleha and S.W. Tam, Effects of linopirdine, HP 749, and glycyl-prolyl-glutamate on transmitter release and uptake, *Drug Dev Res* 29:203–208 (1993).
30. J.A. Saydoff and R. Zaczek, Linopirdine enhances KCl evoked release, but not basal release, of endogenous dopamine in superfused rat striatum, *FASEB Abstr* 1521 (1993).
31. R. Zaczek, W.J. Tinker and S.W. Tam, Unique properties of norepinephrine release from terminals arising from the locus coeruleus: high potassium sensitivity and lack of linopirdine (DuP 996) enhancement, *Neurosci Lett* 155:107–111 (1993).
32. M. Marynowski, C. Maciag, C.M. Rominger, S.W. Tam and R. Zaczek, Effects of linopirdine (DuP 996) on hippocampal extracellular levels of acetylcholine in freely moving animals, *Soc Neurosci Abstr* 19:1040 (1993).
33. T.M. Smith, A.D. Ramirez, S.D. Heck, V.J. Jasys, R.A. Volkmann, J.T. Forman and D.R. Liston, *In vivo* microdialysis and pharmacokinetic studies with DuP 996, *Soc Neurosci Abstr* 19:1041 (1993).
34. L. Cook, G.F. Steinfels, K.W. Rohrbach and V.J. Denoble, Cognition enhancement by the acetylcholine releaser DuP 996, *Drug Devel Res* 19:301–314 (1990).
35. J.D. Brioni, P. Curzon, M.J. Buckley, S.P. Arneric and M.W. Decker, Linopirdine (DuP 996) facilitates the retention of avoidance training and improves performance of septal-lesioned rats in the water maze, *Pharmacol Biochem Behav* 44:37–43 (1993).
36. V.J. DeNoble et al., Comparison of DuP 996, with physostigmine, THA and 3,4-DAP on hypoxia-induced amnesia in rats, *Pharmacol Biochem Behav* 36:957–961 (1990).
37. M.G. Baxter, K.W. Rohrbach, S.W. Tam, R. Zaczek and D.S. Olton, Effects of linopirdine and DuP 921 on age-related impairments in memory and on the cholinergic system, *Soc Neurosci Abstr* 19:1041 (1993).
38. G.W. Dent, B.L. Rule, S.W. Tam and E.B. De Souza, Effects of the memory enhancer linopirdine (DuP 996) on cerebral glucose metabolism in native and hypoxia-exposed rats, *Brain Res* 620:7–15 (1993).
39. G. Dent, S.W. Tam and R. Grzanna, The memory enhancer linopirdine increases c-fos expression in cerebral cortex of aged rats, *Soc Neurosci Abstr* 19:1040 (1993).
40. B. Saletu, A. Darragh, P. Salmon and R. Coen, EEG brain mapping in evaluating the time-course of the central action of DuP 996 - a new acetylcholine releasing drug, *Br J Clin Pharmacol* 28:1–16 (1989).
41. B. Saletu, A. Darragh, H.P. Breuel, W. Herrmann, P. Salmon, R. Coen and P. Anderer, EEG mapping central effects of multiple doses of linopirine -- a cognitive enhancer -- in healthy elderly male subjects, *Human Psychopharmacology* 6:267–275 (1991).
42. R. Zaczek, C. Maciag and W.J. Tinker, Effects of linopirdine (DuP 996) on the KCl, veratridine, NMDA and electrically induced release of [^{3}H]acetylcholine from superfused brain slices, *Soc Neurosci Abstr* 19:1040 (1993).

43. C.P. Smith, L.R. Brougham and H.M. Vargas, Linopirdine (DuP 996) selectively enhances acetylcholine release by high potassium, but not electrical stimulation, in rat brain slices and guinea pig ileum, *Drug Dev Res*, in press.
44. B.W. Lampe and B.S. Brown, Electrophysiological effects of DuP 996 on hippocampal CA1 neurons, *Soc Neurosci Abstr* 17:632.19 (1991).
45. J.M. Frey, P.A. Murphy and B.S. Brown, DuP 996, a novel neurotransmitter releaser, blocks voltage-activated potassium currents in cultured neocortical neurons, *Soc Neurosci Abstr* 17:632.20 (1991).
46. W.J. Tinker, C. Maciag, S.W. Tam and R. Zaczek, Effects of linopirdine (DuP 996) on KCl and $CaCl_2$ dose response of potassium evoked release of [^{3}H]acetylcholine from superfused hippocampal slices, *Soc Neurosci Abstr* 18:1245(1992).
47. T. W. Vickroy, Presynaptic cholinergic actions by the putative cognitive enhancing agent DuP 996, *J Pharmacol Exp Ther* 264 (1993).
48. S.W. Tam, D. Rominger and V.J. Nickolson, Novel receptor site involved in enhancement of stimulus-induced acetylcholine, dopamine and serotonin release, *Mol Pharmacol* 40:16–21 (1991).
49. E.B. De Souza, B.L. Rule and S.W. Tam, [^{3}H]Linopirdine (DuP 996) labels a novel binding site in rat brain involved in the enhancement of stimulus-induced neurotransmitter release: autoradiographic localization studies, *Brain Res* 582:335–341 (1992).
50. J.A. Saydoff and R. Zaczek, The role of Ca^{2+} channels, adenosine, and Ca^{2+} stores on KCl evoked acetylcholine release and linopirdine (DuP 996) release enhancement in rat hippocampal slices, *Soc Neurosci Abstr* 19:423.9 (1993).

DIFFERENTIAL CHANGES IN REGIONAL BRAIN GANGLIOSIDE AND NEUTRAL GLYCOSPHINGOLIPID CONTENTS IN ALZHEIMER'S DISEASE

N.M.K. Ng Ying Kin, L.H. Pan, J.H. Louvaris, Y. Robitaille and N.P.V. Nair

Douglas Hospital Research Centre
6875 LaSalle Blvd
Verdun, Quebec
Canada, H4H 1R3 and
Department of Psychiatry
McGill University
Montreal, Quebec, Canada

INTRODUCTION

Dementia of the Alzheimer type or Alzheimer's disease (AD) is a chronic organic disease of the brain affecting primarily the elderly population.[1] It is characterized histopathologically by the intraneuronal accumulation of neurofibrillary tangles, deposits of neuritic plaques in the neuropil, and neuronal loss in specific regions of the forebrain.[2] Neurochemically, multiple transmitter deficits in autopsy brain tissues have been reported.[3] Since the sialic acid-containing glycosphingolipids, ganglyosides, have been found to be associated with various transmitter functions, e.g., cholinergic,[4] serotonergic[5] and noradrenergic[6] and they are also important neuronal membrane components with demonstrable neuronotrophic properties,[7] changes in brain ganglioside concentrations would probably occur. We have recently analysed the ganglioside contents of cortical tissues from different regional lobes of post-mortem brain specimens from AD and control subjects. For comparison purposes, we have also quantified the levels of neutral glycosphingolipids in these tissues, such neuroconstituents being associated with different brain cells, notably oligodendrocytes. Asialogangliosides present in these glycolipid fractions are also substrates for ganglioside synthesis.

MATERIAL AND METHODS

Human Materials

Patients were diagnosed using standard clinical criteria for senile dementia of the Alzheimer's type, Alzheimer's disease (AD) (DSM III, Reisberg Global Deterioration Scale,[8] and NINCDS-ADRDA CRITERIA[9]).

Brain specimens were obtained between 5–12 hours post-mortem, frozen in pentane and stored at −80°C at the brain bank of the Douglas Hospital Research Center. AD brains, when examined neuropathologically, exhibited characteristic neurofibrillary tangles in the perikarya and neuritic plaques in the neuropil. For ganglioside and neutral glycosphingolipid analyses, discrete areas of the cortex were disected from the fronta, temporal, occipital, parietal, and hippocampal regions of the brain. The age range of patients (total,11) was 47 to 96 years (mean 73 ± 4.3 S.E.M.) and that from control (total,6) 47 to 80 (mean 68 ± 5.1).

Isolation of Gangliosides and Neutral Glycosphringolipids

The double extraction procedures for ganglioside isolation as described by Ledeen and Yu[10] was used. Typically 100 to 300 mg of cortical brain tissues from various brain regions was homgenized with 19 volumes of chloroform-methanol (CM) solvent mixture, 2:1 by volume. The homogenate was centrifuged and the resulting CM supernatant kept. The residual pellet was re-extracted with 10 volumes of CM 1:2 by volume and containing 5% water. After centrifugation, this CM supernatant was combined with the first CM 2:1 supernatant and chloroform added to obtain a final CM solution with a ratio of 2:1 by volume. The combined CM supernatants were than subjected to Folch washing with 0.1 M KCl solutions[11] and the combined upper phases containing gangliosides separated from the non-ganglioside lipids in the lower phase. Gangliosides in the upper phase were purified by dialysis against 0.1 M $CaCl_2$ solution overnight at 4°C. Neutral glycosphingolipids were isolated from lower phase lipids by silicic acid chromatography. Lipids were loaded onto silica gel mini-columns and washed with choloform (4 vol.). Neutral glycosphingolipids were eluted with 4 volumes of acetone-methanol (9:1, by vol.).

Assays of Gangliosides and Neutral Glycosphingolipids

Brain gangliosides were determined using the modified resorcinol assay of Svennerholm[12,13]. N-acetylneuraminic acid (Sigma Chemical Co) was used as standard and spectrophotometric measurements carried out at 580 nm. Neutral glycosphingolipids were estimated by reactions with orcinol-sulphuric acid reagents,[14] with galactose as standard. Spectrophotometric measurements were carried out at 425 nm.

Thin Layer Chromatography (TLC)

Gangliosides, isolated by the method described above, were analysed by TLC. Samples containing 50–90 μgram were freeze-dried and dissolved in 100 xl CM 2:1. They were applied as 1 cm band on a 20 by 20 cm silica gel G coated plate (Brinkmann Instrument, Ont., Canada) which was previously activated at 100°C. Bovine brain gangliosides (Sigma Chemical Company) were used as standards. The TLC plate was developed in CM: 0.02% $CaCl_2$ in water (60:40:9, by vol)[10] and the gangliosides visualized with resorcinol-HCl reagent.[12]

RESULTS

A total of eleven subjects with clinically and neurohistopathologically diagnosed Alzheimer's disease (AD) and an age range of 46 to 96 years together with five age-matched controls of between 47 and 79 years of age were studied for regional brain cortical contents of gangliosides and neutral glycosphingolipids.

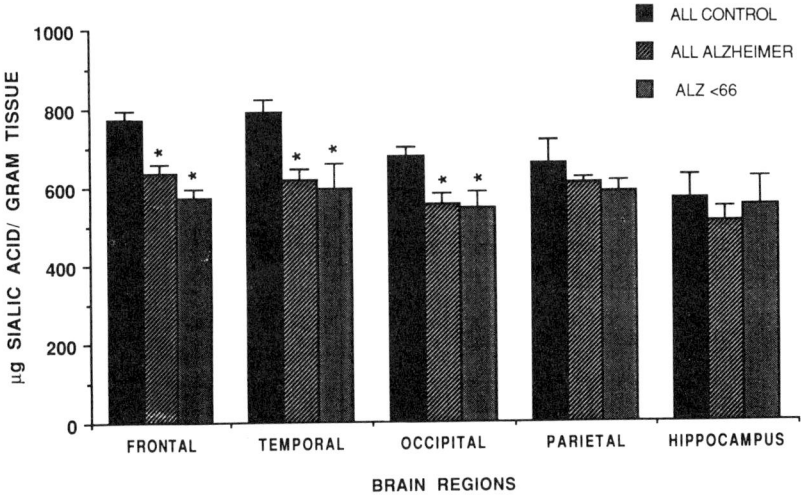

Figure 1. Histogram showing the concentrations of gangliosides, expressed in μg lipid bound sialic acid per gram wet weight of post-mortem cortical tissues, in five different brain regions of clinically and histologically diagnosed subjects with Alzheimer's disease (AD) (age-range 47–96 and less than 66 years). Data are compared with control subjects (CTRL), age-range 47–80 years. Each bar shows the means ± S.E.M. *$p < 0.005$ Student's t-test.

Brain Gangliosides

In general, when compared to controls, total ganglioside contents, in μg lipid bound sialic acid per gram wet weight, were found to be decreased in all the regions analysed in the case of AD subjects but were only significant in frontal, temporal and occipital ones (Fig. 1). Thus, in the frontal area, ganglioside contents in AD were 636 ± 20.7 versus 772 ± 22.3 ug/g in controls, a significant reduction of 18% ($p < 0.001$, 2-tailed t-test). In the temporal region, the values for AD and controls were 619 ± 24.7 and 792 ± 31.9 μg/g respectively, decrease of 22% ($p < 0.001$). Ganglioside contents in the occipital area were also significantly decreased ($p < 0.005$), with a value of 557 ± 23.8 versus 675 ± 23.1 μg/g in controls, a reduction of 18%. Similarly, decreases in ganglioside contents of 7 and 3% were found for the hippocampal and parietal regions respectively, but were not statistically significant. Thus, in the hippocampus, ganglioside contnents in AD were 507 ± 37.1 versus 568 ± 57.0 μg/g in controls. In the parietal region, the respective values were 609 ± 12.2 and 661 ± 57.2 μg/g.

In the four patients under 66 years of age, decreases in brain ganglioside contents were more pronounced particularly for the frontal and occipital regions, 572 ± 23.3 μg/g representing a reduction of 26% (Fig.1). Differences were not as marked for the other regions, although in general there was a tendency towards lower ganglioside contents in young when compared to older patients. In contrast, in the control group, brain cortical contents of gangliosides in all five regions studied, showed little variation with age, a finding consistent with earlier reports by Suzuki.[15]

Thin-layer chromatographic analysis of brain cortical gangliosides showed normal patterns for all regions of AD brains and no apparent difference when compared with those of control brains.

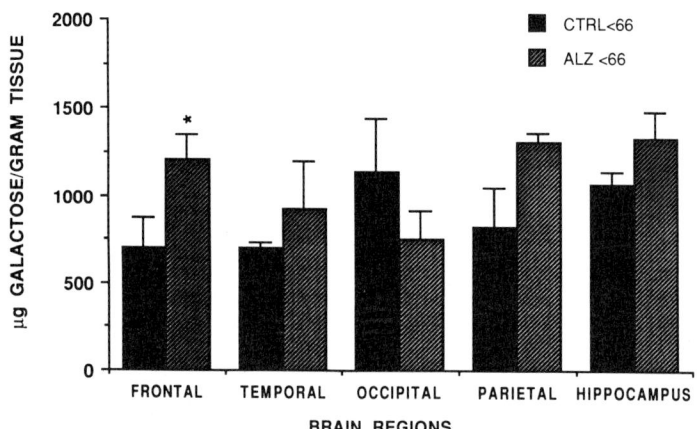

Figure 2. Histogram showing the concentrations of neutral glycospingolipids, expressed in ug galactose per gram wet tissue weight, isolated from the brain cortical tissues used for the qualification of gangliosides (*see Materials and Methods and Fig. 1*). Data are compared between Alzheimer's Disease subjects (AD) and controls (CTRL), age less than 66 years. Each bar shows the mean ± S.E.M. *p < 0.05 Student's t-test.

Neutral Glycosphingolipids

Neutral glycosphingolipids, isolated from the lower phase extracts of brain cortical tissues, after Folch partitioning and silica gel chromatography were estimated using the orcinol-H_2SO_4 reagents. The results are expressed as µg galactose per gram wet tissue weight. The values obrained from all regions analysed were found to be highly variable with respect to age and ranged between approximately 400 to 1800 µg/g in controls as well as in AD subjects. Thus, any comparison between the two groups of subjects must take into account their age. For subjects > 66 years, there was no significant difference between AD subjects and controls. On the other hand, there is a discernible trend for higher glycolipid contents in frontal, temporal, parietal and hippocampal regions of younger AD patients when compared to controls (Fig. 2). For the frontal cortical tissues, this increase was statistically significant (t-test,p < 0.05). In contrast, the occipital contents were lower in this same group of patients.

DISCUSSION

The sialic acid-containing glycolipids, gangliosides, are highly concentrated in neuronal membranes[7] and particular species of the compounds have been found to associated with the following neurotransmitter functions, cholinergic[4] and probably serotonergic[5] and noradrenergic,[6] and also with neurofibrillary tangles found in AD patients.[16] Since AD is a neurodegenerative disorder, characterized by accumulation of intraneuronal neurofibrillary tangles and multiple neurotransmitter deficits,[3] brain ganglioside contents would probably be altered. Indeed, several authors have reported decreased concentrations of these substances in tissues from different cortical and subcortical regions of post-mortom AD brains.[17–20] In the present studies on eleven AD subjects, changes in brain ganglioside contents have also been observed, although the reductions were statistically significant for the frontal, temporal and occipital regions only and not hippocampal and parietal ones. In the four patients aged under 66 years, ganglioside deficit was more pronounced than that found in older patients, particularly for the frontal area. Such findings could only be partially attributed to accelerated neuronal loss in some of the tissues analysed, since in cortical areas with only small neuronal loss, ganglioside decrease of larger order of magnitude was detected. Similarly, there does not appear to be any direct correlation between the number of neurofibrillary tangles or senile plaques and the degree of ganglioside loss. Indeed, the occipital lobes, which are mostly devoid of such abnormal deposits, ganglioside contents are greatly decreased, whereas in both the hippocampus and parietal regions, reductions in ganglioside contents were not significant, despite the high concentrations of plaques and tangles. The more severe deficits found in younger patients would appear to reflect abnormal ganglioside metabolism in AD.

Recently, we carried out studies on rats with quinolinic adid-induced lesions of the forebrain nucleus magnocellularis which resulted in central cholinergic hypofunctions, a neurochemical hallmark of AD.[21] Brain cortical ganglioside contents were found to be decreased despite normal cell counts in these tissues.[22] Based on labelling studies with radioactive sugar precursors, we suggested that these changes due to decreased sialic acid incorporation into the brain gangliosides as a result of deafferentation of the brain cortex.[23,24] Similar mechanisms are probably operative in human AD, since it has been well documented that in patients with early onset AD, loss of neurons in the nucleus basalis of Meynert (NBM) occured at a much faster rate than in older patients.[25,26] A dificiency in sialic acid incorporation would not only result in decreased ganglioside concentrations, as measured by their sialic acid contents, but would also account, in part, for the accompanying increase in neutral (asialo) glycosphingolipid contnents observed, particularly in the younger patients (Fig. 2). It is noteworthy that statistically significant changes occur in the frontal area, which is also the brain region most affected with respect to its ganglioside contents.

Differential changes in ganglioside levels observed in different regions of the brain may reflect regional variations on the compositions of different molecular species of the gangliosides.[15] The methods we used for the quantification of gangliosides as well as their visualisation on TLC are dependent on the contents of sialic acid. In the particular molecular types of gangliosides found in the hippocampus and parietal regions, the sialic acid contents may well be lower than those found in other regions analysed and would therefore appear to be less affected in AD. However, even for the frontal, temporal, and occipital regions, where ganglioside contents were found to significantly reduced, we were unable to demonstrate any compositional difference between AD and controls. A more sensitive method such as high performance liquid chromatography of the 2,4-dinitrophenylhydrazide derivatives of gangliosides[27] would probably be informative. On the other hand, development of antibodies to specific molecular species of gangliosides, such as those associated with cholinergic neurons (chol-1 antigens) have been used by Derrington et al[4] to demonstrate differences in the TLC patterns of ganliosides from hypocholinergic and normal rat brains. It is noteworthy that chol-1 gangliosides have very low mobilities on TLC and are presumably rich in sialic acid. Deafferentation of the brain cortex in AD, as a result of the loss of projections fron the NBM, would probably account for the decreased synthesis of sialic acid-rich gangliosides including chol-1, with conconmitant increase in the concentration of asialogangliosides found in the neutral glycosphingolipid fractions. The status of these gangliosides in AD is being investigated.

ACKNOWLEDGEMENTS

This research was supported by a grant from the Douglas Hospital Research Center.

REFERENCES

1. Katzman R.: (1986) Alzheimer's disease. N.Eng.J.Med. 314:964–973.
2. Tomlinson S.E.,Blessed G.,and Roth M.: (1970) Observations on the brain of demented old people. J.Neurol.Sci. 11:205–242.
3. Francis P.T.,Palmer A.M.,Sims N.R.,Bowen D.M.,Davison A.N.,Esiri M.N.,Neary D.,Snowden J.S.and Wilcock G.K.: (1985) Neurochemical studies of early onset Alzheimer's disease. N.Engl. J.Med. 313:7–11.
4. Derrington E.A.,Masco D., and Whittaker P.V.: (1989) Confirmation of the cholinergic specificity of the chol-1 gangliosides in mammalian brain using affinity-purified antisera and lesions affecting the cholinergic input to the hippocampus. J.Neurochem. 53:1686–1692.
5. Tamir H.,Brunner W.,Casper D., and Rapport M.M.: (1980) Enhancement by gangliosides of the binding of serotonin to serotonin binding protein. J. Neurochem. 34:1719–1724.
6. Molina V.A.,Keller E.A. and Orsingher O.A.: (1989) Gangliosides enhance behavioral and neurochemical effects induced by chronic desipramine (DMI) treatment. Eur.J.Pharmacol. 6012:247–52.
7. Ledeen R.W.: (1985) Gangliosides of the neuron. Trends Neurosci 3:169–175.
8. Reisberg B.,Ferris S.H.,DeLeon M.J. and Crook T.: (1982) The global deterioration scale for assessment of primary degenerative dementia. Amer J. of Psychiatry 139:1136–1139.
9. McKhann G.,Drachman D.,Folstein M.,Katzman R.,Price D. and Stadlan E.M.: (1984) Clinical diagnosis of Alzheimer's disease: report of the NINCDS-ADRDA work group under the auspices of department of health and human services task force on Alzheimer's disease. Neurology 34:939–944.
10. Ledeen R.W. and Yu R.K.: (1982) Ganglioside structure isolation and analysis, in *Methods in Enzymology* (Ginsburg V.,ed.) 83:139–191, Academic Press Inc. N.Y.
11. Folch J.,Lees M. and Sloane-Stanley G.H.: (1957) A simple method for the isolation and purification of total lipids from animal tissue. J.Biol.Chem. 226:497–509.
12. Svennerholm L.: (1957) Quantitative estimation of sialic acids. Biochim. Biophys. Acta 24:604–611.
13. Miettinen T. and Takki-Luukkainen C.T.: (1959) Use of butylacetate on the determination of sialic acid. Acta Chem. Scand. 13:856–858.
14. Hess H.H. and Lewin E.: (1965) Microassay of biochemical structural components in nervous tissue-II. Methods for cerebrosides, proteolipid proteins and residue proteins. J. Neurochem. 12:205–211.
15. Suzuki K.: (1965) The pattern of mammalian brain gangliosides,III: regional and developmental differences. J.Neurochem. 12:969–979.
16. emory R.E.,Ala T.A., and Frey II W.H.: (1987) Ganglioside monoclonal antibody (A2B5) labels Alzheimer's neurofibrillary tangles. Neurology 37:768–772.
17. Cherayil G.D.: (1969) Estimation of gylcolipids in four selected lobes of human brain in neurological diseases. J.Neurochem. 16:913–920.
18. De Kosky S.T. and Bass N.H.: (1982) Aging, senile dementia, and the intralaminar microchemistry of cerebral cortex. Neurology 32:1227–1233.
19. Sorbi S.,Piacentini S., and Andemer L.: (1987) Intralaminar distribution of neurotransmitter-related enzymes in cerebral cortex of Alzheimer's disease. Geronotology 33:197–202.
20. Crino P.B.,Ullman M.D.,Vogt B.A.,Bird E.D., and Volicer L.: (1989) Brain gangliosides in dementia of the Alzheimer's type. Arch Neurol. 46:398–401.
21. Coyle J.T.,Price D.L. and DeLong M.R.: (1983) Alzheimer's disease: A disorder of cortical cholinergic innervation. Science 219:1184–1190.
22. Ng Ying Kin M.M.K.,Pan L.H.,Louvaris J.H.,Robitaille Y. and Nair N.P.V.: (1989) Brain gangliosides in Alzheimer's disease and in rats with forebrain lesions. Soc. NeuroSci 15,PartI,PP859 (Abs).
23. Ng Ying Kin N.M.K. and Chung D.: (1990) In vivo incorporation of N-acetyl-D-[U-C14] mannosamine into brain gangliosides of rats with quinolinic acid-induced lesions of the forebrain nucleus basalis magnocellularis. Mol. Chem Neuropathol. 13:233–241.
24. Ng Ying Kin N.M.K. and Chung D.: (1991) Incorporation of radiolabelled sugar precursors into brain gangliosides of hypocholinergic rats. (Abstr.) J. Neurochem. 57(suppl),S103D.
25. Whitehouse P.T.,Strubble R.G.,Clark A.W.,Coyle J.T., and Delong M.R.: (1982) Alzheimer's disease and senile dementia: loss of neurons in the basal forebrain. Science 215:1237–1239.

26. Mann D.M., Yates P.O., and Marcyniak B.: (1984) A comparison of changes in the nucleus basalis and locus coeruleus in Alzheimer's disease. J. Neurol. Neurosrg. Psychiatry 47:201–203.
27. Miyazaki K., Okamura N., Kishimoto Y., and Lee Y.C.: (1986) Determination of gangliosides as 2,4-dinitrophenylhydrazides by high performance liquid chromatography. Biochem J. 235:755–761.

COMMENTS

Dr. E.H.F. Wong: Dr. Ng, have you studied brain tissues from other neurodegenerative disease such as Parkinson Disease (PD) to determine whether the changes in brain gangliosides observed were specific for Alzheimer's disease (AD)?

Dr. Ng: No, we have not undertaken such studies. To my knowledge, none has been reported either, although in brain areas where neuronal cell death occurs, there would no doubt be decreased ganglioside contents. However, the point I wish to make here is that changes in the different molecular types of gangliosides are much more relevant than those in the total concentrations of these compounds. As I have mentioned in my talk, specific gangliosides are associated with individual neurotransmitters. Since neurotransmitter deficits in AD and PD, for example, are of different nature, changes in specific ganglioside types would be expected for the two disorders. We are now undertaking studies on animal models with lesions in various neurotransmitter systems so as to identify these gangliosides.

ACTION OF ORGANOPHOSPHATE ANTICHOLINESTERASES ON THE THREE CONFORMATIONAL STATES OF NICOTINIC RECEPTOR

Mugen Chi and Manji Sun

Institute of Pharmacology and Toxicology
Beijing 100850, China

ABSTRACT

Organophosphate and other ligands were examined for binding on the membrane-bond nicotinic receptor at three conformational states. Soman (pinacolyl methylphosphonofluoridate), sarin (isopropyl methylphosphonofluoridate, tabun (ethyl N-dimethylphosphoramidocyanidate) and phencyclidine did not show any effect on the binding of $[^{125}I]\alpha$-cobrotoxin to the nicotinic receptor. However, VX, O-ethyl-S-(2-diisopropylaminoethyl) methylphosphonothiolate, at concentrations higher than 10 umol/L exhibited profound inhibition on the equilibrium binding rates in a concentration-dependent manner. Agonist nicotine and antagonist d-tubocurarine also showed significant inhibitions.

INTRODUCTION

It is well known that acetylcholinesterase is the primary target of organophosphate poison in the intoxication. The inhibition of acetylcholinesterase in the nervous system induces a series of serious signs and symptoms. However, there are evidence showing that cholinesterase might not be the sole target in the body. Blockage of the neuromuscular transduction of the skeletal muscle of cat induced by diisopropyl flurophosphate (DFP) could be eliminated by washing, whereas acetylcholinesterase was still inhibited.[1] Exposure of satorius of frog to DFP caused attenuation of the current amplitude at the end-plate and the half-decay period which fairly recovered after washing while the activity of acetylcholinesterase remained inhibitory.[2] It implies that the nicotinic receptor might be another target of organophosphate anticholinesterases. Indeed, direct actions of organophosphates, such as soman and VX caused significant reductions in the initial rates of binding of $[^{125}I]\alpha$-bungarotoxin to the nicotinic receptors from Torpedo.[3] In this paper, evidences of direct actions of various ligands including organophosphates on the equilibrium binding rates of three conformational states of the nicotinic receptor[4] was shown by using radiolabelled ligand displacement tests.

MATERIALS AND METHODS

^{125}I-labelled α-Cobrotoxin

The postsynaptic neurotoxin was seperated from the venom of Naja naja atra.[5] Thirty μl of α-cobrotoxin (105 μg) was added to a glass test tube previously coated with 50 μg of iodo-gen and then 30 μl of [^{125}I] NaI (77.7 MBq) was added. After 30 min at 4°C with shaking, the reaction mixture was applied onto a Sephadex G–25 column previously equilibrated and then eluted with 50 mmol/L sodium phosphate buffer (pH 7.6)–0.02% NaN$_3$–140 mmol/L NaCl monitoring at 280 nm. The α-cobrotoxin fraction with a protein concentration of 0.05 mg/ml and a specific radioactivity of 3.589 TBq/mmol was collected and stored at −20°C in small quantities.

Membrane-Bound Nicotinic Receptor

Electric organs from torpediniforms Nacline timilei were minced and homogenized in a Waring blender with 0.02% NaN$_3$–50 mmol/L Tris-HCl buffer (pH 7.4). The supernatant after centrifugation at 5000 × g for 10 min was collected. the pellet was extracted once again. the pooled supernatant was centrifuged at 30,000 × g for one hour and the sediment fraction was resuspended in the same buffer. Then the membrane-bound nicotinic receptor preparation was stored at −20°C in divided samples. the specific binding capacity was 1170 + 115 pmol [^{125}I]α-toxin/mg protein with a protein concentration of 8.350 + 2.258 mg/ml and a cholinesterase activity of 16 μmol acetylcholine/30min/mg protein.

Chemicals

Carbamylcholine, d-tubocurarine and nicotine were perchased from Sigma,. Acetylcholine bromide from Beijing Chemical factory. Organophsophates of more than 95% purity, phencyclidine and iodo-gen were synthesized in our laboratory. [^{125}I]NaI with a specific radioactivity of 3.7 nBq/ml was a product of the Institute of Atomic Energy, Academia Sinica. All the other reagents were chemical pure.

Glass fiber micropore membrane type 49 was purchased from shanghai Yuguang Company with a non-specific absorption of 8% for [^{125}I]α-cobrotoxin. The autogamma counter is a product of No. 262 Factory in Xian, china.

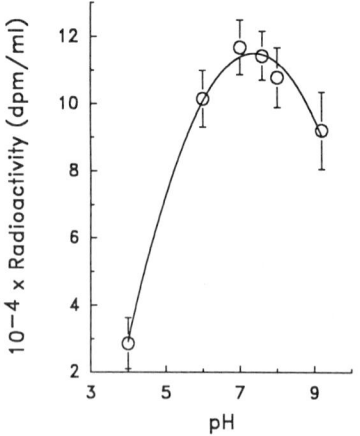

Figure 1. Effects of pH on the binding of [^{125}I]a -cobrotoxin to the membrane-bound nicotinic receptor.

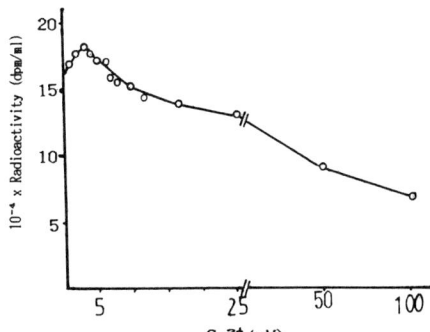

Figure 2. Influence of calcium ion on the binding of [^{125}I]α-cobrotoxin to the membrane-bound nicotinic receptor. Deviation coefficient, 1–9%

Option for Conditions of Ligand-Receptor Binding

Aliquots of 25 μl (1 μg protein) membrane preparation were added to 80 μl (27 ng protein) of [^{125}I]α-cobrotoxin. Torpedo saline (250 mmol/L, NaCl-5 mmol/L, KCl-4 mmol/L, CaCl$_2$-2 mmol/L, MgCl$_2$-5mmol/L phosphate buffer, pH 7.0) was then added to make a total volume of 200 μl. The mixtures were incubated at 30°C for 30 min, and then filtered through glass-fiber membranes. the filter membranes were washed with 3 ml of Torpedo saline for ten times., dried in an oven and counted for radioactivity. In the non-specific control one hundred-fold of unlabelled α-cobrotoxin was added. The optimum pH of the reaction was 7.0 (Fig. 1). the highest binding was shown in presence of Ca^{++} at 2–4 mmol/L concentrations (Fig.2). The time course showed that the reaction approached to a plateau during 15 min (Fig.3). K_D and B_{max} were estimated to be 13.3 nmol/L and 1.2 nmol/mg protein respectively from the scatchard plot (Fig.4) with a Hill coefficient of 1 (Fig. 5).

Binding Assay of Organophosphates to the Nicotinic Receptor at the Resting State

Aliquots (80 μl) of [^{125}I]α-cobrotoxin were mixed with 20 μl of organophosphate in various concentrations and 75 μl of topedo saline. Then 25 μl of membrane-bound nicotinic receptor preparation was added. The reaction was carried out at 30°C for 30 min. Fifty μl of the reaction solution was filtered, washed and counted as were stated above. The binding rates were calculated as the percentage of the control without the adition of organophosphate.

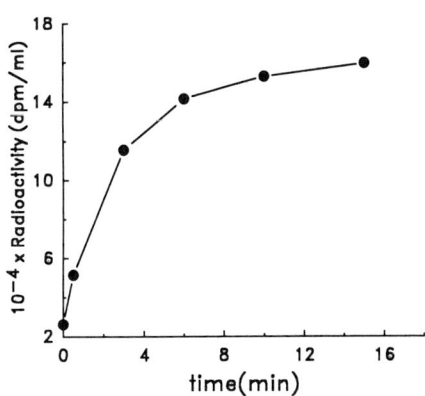

Figure 3. Time course of the binding reaction.

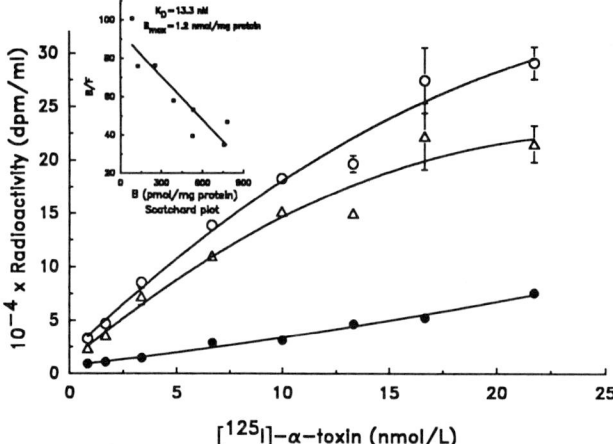

Figure 4. The saturation curve of binding of [^{125}I] α-cobrotoxin to the membrane-bound nicotinic receptor. O–O, total binding; △–△, specific binding; ●–●, non-specific binding.

Non-specific binding was estimated in the presence of 108-fold of unlabelled α-cobrotoxin instead of the organophosphate.

Binding Assay of Organophosphate to the Nicotinic Receptor at the Putative Activated State

Aliquots (20 µl, final concentration 100 µmol/L) of carbamylcholine, organophosphate (20 µl) in various concentrations, [^{125}I]α-cobrotoxin (80 µl) and 55 µl of topedo saline were added in order. Then 25 µl of membrane-bound receptor preparation was added to make a volume of 200 µl. The following procedures were as above mentioned. The binding rates were expressed as the percentages of the control without organophosphate. The non-specific binding was set and substracted.

Binding Assay of Organophosphate to the Nicotinic Receptor at the Putative Desensitized State

Alquots (20 µl) of carbamylcholine and 25 µl membrane-bound nicotinic receptor preparation were incubated at 30°C for 30 min. The, 20 µl of organophosphate in various

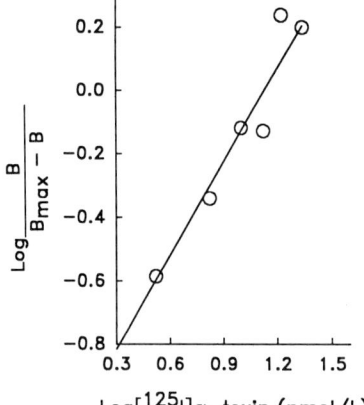

Figure 5. Hill plot Log[^{125}I] α-toxin (nmol/L).

concentrations, 80 µl of [^{125}I]α-cobroyoxin and 55 µl of Topedo saline were sequentially added. The following processes were identical as mentioned above. The binding percentage in terms of the control without organophosphate were calculated. the non-specific binding was established and corrected.

RESULTS AND DISCUSSION

Actions of Nicotinie, d-Tubocurarine and Phencyclidine on the Three Conformation States of Nicotinic Receptor

The classical agonist nicotine and antagonist d-tubocurarine caused significant inhibitions on the binding of [^{125}I]α-cobortoxin to the nicotinic binding site in a dose-dependent manner (Table 1, 2). However, the phencyclidine being not expected to direct to the recognition site on the nicotinic receptor molecule showed no influence on the binding rate (Table 3).

Effects of Organophosphates on the Three Conformational States of Nicotinic Receptor

Soman, sarin and tabun have been examined for their influences on the binding of [^{125}I]α-cobrotoxin to the three states of membrane-bound receptor preparation. As shown in Table 4, 5 and 6. None of them has any effect on the binding rate.

VX at concentration higher than 10 µmol/L exhibited profound inhibitions on the binding rates in a concentration-dependent manner. One mmol/L VX showed an inhibition about 30%. However, there was no significant difference among those three states of nicotinic receptor (Table 7). The chemical structure of G-type organophosphates are different from that of the

Table 1. Effect of Nicotine on the Binding of [^{125}I] α -Cobrotoxin to the Membrane-Bound Nicotinic Receptor

	Concentration of nicotine (µ mol/L)	Total binding (cpm)	Specific binding# (cpm)	Binding rate (%)
Resting state	0	5511 ± 200	3001 ± 249	100
	1000	3462 ± 150	952 ± 194**	31 ± 3.5
	100	4003 ± 114	1493 ± 163**	49 ± 1.5
	10	4551 ± 128	2041 ± 170**	67 ± 1.5
	1	4659 ± 95	2149 ± 144**	71 ± 3.0
	0.1	4898 ± 157	2388 ± 190*	79 ± 5.9
Activated state	0	4643 ± 211	1923 ± 272	100
	1000	3532 ± 183	813 ± 216**	41 ± 6.2
	100	3906 ± 106	1187 ± 90*	62 ± 4.8
	10	3972 ± 96	1253 ± 74*	65 ± 5.9
	1	4017 ± 38	1297 ± 73*	68 ± 7.2
	0.1	4358 ± 76	1638 ± 181	85 ± 8.2
Desensitized state	0	4501 ± 301	2121 ± 299	100
	1000	3035 ± 42	655 ± 315**	30 ± 13.6
	100	3073 ± 24	692 ± 308**	31 ± 13.6
	10	3350 ± 110	969 ± 339*	44 ± 12.5
	1	3501 ± 45	1120 ± 346*	52 ± 13.9
	0.1	4252 ± 496	1871 ± 449	87 ± 10.3

# Substraction of the non-specific binding (for resting state, 2510 ± 50; activated state, 2719 ± 106; desensitized state, 2381 ± 333) from the total binding.
n=3, x̄ ± SD, *p<0.05, **p<0.01.

Table 2. Effect of d-Tubocurarine on the Binding of $[^{125}I]$ α-Cobrotoxin to the Membrane-Bound Nicotinic Receptor

	Concentration of d-tubocurarine (μmol/L)	Total binding (cpm)	Specific binding# (cpm)	Binding rate (%)
Resting state	0	6101 ± 277	3352 ± 126	100
	1000	3434 ± 105	686 ± 207**	20 ± 5.7
	100	4027 ± 126	1278 ± 409**	37 ± 11.3
	10	4056 ± 70	1307 ± 363**	38 ± 10.6
	1	5093 ± 126	2344 ± 267**	70 ± 8.3
	0.1	5543 ± 77	2794 ± 302*	83 ± 7.3
Activated state	0	4814 ± 120	2126 ± 102	100
	1000	3704 ± 9	1017 ± 49**	47 ± 3.6
	100	3771 ± 137	1083 ± 148**	50 ± 5.4
	10	3916 ± 183	1228 ± 202**	57 ± 8.3
	1	3967 ± 241	1279 ± 222**	59 ± 7.4
	0.1	4343 ± 317	1655 ± 295	77 ± 10.0
Desensitized state	0	4521 ± 196	2014 ± 235	100
	1000	2678 ± 113	172 ± 82**	8 ± 3.5
	100	2972 ± 117	465 ± 52**	23 ± 5.5
	10	2999 ± 36	492 ± 92**	24 ± 4.3
	1	4373 ± 139	1866 ± 171	92 ± 2.4
	0.1	4670 ± 411	2164 ± 479	106 ± 14.1

\# Substraction of the non-specific binding (for resting state, 2748 ± 300; activated state, 2687 ± 40; desensitized state, 2506 ± 71) from the total binding.
n=3, x ± SD, *p<0.05, **p<0.01.

Table 3. Effect of Phencyclidine on the Binding of $[^{125}I]$ α-Cobrotoxin to the Membrane-Bound Nicotinic Receptor

	Concentration of phencyclidine (μmol/L)	Total binding (cpm)	Specific binding# (cpm)	Binding rate (%)
Resting state	0	4748 ± 283	2485 ± 283	100
	1000	4553 ± 63	2291 ± 99	93 ± 12.2
	100	4617 ± 179	2354 ± 146	95 ± 15.8
	10	5106 ± 96	2844 ± 67	115 ± 12.2
	1	4785 ± 146	2523 ± 137	102 ± 6.6
	0.1	4861 ± 66	2598 ± 31	105 ± 13.6
Activated state	0	4220 ± 100	1724 ± 171	100
	1000	4133 ± 96	1637 ± 160	95 ± 5.7
	100	4305 ± 115	1809 ± 185	105 ± 5.0
	10	3829 ± 63	1332 ± 112	77 ± 5.7
	1	4295 ± 190	1798 ± 238	104 ± 11.6
	0.1	4318 ± 67	1822 ± 131	105 ± 6.1
Desensitized state	0	3447 ± 170	1131 ± 177	100
	1000	3492 ± 71	1176 ± 98	105 ± 17.8
	100	3530 ± 291	1214 ± 282	107 ± 13.8
	10	3759 ± 183	1443 ± 211	129 ± 26.7
	1	3848 ± 45	1532 ± 67	137 ± 16.3
	0.1	3693 ± 161	1276 ± 119	115 ± 28.3

\# Substraction of the non-specific binding (for resting state, 2262 ± 36; activated state, 2496 ± 74; desensitized state, 2316 ± 30) from the total binding.
n=3, x ± SD.

Table 4. Effect of Soman on the Binding of [^{125}I] α-Cobrotoxin to the Membrane-Bound Nicotinic Receptor

	Concentration of soman (μ mol/L)	Total binding (cpm)	Specific binding# (cpm)	Binding rate (%)
Resting state	0	6151 ± 259	5507 ± 193	100
	1000	6229 ± 849	5586 ± 793	101 ± 11.2
	100	6330 ± 858	5687 ± 807	103 ± 13.1
	10	6185 ± 610	5541 ± 543	100 ± 6.5
	1	5634 ± 411	4990 ± 360	91 ± 4.2
Activated state	0	6438 ± 488	5829 ± 475	100
	1000	6746 ± 27	6137 ± 18	105 ± 8.5
	100	6535 ± 602	5927 ± 589	101 ± 1.7
	10	6853 ± 617	6244 ± 607	106 ± 2.1
	1	6391 ± 287	5782 ± 309	99 ± 8.4
Desensitized state	0	6434 ± 50	5763 ± 54	100
	1000	6942 ± 687	6272 ± 648	108 ± 11.6
	100	6444 ± 403	5773 ± 400	100 ± 6.1
	10	6513 ± 542	5843 ± 502	101 ± 9.1
	1	6050 ± 306	5379 ± 292	93 ± 4.5

# Substraction of the non-specific binding (for resting state, 643 ± 67; activated state, 608 ± 29; desensitized state, 670 ± 39) from the total binding.
n=3, x ± SD.

Table 5. Effect of Sarin on the Binding of [^{125}I] α-Cobrotoxin to the Membrane-Bound Nicotinic Receptor

	Concentration of sarin (μ mol/L)	Total binding (cpm)	Specific binding# (cpm)	Binding rate (%)
Resting state	0	4296 ± 202	2308 ± 127	100
	1000	4347 ± 146	2358 ± 122	102 ± 10.4
	100	4416 ± 427	2428 ± 261	105 ± 15.3
	10	4367 ± 160	2378 ± 187	103 ± 13.0
	1	4320 ± 177	2332 ± 22	101 ± 6.7
	0.1	4285 ± 188	2297 ± 144	99 ± 11.7
Activated state	0	4031 ± 78	1537 ± 98	100
	1000	4120 ± 160	1626 ± 160	105 ± 9.8
	100	3950 ± 331	1456 ± 323	95 ± 22.4
	10	4326 ± 150	1832 ± 137	119 ± 15.6
	1	4184 ± 182	1689 ± 193	109 ± 8.0
	0.1	4040 ± 128	1545 ± 107	100 ± 12.1
Desensitized state	0	3874 ± 174	1741 ± 200	100
	1000	3805 ± 74	1672 ± 106	96 ± 7.4
	100	3771 ± 58	1638 ± 32	94 ± 10.9
	10	3905 ± 242	1772 ± 265	102 ± 17.1
	1	3670 ± 174	1495 ± 148	86 ± 15.4
	0.1	3915 ± 140	1782 ± 169	102 ± 2.5

# Substraction of the non-specific binding (for resting state, 1988 ± 175; activated state, 2494 ± 21; desensitized state, 2133 ± 33) from the total binding.
n=3, x ± SD.

Table 6. Effect of Tabun on the Binding of [^{125}I] α-Cobrotoxin to the Membrane-Bound Nicotinic Receptor

	Concentration of tabun (μmol/L)	Total binding (cpm)	Specific binding# (cpm)	Binding rate (%)
Resting state	0	4271 ± 155	1752 ± 156	100
	1000	4227 ± 355	1708 ± 350	96 ± 13.1
	100	4298 ± 342	1779 ± 335	101 ± 12.6
	10	4393 ± 568	1875 ± 569	105 ± 24.2
	1	4190 ± 62	1672 ± 54	95 ± 6.5
	0.1	4241 ± 153	1723 ± 148	98 ± 3.6
Activated state	0	4357 ± 144	1812 ± 198	100
	1000	3638 ± 155	1089 ± 256	60 ± 7.5
	100	4266 ± 286	1717 ± 197	96 ± 21.0
	10	4324 ± 109	1775 ± 125	98 ± 7.0
	1	4217 ± 68	1668 ± 143	92 ± 6.7
	0.1	4289 ± 97	1740 ± 118	96 ± 6.4
Desensitized state	0	4190 ± 133	1831 ± 248	100
	1000	4053 ± 257	1693 ± 273	93 ± 17.0
	100	3750 ± 41	1390 ± 81	76 ± 6.3
	10	4262 ± 88	1902 ± 203	103 ± 3.1
	1	4109 ± 97	1749 ± 135	97 ± 11.3
	0.1	4215 ± 219	1855 ± 99	102 ± 20.0

# Substraction of the non-specific binding (for resting state, 2518 ± 15; activated state, 2549 ± 100; desensitized state, 2359 ± 120) from the total binding.
n=3, x ± SD.

V-type organophosphates in the leaving group of the molecule. The recognition site of the nicotinic receptor composed of esteristic, anionic and hydrophobic structures. It is possible that the interaction between thiocholine group of VX and the hydrophobic domain of the receptor underlies the molecular basis of the inhibition which is absent in cases of the G-type organophosphates.

Table 7. Effect of VX on the Binding of [^{125}I] α-Cobrotoxin to the Membrane-Bound Nicotinic Receptor

	Concentration of VX (μmol/L)	Total binding (cpm)	Specific binding# (cpm)	Binding rate (%)
Resting state	0	6824 ± 159	5658 ± 187	100
	1000	5003 ± 308	3838 ± 343**	67 ± 6.4
	100	6080 ± 446	4915 ± 430	86 ± 6.7
	10	6593 ± 492	5427 ± 491	96 ± 6.6
Activated state	0	6758 ± 163	5586 ± 184	100
	1000	5082 ± 190	3910 ± 185**	70 ± 1.9
	100	6099 ± 151	4926 ± 194*	88 ± 3.7
	10	6307 ± 160	5134 ± 178*	92 ± 0.3
Desensitized state	0	6952 ± 193	5747 ± 186	100
	1000	5313 ± 170	4107 ± 141**	71 ± 2.1
	100	6635 ± 302	5429 ± 268	94 ± 5.5
	10	6904 ± 260	5698 ± 268	99 ± 2.2

# Substraction of the non-specific binding (for resting state, 1165 ± 47; activated state, 1172 ± 44; desensitized state, 1205 ± 33) from the total binding.
n=3, x ± SD, *p<0.05, **p<0.01.

The two types of organosphosphate are similar in the mechanism of actions on the cholinesterases. Their modes of metabolism and interactions with the icotinic receptor are entirely different.

REFERENCES

1. McNamara, BP et al: (1954) Studies on the mechanism of action of DFP and TEPP. J. Pharmacol. Exp. Ther. 110:232-334.
2. Kuba, K and Albuquerque, EX: (1973) Diisopropy fluorophosphate: suppression of ionic conductance of the cholinergic receptor. Science 181:853-856.
3. Bakry, NMS et al: (1988) Direct actions of organophosphate anticholinesterases on nicotinic and muscarinic acetylcholine receptor by diisopropylfluophophate. J. Biochem. Toxicol. 3:235-259.
4. Eldefrawi, ME et al: (1988) Desensitization of the nicotinic acetylcholine receptor by diisopropylfluophosphate. J. Biochem. Toxicol. 3:21-32.
5. Xu, HP et al: (1984) Purification of nicotinic cholinergic receptor protein from the electric organ of Torpediniforms Nacline timilei. Acta Biochem. Biophys. Sinica 16:50-55.

PURIFICATION AND CHARACTERIZATION OF A NOVEL NEUROTOXIN-KAPPA BUNGAROTOXIN

Hu Ben-rong and Zhang Ze

Sun Yat-sen University of Medical Sciences
Guangzhou, China

Alpha-neurotoxins such as alpha-bungarotoxin and cobrotoxin are now used frequently to characterize the nicotinic receptors in vertebrate skeletal muscle and in electric tissue of some kinds of electric fish. Although they also bind to a variety of neuronal preparations which contain nicotinic receptors, but in most cases, this binding fails to block the nicotinic transmission in neuro-muscular junction. So the physiological significance of alpha-neurotoxins binding sites in neural tissue remains unclear.

Now, we report the works for purifying the kappa Bungarotoxin, a noval long-chain neurotoxin which inhibits neuronal acetylcholine receptor (AChR) but does not block neuromuscular AChR, from three sources of Bungarus multicinctus venoms. Six active fractions (Frac.4-C, Frac.4-D_2 from Guangdong bungarotoxin venom, Frac.4-B_2 from Jiangxi bungarotoxin venom, Frac.5-C_2, Frac. 4-B_3 and Frac. 6-B from Hunan bungarotoxin venom) have been purified from the crude venoms by cation exchange chromatography. They can produce a reversible blockade on nicotinic transmission in chick ciliary ganglion at the concentration of 1–2 µg/ml or lower. On SDS-gradient PAGE, they run as a single band with the molecular weight of 7000–7500 daltons except the FRAC.4-B_3 and Frac.6-B of Hunan bungarotoxin multicinctus venom. the name of kappa-bungarotoxin (kx-Bgt) has been given to the Frac.5-C_2 of Hunan bungarotoxin multicinctus venom, because the fraction is the most potent one at blocking ciliary ganglion transmission, and has a very similar biochemical properties comparison to kx-Bgt or toxin F which have been reported. It consists of 66 amino acids and has a isoelectric point of 8.70 as well as a molecular weight of 6978 daltons. It blocks the chicken ciliary ganglion at 0.25 µg/ml within 20–30 min incubation. Otherwise, it can block the nicotinic transmission through guinea pig hypogastric ganglion at 0.5–1.0 µg/ml. In SD rats, kx-Bgt can induce a reversible reducing of blood pressure at very low concentration (80 µg/ml * 0.1 ml/100g body weight). The blockade of kx-Bgt on chick ciliary ganglion is reversible, but the recovery is slow and difficult (overall several hours). However, pre-exposure of the ganglion to d-TC, nicotine or Co was able to protect against subsequent addition of kx-Bgt, suggesting that kx-Bgt is acting via the same mechanism as these ganglionic antagonists. Our subsequent results show that the action of kx-Bgt is selective to neuronal nicotinic receptors, but not to neuromuscular AChR. The homogenus of kx-Bgt have been proven by electrophoresis, gel-filtration and high pressure liquid chromatography (HPLC).

Binding of [^{125}I] kx-Bgt to rat brain suggests two binding sites: one which can be recognised by alpha-Bgt and another which is identified by nicotine and d-TC, but not by alpha-Bgt. This binding site is about 3.47 fmole/mg tissue. The kd value for kx-Bgt to bind is about 2.53 nM, revealing very high affinity.

These results suggest that kx-Bgt may be a new kappa-neurotoxin. It selectively blocks the ganglionic transmission by acting on neuronal nicotinic receptor, and it may act on some kinds brain nicotinic receptors. We are confident that kx-Bgt can be used as a selective probe for neuronal, especially for mammalian brain nicotinic receptors.

DEVELOPMENT OF ANTIDEPRESSANT DRUGS

Fluoxetine (Prozac) and Other Selective Serotonin Uptake Inhibitors

David T. Wong and Frank P. Bymaster

Lilly Research Laboratories
Eli Lilly and Company
Lilly Corporate Center
Indianapolis, Indiana 46285

INTRODUCTION

Active uptake processes have been described for the monoaminergic neurotransmitters including serotonin (5-hydroxytryptamine, 5-HT), norepinephrine (NE) and dopamine (DA) in nervous tissues (Whitby, Axelrod, and Weil-Malherbe, 1961; Burgen and Iverson, 1965; Iverson, 1971). The uptake of 5-HT and NE in brain tissues has been the target sites in our search of selective inhibitors, which may have therapeutic potential for treatment of depressive disorders (Wong et al., 1974; 1975a; 1993a). Fluoxetine [Prozac], a member of the substituted phenoxyphenylpropylamines, was the first selective 5-HT uptake inhibitor to appear in the scientific literature (Wong et al., 1974) and has served as a useful tool to establish the physiological role of 5-HT neurons and a reference of pharmacological responses indicative of an enhanced transmission of 5-HT neurons (Fuller and Wong, 1977; Wong and Fuller, 1987; Wong and Murphy, 1989; Fuller and Wong, 1990). After its introduction in 1988, fluoxetine (Prozac) has become a major antidepressant drug in the United States and in many countries (Beasley et al., 1990; Boyer and Feighner, 1991a). In the U.S., two other selective 5-HT uptake inhibitors, sertraline and paroxetine, have been introduced as antidepressant drugs. In this article, we present some of the background which attracted our attention, and some studies on the chemical series of phenoxyphenylpropylamines and the enantiomers of fluoxetine and its major metabolite norfluoxetine as inhibitors of 5-HT and NE uptake in vitro. We also present the consequences of 5-HT reuptake inhibition in vivo, and the contrast between the classical tricyclic antidepressant drugs and the newly developed selective 5-HT reuptake inhibitors in terms of their interaction with receptors of neurotransmitters.

MECHANISMS OF TRICYCLIC ANTIDEPRESSANT DRUGS

After imipramine was realized to have antidepressive property (Kuhn, 1957), imipramine and its congeners have become the major therapeutic agents. The three fused-ring system in chemical structure and their therapeutic effectiveness are generically described as the tricyclic antidepressant drugs (Table 1). Studies in the early 1960s showed that imipramine potently inhibited neuronal NE uptake in various tissues including the heart and brain (Dengler and Titus, 1961; Hertting et al., 1961; Glowinski and Axelrod, 1964; Iversen, 1971). The desmethylated metabolite of imipramine or desipramine was isolated from rat brain (Gillette et al., 1961) and proved to be a potent inhibitor of NE uptake in brain (Glowinski and Axelrod, 1964; Carlsson et al., 1966). Antagonism of the behavioral effects of reserpine and tetrabenazine by the tricyclic antidepressant drugs might be associated with the inhibition of NE uptake in brain (Sulser et al., 1962; Ross and Renyi, 1967) and these behavioral tests are considered predictive of antidepressive activity (Jesberger and Richardson, 1985; Bourin, 1990). These findings are consistent with the explanation that uptake is a major mechanism for terminating the physiological actions of NE after its release from presynaptic nerve terminals, and blockade of reuptake increases synaptic availability of the neurotransmitter thereby enhancing receptor activation.

In the late 1960s, imipramine was found to block the reserpine-resistant accumulation of 5-HT in brain (Corrodi and Fuxe, 1968; Carlsson, Fuxe and Ungerstedt, 1968). Imipramine and other tertiary amine containing congeners are potent inhibitors of 5-HT uptake and NE uptake in brain, while the major metabolite of imipramine, desipramine and its secondary amine congeners mainly inhibit NE uptake (Carlsson et al., 1969 a, b; Ross and Renyi, 1969). Although the only difference of a single N-methyl group, imipramine is an order of magnitude more potent than desipramine as an inhibitor of 5-HT uptake in brain slices (Ross and Renyi, 1969; Carlsson 1970) and in hypo-thalamic slices (Shaskan and Snyder, 1970). Moreover, clomipramine, which contains a chloro-group in the fused-tricyclic rings of imipramine, showed a further increase in potency to inhibit 5-HT uptake in brain slices (Carlsson, 1970) and in hypothalamic slices (Shaskan and Snyder, 1970). Using synaptosomal preparations of

Table 1. Tricyclic Antidepressant Drugs Inhibition of 5-HT and NE Uptake in Cerebral Slices

Drug	Inhibition of Uptake	
	a. 5HT	b. NE
Desipramine $CH_2CH_2CH_2NHCH_3$	5000	1
Imipramine $CH_2CH_2CH_2N(CH_3)_2$	500	20
Clomipramine $CH_2CH_2CH_2N(CH_3)_2$	100	24

Values are expressed as a. IC_{50}, nM (Carlsson, 1970) and b. K_i, nM (Hyttel, 1982).

rat brain (Snyder and Coyle, 1969; Wong et al., 1973), these differences in potency for inhibiting 5-HT uptake by imipramine, clomipramine and nortriptyline, another secondary amine tricyclic antidepressant drug, were confirmed (Wong et al., 1974). The tricyclic antidepressant drugs are relatively ineffective inhibitors of DA uptake (Carlsson et al., 1966; Ross and Renyi, 1967; Wong et al., 1974). Thus, inhibition of both NE and 5-HT uptake in their respective neurons is linked to the antidepressive mechanism of the tricyclic drugs (Gillette et al., 1961; Glowinski and Axelrod, 1964; Carlsson et al., 1968; Ross and Renyi, 1969; Weil-Malherbe and Szara, 1971). Furthermore, it has been suggested that inhibition of 5-HT uptake may be responsible for the mood-elevating effect of tricyclic antidepressant drugs (Carlsson, 1969a; 1969b; Lapin and Oxenkrug, 1969).

Although imipramine and clomipramine in vitro preferentially inhibit 5-HT uptake and amitriptyline equipotently inhibits both 5-HT and NE uptake in vitro, these tertiary amine containing tricyclic drugs are metabolically N-demethylated in vivo to the corresponding secondary amines, desipramine, chlorodesipramine and nortriptyline. Consequently, NE uptake inhibition in rat brain was mainly observed when the tertiary amine drugs were administered in vivo (Ross and Renyi, 1967; Wong and Bymaster, 1976). Therefore, the extent of inhibition of 5-HT and NE uptake by the tertiary amines depends on their relative degrees of metabolic N-demethylation. Human blood levels of imipramine, clomipramine and amitriptyline and their demethylated metabolite levels vary widely (Muscettola et al., 1978; Narasimhachari and Landa, 1991). The tertiary amine containing tricyclic drugs, therefore, may not consistently provide the 5-HT concentrations needed to adequately activate the serotonergic pathway in antidepressive therapy.

INVOLVEMENT OF SEROTONIN NEURON IN DEPRESSION

Beginning in the 1960s, medical evidence provides a link of the neurotransmitter 5-HT to depression (Weil-Malherbe and Szara, 1971; Coppen et al., 1972; Asberg and Wagner, 1986). By this time, basic research on 5-HT satisfied the neurotransmitter role of 5-HT (Page, 1969) and the synaptic events together with the multiple receptors of 5-HT are illustrated in Figure 1. Concentrations of 5-HT and its metabolite 5-hydroxy-indole acetic acid (5-HIAA) were found to be lower in hindbrains of depressed victims of suicide than in hindbrains killed in sudden deaths (Shaw et al., 1967; Lloyd et al., 1974), coronary occlusion (Broune et al., 1968) or schizophrenic patients (Korpi et al., 1968). The concentration of 5-HIAA in cerebrospinal fluid was lower in depressed patients (Ashcroft et al., 1966; Dencker et al., 1966) particularly among those who made attempts of suicides (Asberg and Wagner, 1986; Asberg et al., 1986). Moreover, treatment with amino acid precursors of 5-HT, tryptophan (Coppen et al., 1963; 1967; Pare, 1963; Hertz and Sulman, 1968) and 5-hydroxytryptophan (5-HTP) (Kline and Sacks, 1963; Sano, 1977), and in combination with an inhibitor of monoamine oxidase (Coppen et al., 1963; Pare, 1963, Kline and Sacks, 1963) showed antidepressive effects. Selective 5-HT reuptake inhibitors, which can retain the selectivity in vivo, would be highly desirable in order to ascertain the importance of 5-HT neurons in treatment of depression (Asberg et al., 1973; Wong et al., 1974).

SYNAPTOSOMAL 5-HT UPTAKE, A CONVENIENT TARGET FOR DRUG DEVELOPMENT

A high affinity uptake of 5-HT (Figure 1) was demonstrated in brain slices (Ross and Renyi, 1969; Shaskan and Snyder, 1970), and showed kinetics distinctly different from those

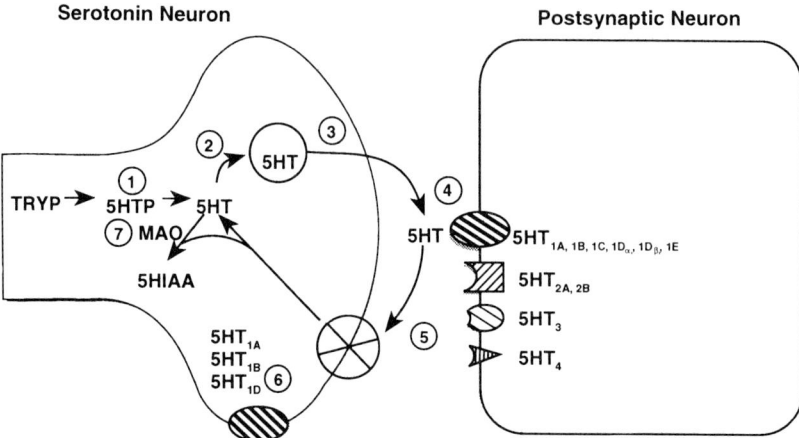

Figure 1. Molecular mechanism of serotonergic synapses. 1. Synthetic pathway is from tryptophan, TRYP to 5 hydroxytryptophan, 5HTP, (catalyzed by tryptophan hydroxylase) and to 5 hydroxytryptamine, 5HT or serotonin (catalyzed by 5-hydroxytryptophan decarboxylase). 2. Storage. 3. Release. 4. Activation of postsynaptic receptors. 5. Reuptake that terminates 5HT action. 6. Possible activation of presynaptic receptors. 7. Degradation by monoamine oxidase, MAO.

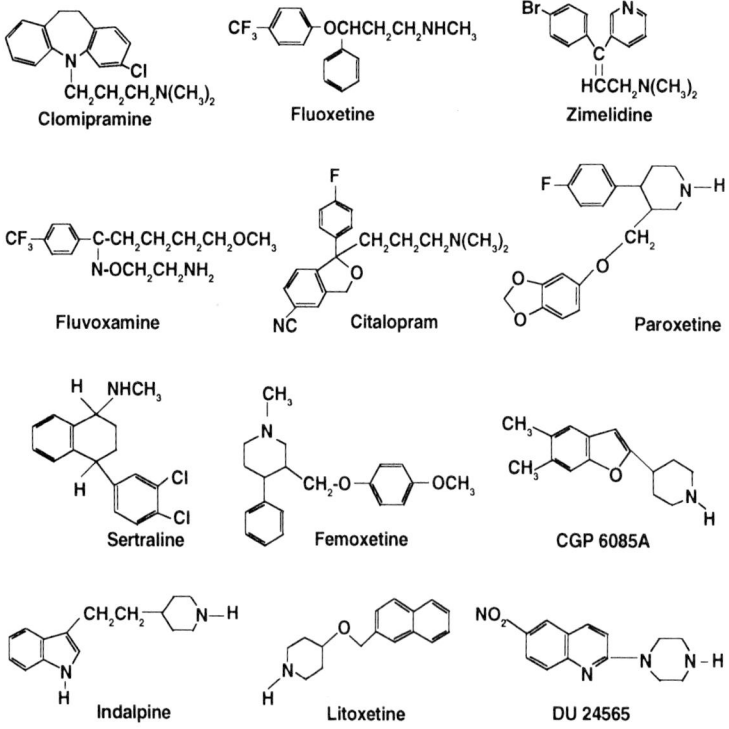

Figure 2. Chemical structures of selective 5-HT uptake inhibitors.

for catecholamines (Shaskan and Snyder 1970; Snyder and Coyle, 1969). Distribution of 5-HT uptake varies in brain regions (Shaskan and Snyder, 1970; Wong et al., 1975a). Endogenous 5-HT and accumulation of exogenous 5-HT are concentrated in the nerve-ending fraction (synaptosomes) of brain homogenates (Kuhar et al., 1971). The unique kinetics of high affinity 5-HT uptake were confirmed with the synaptosomal preparations of rat brain by us in late 1971 at the Lilly Research Laboratories and published in the literature later (Wong et al., 1973). Uptake of 5-HT was significantly reduced after a single administration of the neurotoxin p-chloroamphetamine (p-CA) in vivo (Wong et al., 1973) or by electrolytic lesions in midbrain raphe nuclei (Kuhar et al., 1972), where 5-HT cell bodies localize (Dahlstrom and Fuxe, 1964). Thus, the high affinity uptake of 5-HT in synaptosomes is associated with the 5-HT neurons. The tertiary amine containing tricyclic antidepressant drugs were more potent than the secondary amine containing tricyclic antidepressant drugs as inhibitors of the high affinity 5-HT uptake in synaptosomal preparations of rat forebrain (Kuhar et al., 1972) or whole brain (Wong et al., 1974) as previously demonstrated in brain slices (Ross and Renyi, 1969; Carlsson, 1970) and in vivo (Carlsson et al., 1968; 1969a, b; and Ross and Renyi, 1969). Thus, the specific 5-HT uptake is considered a mechanism for terminating the physiological function of 5-HT in the synapse. Blockade of this reuptake process would increase synaptic availability of 5-HT, which in turn activates the postsynaptic receptors located on the adjacent neuron (Figure 1). In our search for selective 5-HT uptake inhibitors, we chose to investigate the non-tricyclic chemical entities and used the high affinity 5-HT uptake in synaptosomal preparations as a convenient target (Wong, et al., 1974; 1988; 1990b; 1993a).

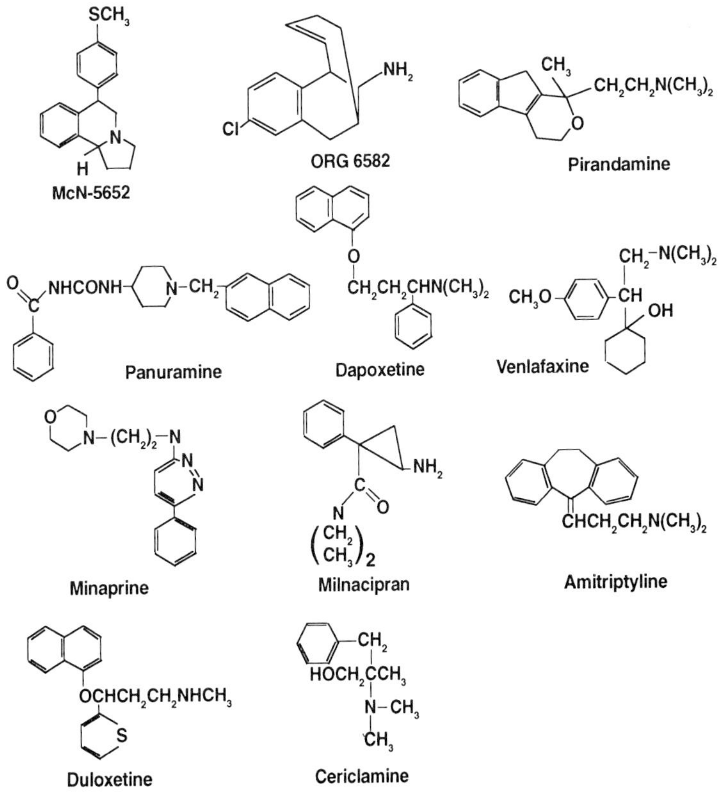

Figure 2. Continued.

Fluoxetine, or previously known as Lilly 110140 (p-trifluoro-phenoxy-phenyl-N-methyl-propylamine, Figure 2), was first reported to have a biochemical and pharmacological property of a selective 5-HT uptake inhibitor in vitro and in vivo (Wong et al., 1974; 1975a; Fuller et al., 1974; 1975; Fuller and Wong, 1990). Among a series of phenoxyphenylpropylamines tested for the reversal of reserpine-induced responses, fluoxetine was found relatively ineffective (Slater et al., 1979).

STRUCTURE-ACTIVITY RELATIONSHIP OF PHENOXYPHENYLPROPYL-AMINES

In micromolar concentrations para-trifluoromethyl-phenoxy-N-methyl-alkylamines, which have a chain length of either 2, 3, or 4 carbon atoms, half the 5-HT uptake in synaptosomes of rat brain and at 10 μM weakly inhibited NE uptake. The alkylamine-side chain having three carbon atoms in length appeared to inhibit 5-HT uptake more effectively. The presence of a propylamine side chain is shared with the chemical structures of the tricyclic drugs, clomipramine, imipramine and amitriptyline.

Adding a phenyl ring to the same carbon atom connected to the phenoxy ring of the p-trifluoromethylphenoxy-N-methyl-propylamine enhanced the potency almost 41-fold to inhibit 5-HT uptake with a K_i of 17 nM (Table 2). Moving the CF_3 group from the para-position of the phenoxy ring to its meta and ortho position drastically reduced the potency to inhibit 5-HT uptake and the relative selectivity for the 5-HT uptake site over that of the NE uptake site. The uptake site for 5-HT prefers the larger substituent of CF_3 over other substituents (H, F, Cl, CH_3 and para-OCH_3) in the para-position of the phenoxy ring. In contrast to the tertiary amine containing tricyclic drugs, the primary amine compound (norfluoxetine) was as potent and selective as fluoxetine. Thus, the position and size of the substituent in the phenoxy ring

Table 2. Inhibition of 5-HT and NE Uptake in Synaptosomal Preparations of Rat Whole Brain by Derivatives of Phenoxyphenylpropylamines.

Substitution				Inhibition of uptake	
R_1	R_2	R_3	R_4	5-HT (K_i, μM)	NE/5-HT (K_i ratio)
H	CH_3	H	H	0.10	2
H	CH_3	o-CF_3	H	1.49	3
H	CH_3	m-CF_3	H	0.17	8
H (fluoxetine)	CH_3	p-CF_3	H	0.02	159
H (norfluoxetine)	CH_3	p-CF_3	H	0.02	128
CH_3	CH_3	p-CF_3	H	0.05	53
CH_3	CH_3	p-CF_3	p-CF_3	0.85	>3
H	CH_3	H	p-CF_3	0.62	3
H	CH_3	p-F	H	0.64	2
H	CH_3	p-Cl	H	0.14	4
H	CH3	p-CH_3	H	0.10	6
H	CH_3	p-OCH_3	H	0.07	17

Wong et al., 1975.

Development of Antidepressant Drugs

confer the selectivity and potency of the phenoxyphenylpropylamines as inhibitors of 5-HT uptake. The spatial requirements of fluoxetine have been discussed in detail based on x-ray crystallographic findings (Robertson et al., 1987) and compared with those of other inhibitors of monoamine uptake (Fuller et al., 1991).

Substitution at the ortho position of the phenoxy ring appears to confer selectivity toward inhibition of NE uptake. For example, nisoxetine and tomoxetine having an ortho-OCH$_3$ and ortho-CH$_3$, respectively, inhibited NE uptake in nM concentrations (Wong et al., 1975a; b; 1982) and were comparable to the potency of desipramine (Table 3). The ortho-substituted compounds were 2 to 3 orders of magnitude weaker inhibitors of DA and 5-HT uptake. Inhibition of NE appears to be stereospecific for tomoxetine, a R enantiomer, which is almost 10 times more potent than its S enantiomer as inhibitors of NE uptake (Wong et al., 1982).

ENANTIOMERS OF FLUOXETINE AND NORFLUOXETINE

Fluoxetine is a racemic mixture containing equal amounts of R and S enantiomers (Figure 3). Both enantiomers inhibited 5-HT uptake in synaptosomal preparations of cerebral cortex with comparable potencies (Figure 4A) as indicated by the K_i value of 21 nM for R-fluoxetine and 16 nM for S-fluoxetine. The demethylated compounds (Figure 3), R- and S-norfluoxetine are the major metabolites of the corresponding enantiomers of fluoxetine. Unexpectedly, S-norfluoxetine is 14 times more potent than R-norfluoxetine (Figure 4B) with K_i value of 20 and 268 nM, respectively. Comparable data have been reported on the enantiomers of fluoxetine (Wong et al., 1985a; 1991) and norfluoxetine (Wong et al., 1993b).

Table 3. Inhibition of NE, DA and 5-HT Uptake by Other Substituted Phenoxyphenyl-Propylamines.

Substitution					Inhibition of uptake (K_i, M)		
R_1	R_2	R_3	R_4	Isomer	NE	DA	5-HT
H	H	H	H	R,S	0.006	0.46	0.54
H (nisoxetine)	H	o-CH$_3$	H	R,S	0.002	0.39	1.37
H	H	o-F	H	R,S	0.005	0.58	0.90
H	H	o-CH$_3$	H	R,S	0.003	1.75	0.39
H (tomoxetine)	H	o-CH$_3$	H	R	0.002	1.60	0.75
H	H	o-CH$_3$	H	S	0.017	2.45	0.69
Desipramine					0.003	6.70	1.70

*Synaptosomal preparations of brain regions used to measure uptake of NE (hypothalamus), DA (striatum) and 5-HT (cerebral cortex) of rat.

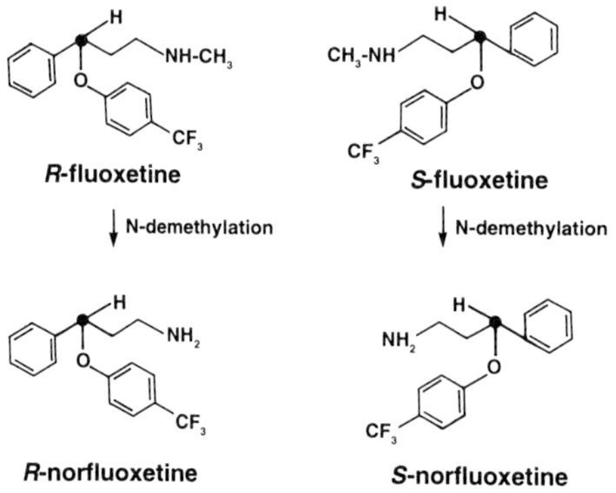

Figure 3. Enantiomers of fluoxetine and norfluoxetine.

SELECTIVITY OF UPTAKE INHIBITORS

Since the mid 1970s, a large number of uptake inhibitors with diverse chemical structures has appeared in the literature (Figure 2). These compounds provide a wide spectrum of selectivity toward the inhibition of the 5-HT uptake carrier relative to their abilities to inhibit the NE and DA uptake carriers (Table 4). The uptake carriers of the three monoamines in their respective neurons can amazingly discern the diverse differences in chemical structures of these inhibitors even though the affinities indicated by the dissociation constants (K_m) for 5-HT, NE and DA are comparable (Shaskan and Snyder, 1970; Snyder and Coyle, 1969). Paroxetine is most potent with a K_i of 0.0003 µM (Hyttel, 1982), which reflects an affinity 2–3 order of magnitude higher than the K_m of 5-HT for its uptake carrier (Shaskan and Snyder, 1970; Wong et al., 1973). Moreover, the selectivity of inhibitors (Table 4) for the 5-HT uptake carrier over that of the NE carrier can achieve a remarkably high ratio of 4000 in citalopram, followed by 800 (indalpine), 300 (paroxetine), 171 (litoxetine) and 113 (dapoxetine). Other inhibitors show selectivity ratios between one and two order of magnitude differences including zimelidine, fluoxetine, femoxetine, sertraline and clomipramine.

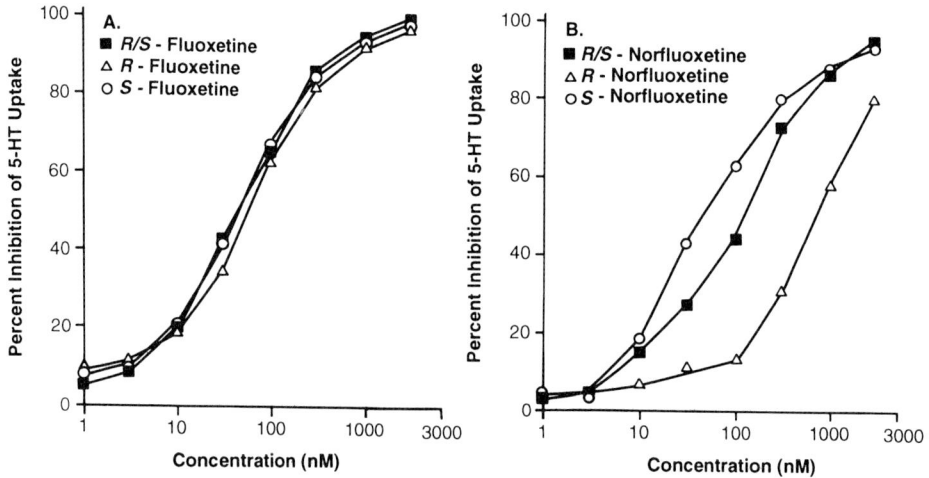

Figure 4. Inhibition of 5-HT uptake by enantiomers of fluoxetine (A) and norfluoxetine (B).

Table 4. Inhibitors of 5-HT, NE and DA Uptake

Inhibitor	Inhibition of uptake			Selectivity Ratio	Ref
	a. 5-HT	b. NE	c. DA		
Paroxetine	0.0003	0.09	5.9	300	a
Clomipramine	0.002	0.024	4.3	12	a
Citalopram	0.002	8	> 10	4000	a
Indalpine	0.003	2.4	1.3	800	a
Duloxetine	0.005	0.016	0.37	3	b
Fluoxetine	0.007	0.38	5	54	a
Litoxetine	0.007	1.2	1.4	171	c
Femoxetine	0.008	0.4	1.4	51	a
Dapoxetine	0.008	0.9	0.9	113	d
Sertraline	0.014	0.7	0.5	51	e
Venlafaxine	0.04	0.2	5.3	5	f
Amitriptyline	0.04	0.02	5.4	0.6	a
Zimelidine	0.06	3.2	> 10	57	a
Fluvoxamine	0.13	1.1	> 10	9	g

K_i values in μM were derived from the following sources: a. Hyttel, 1982; b. Wong et al., 1993a; Scatton et al., 1988; d. Wong et al., 1990b; e. Koe et al., 1983; f. Bolden-Watson and Richelson, 1993; g. Claassen et al., 1977.

There are also inhibitors which potently inhibit the uptake of both 5-HT and NE with a reduced selectivity for the 5-HT uptake carrier (Table 4). Venlafaxine inhibited 5-HT uptake with equipotency as amitriptyline, but was 1/10 as potent inhibiting NE uptake, having a selectivity ratio of 5. On the other hand, amitriptyline inhibited 5-HT and NE uptake with comparable potencies or selectivity ratio approaching unity, but as described in an earlier section amitriptyline is subjected to metabolic N-demethylation to a major metabolite, nortriptyline, a potent and selective inhibitor of NE uptake. The newest inhibitor, duloxetine (LY248686, Wong et al., 1993a), has potencies approaching those of clomipramine as an inhibitor of 5-HT and NE uptake yielding a selectivity ratio of 3. In contrast to clomipramine, duloxetine in vivo exerted inhibitory effects on the uptake of both monoamines with comparable potencies and durations (Wong et al., 1993a).

MONOAMINE UPTAKE INHIBITION IN VIVO

Inhibition of 5-HT uptake in vivo can be demonstrated by at least the following three methods. For example, inhibition of 5-HT uptake ex vivo in synaptosomes of tissue homogenates was dose-dependent in rats treated with fluoxetine or norfluoxetine, which had comparable ED_{50} of 6 and 7 mg/kg i.p. Fluoxetine contains the R and S enantiomers in equal portions and the two enantiomers inhibited 5-HT uptake with comparable potencies (Wong et al., 1985a). However, in vivo R-fluoxetine is metabolically converted to the less effective metabolite, R-norfluoxetine, whereas S-nor-fluoxetine and its parent drug are equipotent as inhibitors of 5-HT uptake (Wong et al., 1993b). Other selective 5-HT reuptake inhibitors have also been demonstrated to be effective in similarly designed ex vivo studies (Maitre et al., 1982; Nelson et al., 1989).

A second method for demonstrating monoamine uptake inhibition in vivo involves the blockade of toxicity induced by the uptake carrier dependent neurotoxins. Fluoxetine and norfluoxetine appeared to be equally effective in preventing the depletion of 5-HT content in mouse brain caused by the 5-HT neuron-specific toxin, p-CA (Fuller et al., 1975). In order to deplete 5-HT, p-CA apparently gains entry into the 5-HT neurons via the membrane uptake carrier (Wong et al., 1973), which is effectively blocked by uptake inhibitors. At a relatively high dose of 32 mg/kg i.p., fluoxetine and norfluoxetine failed to block the depletion of NE from mouse heart by 6-hydroxydopamine (6-OHDA), a NE neuron-specific toxin which gains entry via the NE uptake carrier (Fuller et al., 1974).

Finally, accumulation of intravenously injected radioactive 5-HT in platelets and NE in heart have been used to demonstrate the inhibition of monoamine uptake in vivo. For example, fluoxetine could dose dependently blocked the accumulation of tritium-labeled 5-HT in rat platelets in vivo (Horng and Wong, 1976). Contrary to the tricyclic drugs, both fluoxetine and norfluoxetine at doses up to 20 mg/kg i.p., failed to inhibit the accumulation of radioactive NE in rat heart in vivo (Wong et al., 1975a). Thus, utilizing these various methods of determining monoamine uptake in vivo, fluoxetine and norfluoxetine were confirmed as potent and selective inhibitors of 5-HT reuptake as observed in vitro.

The above techniques have shown that fluoxetine is long acting as an inhibitor of 5-HT uptake, but is without an effect on NE uptake. After a single dose of fluoxetine, the inhibitory effect on 5-HT uptake ex vivo lasts as long as 24 hours without an effect on NE uptake (Wong et al., 1975a). Norfluoxetine in rat brain (Parli and Hicks, 1974), especially the more active metabolite, S-norfluoxetine (Wong et al., 1993b) probably account for the long-lasting inhibition of 5-HT uptake. In contrast, the N-demethylated derivatives of tricyclic antidepressant drugs, including desipramine, chlorodesipramine, and nortriptyline, are markedly more potent inhibitors of NE uptake and markedly less potent as inhibitors of 5-HT uptake than their parent drugs (imipramine, clomipramine and amitriptyline, respectively, Hyttel et al., 1982). Indeed, higher doses of imipramine and clomipramine were required to block the neurotoxic effects of p-CA (Wong and Bymaster, 1976) than to block the accumulation of radioactive NE in rat heart (Wong, et al., 1975a) or the neurotoxic effects of 6-OHDA (Wong and Bymaster, 1976).

NEUROCHEMICAL EVIDENCE OF INHIBITION OF 5-HT UPTAKE

Several techniques have been applied to demonstrate a greater availability of 5-HT in the synapse, when presynaptic uptake is inhibited. The cytofluorimetric technique called "fading" showed increased extracellular concentrations of 5-HT in the raphe region of rat brain after fluoxetine treatment (Geyer et al., 1978). Using in vivo voltammetry, fluoxetine administration produced a small, but long-lasting, increase in 5-HT current and antagonized the large, acute increase produced by p-CA, a 5-HT releaser, which utilized the fluoxetine sensitive 5-HT uptake carrier (Marsden et al., 1979). By means of a push-pull cannula in the nucleus accumbens of rat brain, a large and significant increase in extracellular 5-HT concentration occurred during the first hour after administration of fluoxetine (Guan and McBride, 1988). It is believed that the increase in the extracellular concentration of 5-HT is a reflection of a greater increase of 5-HT concentration within the synaptic cleft.

More recently, increased extracellular concentrations of brain 5-HT in microdialysis fluid have also been shown following the administration of fluoxetine either locally into the dialysate of brain areas or peripherally in vivo (Table 5). Fluoxetine at 10 µM added to the dialysis solution caused a 5.6-fold increase of 5-HT above the baseline levels of hypothalamus while the elevation of potassium from 3 to 120 mM caused a 4-fold increase of extracellular 5-HT (Auerbach et al., 1989); the difference reflects an avid reuptake of the released 5-HT under the latter conditions. Peripheral administration of fluoxetine also elicited increases of extracellular 5-HT in various brain areas including striatum (Perry and Fuller, 1992), thalamus (Dailey et al., 1992) and diencephalon (Rutter and Auerbach, 1993). Similar studies have been reported on other selective serotonin reuptake inhibitors including indalpine (Kalen et al., 1988),

Table 5. Increase of Extracellular 5-HT by Fluoxetine and Other Uptake Inhibitors Using Microdialysis Technique in Conscious Rats

Uptake inhibitor	Route	Brain area	Increase in 5-HT levels
Fluoxetine, 10 μM	dialysate	hypothalamus	5.6-fold (1)
Fluoxetine, 10 mg *	i.p.	striatum	4-fold (2)
Fluoxetine, 20 μM	dialysate	thalamus	1.5-fold (3)
Fluoxetine, 45 mg	i.p.	diencephalon	2-fold (4)
		striatum	1.5-fold (4)
Indalpine, 1 μM	dialysate	caudate-putamen	6-fold (5)
Indalpine, 5 mg	i.p.	caudate-putamen	3-fold (5)
Citalopram, 10 mg	i.p.	dorsal raphe nucleus	4-fold (6)
		frontal cortex	0.7-fold (6)
Clomipramine, 10 μM	dialysate	frontal cortex	2.2-fold (7)
Clomipramine, 20 mg	s.c.	frontal cortex	1-fold (7)
Clomipramine, 10 μM	dialysate	frontal cortex	4-fold (8)
Clomipramine, 20 mg	i.p.	frontal cortex	no change (8)
Clomipramine, 10 μM	dialysate	raphe	4.6-fold (8)
Clomipramine, 20 mg	i.p.	raphe	2.9-fold (8)

Source of data: 1) Auerbach et al., 1989; 2) Perry and Fuller, 1992; 3) Daily et al., 1992; 4) Rutter and Auerbach, 1993; 5) Kalen et al., 1988; 6) Invernizzi et al., 1992; 7) Carboni and Di Chiara, 1989; 8) Adell and Artigas, 1991.
*Dose in mg/kg.

citalopram (Invernizzi et al., 1992) and clomipramine (Carboni and Di Chiara, 1989; Adell and Artigas, 1991). Thus, selective 5-HT reuptake inhibitors administered in vivo do increase the availability of the endogenous 5-HT concentrations in various brain areas.

Presynaptic autoreceptors, which regulate the synthesis, release or turnover of 5-HT are the sites of feedback. Measuring neurochemical indicators of 5-HT turnover provides evidence that inhibition of reuptake increases activation of 5-HT autoreceptors. Fluoxetine significantly lowered brain levels of 5-HIAA when its efflux was blocked by probenecid (Fuller et al., 1975); accumulation of 5-HTP after decarboxylase inhibition with NSD 1015 (Fuller and Wong, 1977), the rate of decline in 5-HT concentration after inhibition of tryptophan hydroxylation (Fuller et al., 1975) and the conversion of radioactive tryptophan into 5-hydroxy-indoles (Bymaster and Wong, 1977). Furthermore, the duration of lowering 5-HIAA levels in rat brain paralleled the duration of 5-HT uptake inhibition by fluoxetine and lasted up to 24 hours (Wong et al., 1975a; Schmidt et al., 1988). Thus, by a feedback mechanism, 5-HT turnover is reduced when the autoreceptor senses a greater abundance of 5-HT in the synapse upon the blockade of 5-HT reuptake by fluoxetine in vivo. These neurochemical findings agreed with the decreased electrical discharges of 5-HT neurons after treatment with fluoxetine (Clemens et al., 1977) and citalopram (Chaput et al., 1986).

PHARMACOLOGICAL RESPONSES OF 5-HT REUPTAKE INHIBITION

Pharmacological studies of fluoxetine and other inhibitors of 5-HT reuptake have been extensively reviewed (Fuller and Wong, 1990; Wong et al., 1990a). For the sake of completeness, however, some studies of fluoxetine are briefly described. In rats trained to eat during hours of daylight, fluoxetine was first shown to lower food intake and to potentiate and prolong the anorectic effect of 5-HTP (Goudie et al., 1976). No indication of tolerance to the anorectic effect of fluoxetine was observed during daily administration of fluoxetine in normal rats up to 24 days (Wong et al., 1985b) or to the decrease of food intake and weight gain in normal and obese mice up to 21 days of daily fluoxetine treatment (Yen et al., 1987).

Fluoxetine and other selective 5-HT reuptake inhibitors including citalopram and sertraline lowered alcohol intake of rats which showed a preference for drinking ethanol containing fluid (Rockman et al., 1982; Gill et al., 1988). Fluoxetine and fluvoxamine also lowered alcohol intake in rats selectively bred for preference of alcohol drinking (Murphy et al., 1985), and the suppressive effect persisted during daily treatment with fluoxetine up to 7 days (Murphy et al., 1988).

Other pharmacological effects of fluoxetine include increases of serum corticosterone in rats (Fuller, 1981), increases in corticotropin-releasing hormone release and adrenocorticotrophin (Gibbs and Vale, 1983); potentiation of morphine-induced analgesia (Messing et al., 1975; Sugrue, 1979); inhibition of muricidal behavior in rats (Molina et al., 1986); decrease of rapid eye movement sleep in rats and cats (Slater et al., 1978; Pastel and Fernstrom, 1987); enhanced processing of memory in mice (Flood and Cherkin, 1987) and rats (Lee et al., 1992); and improvement in passive avoidance behavior of bulbectomized rats (Joly and Danger, 1986). Fluoxetine also potentiated 5-HTP-induced effects, which include increase of serum prolactin levels in rats while fluoxetine itself had no effect (Krulich, 1975); head twitch in mice (Ortmann et al., 1980) and discriminative cue stimulus properties of 5-HTP in rats (Barrett et al., 1982).

INTERACTIONS WITH RECEPTORS OF NEUROTRANSMITTERS

Tricyclic antidepressants can effectively inhibit radioligand binding to many receptors as antagonists including muscarinic-acetylcholinergic receptors, adrenergic receptors, histaminergic receptors, serotonergic receptors and dopaminergic receptors (Hall and Ogren, 1981; Wong et al., 1983; Thomas et al., 1987). The antagonism to these receptors may relate to the side effects profile of tricyclic antidepressants encountered in antidepressive therapy including the anticholinergic effect of constipation, urinary retention, blurred vision and dry mouth; the anti-adrenergic effect of postural hypotension; and the anti-histaminergic effect of drowsiness and sedation (Snyder and Yamamura, 1977; U'Prichard et al., 1978; Richelson and Nelson, 1984).

The non-tricyclic inhibitors of monoamine uptake including fluoxetine, fluvoxamine, zimelidine, paroxetine, sertraline, tomoxetine and duloxetine exhibit relatively low affinities for receptors of neurotransmitters compared to the tricyclic antidepressant drugs (Hall and Ogren, 1981; Wong et al., 1983; Thomas et al., 1987). Indeed, the non-tricyclic drugs including fluoxetine show relatively few side effects (Beasley et al., 1990) typically associated with the tricyclic antidepressant drugs (Richelson and Nelson, 1984).

AFFINITY FOR 5-HT$_{1C}$

Among selective 5-HT reuptake inhibitors, R-fluoxetine (Wong, et al., 1991) and R-norfluoxetine (Wong et al., 1993b) exhibit moderate affinities for the 5-HT$_{1C}$ receptor in bovine choroid plexus (Table 6), while indalpine, citalopram, paroxetine, sertraline, fluvoxamine and femoxetine showed at least 10 times weaker affinity (Wong et al., 1991; Jenck et al., 1993). However, some tricyclic drugs including amitriptyline, nortriptyline, and clomipramine were recently reported to have relatively high affinities for the subtype of 5-HT$_{1C}$

Table 6. Affinities of Tricyclic Antidepressant Drugs for Subtypes of Serotonin Receptors

Drug	5-HT1A	5-HT1B	5-HT1C	5-HT$_{1D}$	5-HT$_2$	5-HT$_3$
R-Fluoxetine	23.2±4.8	12.2±2.8	0.065±0.002	40.2±9.9	0.98±0.22	3.9; 4.2
S-Fluoxetine	6.51±3.07	3.87 ± 0.47	1.84±0.10	9.50±2.20	2.04±0.36	> 40
R-Norfluoxetine	19; 37	13; 7	0.18±0.04	57.5±16.5	0.57±0.03	5; 3
S-Norfluoxetine	30; 49	10.3±3	3.43±0.6	48.5±11.2	3.8±0.8	15.8; 17.4
Amitriptyline	0.82	3.76	0.014±0.001	3.34	0.024±0.004	585
Nortriptyline	0.63; 0.69	6.40	0.022±0.002	> 1 (36)*	0.048±0.012	> 10 (39)
Clomipramine	> 10 (8)	> 10 (33)	0.045±0.013	> 10 (45)	0.075±0.015	1.40
Maprotiline	> 10 (6)	> 10 (45)	0.165±0.04	> 10 (34)	0.24±0.02	1.52
Imipramine	> 10	> 10	0.67±0.03	3.85	0.26±0.03	2.82
Desipramine	> 10 (7)	> 10 (0)	0.64±0.21	> 10 (0)	0.56±0.09	6.05

receptor in the receptor transfected cell line (Roth et al., 1992) and in pig (Jenck et al., 1993) and bovine (Wong and Threlkeld, 1993) choroid plexus (Table 6). On the other hand, imipramine, desipramine and maprotiline possess only moderate affinities for the 5-HT$_{1C}$ receptor (Table 6). In a behavioral model thought to show functional properties of 5-HT$_{1C}$ receptors, antidepressants including mianserin (a known antagonist of 5-HT$_{1C}$ receptors), amitriptyline, nortriptyline, imipramine, maprotiline, and desipramine were shown to be antagonists, while fluoxetine, clomipramine and other selective 5-HT reuptake inhibitors were inactive (Jenck et al., 1993). Among other subtypes of 5-HT receptors (Table 6), some tricyclic drugs are known to have affinities for 5-HT$_2$ receptors. While the significance for the tricyclic drugs to have affinity and antagonist properties toward the 5-HT$_{1C}$ receptor is unclear, the involvement of 5-HT$_{1C}$ in the pathology of mood, anxiety, psychosis and eating disorders has been implicated (Moreau et al., 1993).

ANTIDEPRESSIVE EFFECTS OF SELECTIVE 5-HT REUPTAKE INHIBITORS

Selective 5-HT reuptake inhibitors represent a new class of antidepressant drugs (Boyer and Feighner, 1991a, a review) and support the involvement of 5-HT neurons in depression (Asberg et al., 1986). The antidepressive effects of zimelidine (Heel et al., 1982), fluvoxamine (Itil et al., 1983; Dominguez et al., 1985), fluoxetine (Bremner, 1984; Feighner, 1985), indalpine (Guelfi et al., 1981), citalopram (Bjerkenstedt et al., 1985), paroxetine (Cohn and Wilcox, 1992) and sertraline (Doogan et al., 1988) are well documented. Moreover, the efficacy of selective 5-HT reuptake inhibitors is comparable to that of the tricyclic antidepressant drugs. The side effect profile, however, is quite different from those described in the previous section on the tricyclic antidepressant drugs. Nausea, diarrhea, insomnia, and agitation are the side effects commonly associ-ated with the selective 5-HT reuptake inhibitors (Beasley et al., 1990; Boyer and Feighner, 1991b).

Moreover, the observation of relapse into depression upon ingestion of a tryptophan limiting amino-acid drink by patients, who were treated successfully with fluoxetine or fluvoxamine, further supports the importance of 5-HT neurons as an antidepressive pathway (Delgado et al., 1990). An earlier onset of antidepressive efficacy was observed when patients were treated with a combination of fluoxetine and desipramine (Nelson et al., 1991). Thus, enhancement of both 5-HT and NE transmission by a single molecule like duloxetine, which inhibits the uptake of both 5-HT and NE without much affinity for receptors of neurotransmitters (LY248686, Wong, et al., 1993a), might be advantageous in the treatment of depression.

ACKNOWLEDGMENTS

In the course of developing fluoxetine as the first selective 5-HT reuptake inhibitor and a major 5-HT selective antidepressant drug, we have benefited from working with many colleagues including Mr. Jong-Sin. Horng, Mr. Kenneth Hauser, Mr. Richard Kattau, Mr. Joseph H. Krushinski, Mr. Kenneth W. Perry, Mr. Leroy R. Reid, Ms. Penny G. Threlkeld, Drs. Ray W. Fuller, Louis Lemberger, Bryan B. Molloy and David W. Robertson of the Lilly Research Laboratories. We also appreciate the encouragement received from Drs. Irwin H. Slater and Russell J. Kraay. The authors also thank Mrs. Ruth Leonard and Mrs. Joan Hager for preparation of the manuscript.

REFERENCES

Adell, A. and Artigas, F. (1991). Differential effects of clomipramine given locally or systemically on extracellular 5-hydroxytryptamine in raphe nuclei and frontal cortex: an in vivo brain microdialysis study. Naunyn-Schmiedeberg's Arch. Pharmacol. 343:237–244.

Asberg, M., Bertilsson, L., Tuck, D. et al. (1973). Indolamine metabolites in the cerebrospinal fluid of depressed patients before and during treatment with nortriptyline. Clin. Pharmacol. Ther. 14:277–286.

Asberg, M, Bertilsson, L., Martensson, B. et al. (1986). Therapeutic effects of serotonin uptake inhibitors in depression. J. Clin. Psychiatry 47 (suppl): 23–35.

Asberg, M. and Wagner, A. (1986). Biochemical effects of antidepressant treatment --studies of monoamine metabolites in cerebrospinal fluid and platelet [^{3}H]imipramine, in "Antidepressants and Receptor Function," Wiley, Chichester, Ciba Foundation Symposium 123:57–83.

Aschroft, G. W., Crawford, T. B. B. and Eccleston, D. (1966). 5-Hydroxyindole compounds in the cerebrospinal fluid of patients with psychiatric or neurological diseases. Lancet II:1049–1050.

Auerbach, S. B., Minzenberg, M. J. and Wilkinson, L. O. (1989). Extracellular serotonin and 5-hydroxyindoleacetic acid in hypothalamus of the unanesthetized rat measured by in vivo dialysis coupled to high-performance liquid chromatography with electrochemical detection: dialysate serotonin reflects neuronal release. Brain Res. 499:281–290.

Barrett, R. J., Blackshear, M. A. and Sanders-Bush, E. (1982). Discriminative stimulus properties of L-5-hydroxytryptophan: Behavioral evidence for multiple serotonin receptors. Psychopharmacology 76:29–35.

Beasley, C. M., Bosomworth, J. C. and Wernicke, J. F. (1990). Fluoxetine: relationship among dose, response, adverse events, and plasma concentrations in the treatment of depression. Psychopharmacol. Bull. 26:18–24.

Bjerkenstedt, L., Edman, G., Flyckt, L. et al. (1985). Clinical and biochemical effects of citalopram, a selective 5-HT reuptake inhibitor-a dose-response study in depressed patients. Psychopharmacol. 87:253–259.

Bolden-Watson, C. and Richelson, E. (1993). Blockade by newly developed anti-depressants of biogenic amine uptake into rat brain synaptosomes. Life Sci. 52: 1023–1029.

Bourin, M. (1990). Is it possible to predict the activity of a new antidepressant in animals with simple psychopharmacological tests? Fundam. Clin. Pharmacol. 4:49–64.

Boyer, W. E. and Feighner, J. D. (1991a). The efficacy of selective serotonin re-uptake inhibitors in depression, in "Selective Serotonin Re-Uptake Inhibitors, The Clinical Use of Citalopram, Fluoxetine, Fluvoxamine, Paroxetine and Sertraline," J. D. Feighner and W. E. Boyer, eds., John Wiley & Son, Chichester, England, pp. 89–108.

Boyer, W. E. and Feighner, J. D. (1991b). Side-effects of the selective serotonin re-uptake inhibitors, ibid, pp. 133–152.

Bremner, J. D. (1984). Fluoxetine in depressed patients: a comparison with imipramine. J. Clin. Psychiatry 45:414–419.

Broune, H. R., Bunney, W. E., Jr., Colburn, R. W. et al. (1968). Noradrenaline, 5-hydroxytryptamine and 5-hydroxyindoleacetic acid in hindbrains of suicidal patients. Lancet II:805–808.

Burgen, A.S.V. and Iversen, L.L. (1965). The inhibition of noradrenaline uptake by sympathomimetic amines in the rat isolated heart. Brit. J. Pharmacol. 25:34–49.

Bymaster, F. P. and Wong, D. T. (1977). Effect of Lilly 110140, 3-(p-trifluoro-methylphenoxy)-N-methyl-3-phenylpropylamine on synthesis of 3H-serotonin from 3H-tryptophan in rat brain. The Pharmacologist 16:244.

Carboni, E. and Di Chiara, G. (1989). Serotonin release estimated by transcortical dialysis in freely-moving rats. Neuroscience 32:637–645.
Carlsson, A. (1970). Structural specificity for inhibition of [14C]-5-hydroxytryptamine uptake by cerebral slices. J. Pharm. Pharmacol. 22:729–732.
Carlsson, A., Fuxe, K., Hamberger, B. and Lindqvist, M. (1966). Biochemical and histochemical studies on the effects of imipramine-like drugs and (+)-amphetamine on central and peripheral catecholamine neurons. Acta Physiol. Scand. 67:481–497.
Carlsson, A., Corrodi, H., Fuxe, K. and Hokfelt, T. (1969a). Effect of antidepressant drugs on the depletion of intraneuronal brain 5-hydroxytryptamine stores caused by 4-methyl-a-ethyl-meta-tyramine. Eur. J. Pharmacol. 5:357–366.
Carlsson, A., Corrodi, H., Fuxe, K. and Hokfelt, T. (1969b). Effects of some antidepressant drugs on the depletion of intraneuronal brain catecholamine stores caused by 4, a-dimethyl-meta-tyramine. Eur. J. Pharmacol. 5:367–373.
Carlsson, A., Fuxe, K. and Ungerstedt, U. (1968). The effects of imipramine of central 5-hydroxytryptamine neurons. J. Pharm. Pharmacol. 20:150–151.
Chaput, H., de Montigny, C. and Blier, P. (1986). Effects of a selective 5-HT reuptake blocker, citalopram, on the sensitivity of 5-HT autoreceptors: electrophysiological studies in the rat brain. Naunyn-Schmiedeberg's Arch. Pharmacol. 333:342–348.
Claassen, V., Davis, J. E., Hertting, G. et al. (1977). Fluvoxamine, a specific 5-hydroxytryptamine uptake inhibitor. Brit. J. Pharmacol. 60:505–516.
Clemens, J. A., Sawyer, B. D. and Cerimele, B. (1977). Further evidence that serotonin is a neurotransmitter involved in the control of prolactin secretion. Endocrinology 100:692–698.
Cohn, J. B. and Wilcox, C. S. (1992). Paroxetine in major depression: a double-blind trial with imipramine and placebo. J. Clin. Psychiatry 53: 2(suppl) 52–56.
Coppen, A. J., Prange, A. K. and Whybrow, P. C. (1972). Abnormalities of indoleamines in affective disorders. Arch. Gen. Psychiatry 26:474–478.
Coppen, A., Shaw, D. M., and Farrell, J. P. (1963). Potentiation of the antidepressive effect of a monoamine oxidase inhibitor by tryptophan. Lancet I:79–81.
Coppen, A., Shaw, D. M., Herzberg, B. and Maggs, R. (1967). Tryptophan in the treatment of depression. Lancet II:1178.
Corrodi, H. and Fuxe, K. (1968). The effects of imipramine on central monoamine neurones. J. Pharm. Pharmacol. 20:230–231.
Dahlstrom, A. and Fuxe, K. (1964). Evidence for the existence of monoamine-containing neurons in the central nervous system. 1. Demonstration of monoamines in the cell bodies of brain stem neurons. Acta Physiol. Scand. 62:suppl. 232: 6–55.
Dailey, J.W., Yan, Q.S., Mishra, P.K. et al. (1992). Effects of fluoxetine on convulsions and in brain serotonin as detected by microdialysis in genetically epilepsy-prone rats. J. Pharmacol. Exp. Ther. 260:533–540.
Delgado, P. L., Charney, D. S., Price, L. H. et al. (1990). Serotonin function and the mechanism of antidepressant action. Arch. Gen. Psychiatry 47:411–418.
Dencker, S. J., Malm, U., Roo, B. E. et al. (1966). Acid monoamine metabolites of cerebrospinal fluid in mental depression and mania. J. Neurochem. 13:1545–1548.
Dengler, H.J. and Titus, E.O. (1961). The effect of drugs on the uptake of isotopic norepinephrine in various tissues. Biochem. Pharmacol. 8:64.
Dominguez, R. A., Goldstein, B. J., Jacobson, A. F. and Steinbook, R. M. (1985). A double-blind placebo-controlled study of fluvoxamine and imipramine in depression. J. Clin. Psychiatry 46:84–87.
Doogan, D. P. and Caillard, V. (1988). Sertraline: A new antidepressant. J. Clin. Psychiatry 49:8 (suppl) 46–51.
Feighner, J. P. (1985). A comparative trial of fluoxetine and amitriptyline in patients with major depressive disorder. J. Clin. Psychiatry 46:69–372.
Flood, J. F. and Cherkin, A. (1987). Fluoxetine enhances memory processing in mice. Psychopharmacology 93:36–43.
Fuller, R. W. (1981). Serotonergic stimulation of pituitary-adrenocortical function in rats. Neuroendocrinology 32:118–127.
Fuller, R. W., Perry, K. W. and Molloy, B. B. (1974). Effect of an uptake inhibitor on serotonin metabolism in rat brain: studies with 3-(p-trifluoromethylphenoxy)-N-methyl-3-phenylpropylamine. Life Sci. 15:1161–1171.
Fuller, R. W., Perry, K. W. and Molloy, B. B. (1975). Effect of 3-(p-trifluoro-methylphenoxy)-N-methyl-3-phenylpropylamine on the depletion of brain serotonin by 4-chloroamphetamine. J. Pharmacol. Exp. Ther. 193:796–803.

Fuller, R. W. and Wong, D. T. (1977). Inhibition of serotonin reuptake. Fed. Proc. 36: 2154–2158.

Fuller, R. W. and Wong, D. T. (1990). Serotonin uptake and serotonin uptake inhibition. Ann. N.Y. Acad. Sci. 600:68–78.

Fuller, R. W., Wong, D.T. and Robertson, D. W. (1991). Fluoxetine, a selective inhibitor of serotonin uptake. Medicinal Res. Rev. 11:17–34.

Geyer, M. A., Dawsey, W. J. and Mandell, A. L. (1978). Fading: a new cytofluorimetric measure quantifying serotonin in the presence of catecholamines at the cellular level in brain. J. Pharmacol. Exp. Ther. 207:650–667.

Gibbs, D. M. and Vale, W. (1983). Effect of the serotonin reuptake inhibitor fluoxetine on corticotropin-releasing factor and vasopressin secretion into hypophysial portal blood. Brain Res. 280:176–179.

Gill, K., Amit, Z. and Koe, B. (1988). Treatment with sertraline, a new serotonin uptake inhibitor, reduces voluntary ethanol consumption in rats. Alcohol 5:349–354.

Gillette, J. R., Dingell, J. V., Sulser, F. et al. (1961). Isolation from rat brain of a metabolic product, desmethylimipramine, that mediates the antidepressant activity of imipramine (Tofranil). Experientia XVII:417–418.

Glowinski, J. and Axelrod, J. (1964). Inhibition of uptake of tritiated-noradrenaline in the intact rat brain by imipramine and structurally-related compounds. Nature 204: 1318–1319.

Goudie, A. J., Thornton, E. W. and Wheeler, T. J. (1976). Effects of Lilly 110140, a specific inhibitor of 5-hydroxytryptamine uptake, on food intake and 5-hydroxy-tryptophan-induced anorexia. Evidence for serotonergic inhibition of feeding. J. Pharm. Pharmacol. 28:318–320.

Guan, X. M., McBride and W. J. (1988). Fluoxetine increases the extracellular levels of serotonin in the nucleus accumbens. Brain Res. Bull. 21:43–46.

Guelfi, J. D., Dreyfus, J. F., Boyer, P. and Pichot, P. (1981). A double-blind controlled multicenter trial comparing indalpine and imipramine. 3rd World Congress of Biological Psychiatry, Stockholm, June 28-July 3, 1981.

Hall, H. and Ogren, S. O. (1981). Effects of antidepressant drugs on different receptors in rat brain. Eur. J. Pharmacol. 70:393–407.

Heel, R. C., Morley, P. A., Brogden, R. N., et al. (1982). Zimelidine: a review of its pharmacological properties and therapeutic efficacy in depressive illness. Drugs 24: 169–206.

Hertting, G., Axelrod, J. and Whitby, L. G. (1961). Effect of drugs on the uptake and mechanism of [3H]-norepinephrine. J. Pharmacol. Exp. Ther. 134:146–153.

Hertz, D. and Sulman, F. G. (1968). Preventing depression with tryptophan. Lancet I: 531.

Horng, J. S. and Wong, D. T. (1976). Effects of serotonin uptake inhibitor, Lilly 110140, on transport of serotonin in rat and human blood platelets. Biochem. Pharmacol. 25: 865–867.

Hyttel, J. (1982). Citalopram--pharmacological profile of a specific serotonin uptake inhibitor with antidepressant activity. Neuro-Psychopharmacol. Biol. Psychiatry 6: 277–295.

Invernizzi, R., Belli, S. and Samanin, R. (1992). Citalopram's ability to increase the extracellular concentrations of serotonin in the dorsal raphe prevents the drug's effect in the frontal cortex. Brain Res. 584:322–324.

Itil, T. M., Shrivastava, R. K., Mukherjee, S. et al. (1983). A double-blind placebo-controlled study of fluvoxamine and imipramine in out-patients with primary depression. Brit. J. Clin. Pharmacol. 15:433S-438S.

Iversen, L. L. (1971). Role of transmitter uptake mechanisms in synaptic neuro-transmission. Brit. J. Pharmacol. 41:571–591.

Jenck, F., Moreau, J.-L., Mutel, V. et al. (1993). Evidence for a role of 5-HT1C receptors in the antiserotonergic properties of some antidepressant drugs. Eur. J. Pharmacol. 231:223–229.

Jesberger, J. A. and Richardson, J. S. (1985). Animal models of depression: Parallels and correlates to severe depression in human. Biol. Psychiatry 20:764–784.

Joly, D. and Danger, D.J. (1986). The effects of fluoxetine and zimelidine on the behavior of olfactory bulbectomized rats. Pharmacol. Biochem. Behav. 24:199–204.

Kalen, P., Strecker, R. E., Rosengren, E. and Bjorklund, A. (1988). Endogenous release of neuronal serotonin and 5-hydroxyindoleacetic acid in the caudate-putamen of the rat as revealed by intracerebral dialysis coupled to high-performance liquid chromatography with fluorimetric detection. J. Neurochem. 51:1422–1435.

Kline, N. S. and Sacks, W. (1963). Relief of depression within one day using an M.A.O. inhibitor and intravenous 5-HTP. Am. J. Psychiatry, 120:274.

Koe, K. K., Weissman, A., Welch, W. M. et al. (1983). Sertraline, 1S,4S-N-methyl-4-(3,4-dichlorophenyl)-1,2,3,4-tetrahydro-1-naphthylamine, a new uptake inhibitor with selectivity for serotonin. J. Pharmacol. Exp. Ther. 226:686–700.

Korpi, E. R., Kleinman, J. E., Goodman, S. I. et al. (1968). Serotonin and 5-hydroxy-indoleacetic acid in brain of suicide victims: Comparison in chronic schizophrenic patients with suicide as cause of death. Arch. Gen. Psychiatry 43:594–600.

Krulich, L. (1975). The effect of a serotonin uptake inhibitor (Lilly 110140) on the secretion of prolactin in the rat. Life Sci. 17:1141–1144.

Kuhar, M.J., Shaskan, E.G. and Snyder, S.H. (1971). The subcellular distribution of endogenous and exogenous serotonin in brain tissue: Comparison of synaptosomes storing serotonin, norepinephrine, and gamma-aminobutyric acid. J. Neurochem. 18: 333–343.

Kuhar, M. J., Roth, R. and Aghajanian, G. K. (1972). Synaptosomes from forebrains of rats with midbrain raphe lesions: selective reduction of serotonin uptake. J. Pharmacol. Exp. Ther. 181:36–45.

Kuhn, R. (1957). Die behandlung depressiver zustande mit einem iminodebenzylderivat (G22355). Schweiz. und Wschr. 87:1135–1140.

Lapin, I. P. and Oxenkrug, G. F. (1969). Intensification of the central serotonergic processes as a possible determinant of the thymoleptic effect. Lancet 1:132–136.

Lee, E. H. Y., Lin, W. R., Chen, H. Y. et al. (1992). Fluoxetine and 8-OHDPAT in the lateral septum enhances and impairs retention of an inhibitory avoidance response in rats. Physiol. Behav. 51:681–688.

Lloyd, K. G., Farley, I. J., Deck, J.H.H. et al. (1974). Serotonin and 5-hydroxy-indolacetic acid in discrete areas of the brainstem of suicide victims and control patients. Adv. Biochem. Psychopharmacol. 11:387–397.

Maitre, A., Baumann, P.A., Jaekel, J. et al. (1982). 5-HT uptake inhibitors; Psycho-pharmacological and neurochemical criteria of selectivity, in: "Serotonin in Biological Psychiatry," B. T. Ho et al (eds), Raven Press; New York, pp 229–246.

Marsden, C. A., Conti, J., Strope, E. et al. (1979). Monitoring 5-hydroxytryptamine release in the brain of the freely moving unanesthetized rat using in vivo voltammetry. Brain Res. 171:85–99.

Messing, R. B., Phebus, L., Fisher, L. A. et al. (1975). Analgesic effect of fluoxetine hydrochloride (Lilly 110140), a specific inhibitor of serotonin uptake. Psychopharmacol. Comm. 1:511–521.

Molina, V.A., Gobaille, S. and Mandel, P. (1986). Effects of serotonin-mimetic drugs on mouse-killing behavior. Aggress. Behav. 12:201–269.

Moreau, J.-L., Jenck, F., Martin, J. R. et al. (1993). Effects of repeated mild stress and two antidepressant treatments on the behavioral response to 5-HT_{1C} receptor activation in rats. Psychopharmacology 110:140–144.

Murphy, J. M., Waller, M. B., Gatto, G. J. et al. (1985). Monoamine uptake inhibitors attenuate ethanol intake in alcohol-preferring (P) rats. Alcohol 2:349–352.

Murphy, J. M., Waller, M. B., Gatto, G.J. et al. (1988). Effects of fluoxetine on the intragastric self-administration of ethanol in the alcohol preferring P lines of rats. Alcohol 5:283–286.

Muscettola, G., Goodwin, F. K., Potter, W. Z. et al. (1978). Imipramine and desipramine in plasma and spinal fluid. Arch. Gen. Psychiatry 35: 621–625.

Narasimhachari, N. and Landa, B. (1991). Incidence of high and low hydroxylator and demethylator phenotypes in patients on antidepressant therapy. Clin. Pharmacol. Ther. Abs. Pi-51:136.

Nelson, D. R., Thomas, D. R. and Johnson, A. M. (1989). Pharmacological effects of paroxetine after repeated administration to animals. Acta Psychiatry Scand. 80 (suppl 350), 21–23.

Nelson, J. C., Mazure, C. M., Bowers, M. B. and Jatlow, P. L. (1991). A preliminary, open study of the combination of fluoxetine and desipramine for rapid treatment of major depression. Arch. Gen. Psychiatry 48:303–307.

Ortmann, R., Waldmeier, P. C., Radeke, E. et al. (1980). The effect of 5-HT uptake and MAO-inhibitors on L-5-HTP-induced excitation in rats. Naunyn-Schmiedeberg's Arch. Pharmacol. 311:185–192.

Page, I. H. (1969). Serotonin and the brain, in "The Structure and Function of Nervous Tissue," G. H. Bourne (ed.), Vol. III, Biochemistry and Disease, Academic Press, NY, pp. 289–307.

Pare, C.M.B. (1963). Potentiation of monoamine oxidase inhibitors by tryptophan. Lancet II:527–528.

Parli, C. J. and Hicks, J. (1974). In vivo demethylation of Lilly 110140: 3(p-trifluoromethylphenoxy)-N-methylphenoxy)-N-methyl-3-phenyl propylamine to an active metabolite - Lilly 103947. Fed. Proc. 33:560.

Pastel, R. H. and Fernstrom, J. D. (1987). Short-term effects of fluoxetine and trifluorophenylpiperazines on electroencephalographic sleep in the rat. Brain Res. 436:92–102.

Perry, K. W. and Fuller R. W. (1992). Effect of fluoxetine on serotonin and dopamine concentration in microdialysis fluid from rat striatum. Life Sci. 50:1683–1690.

Richelson, E. and Nelson, A. (1984). Antagonism by antidepressants of neurotransmitter receptors of normal human brain in vitro. J. Pharmacol. Exp. Ther. 230:94–102.

Robertson, D. W., Jones, N. D., Swartzendruber, J. K. et al. (1987). Molecular structure of fluoxetine hydrochloride, a highly selective serotonin uptake inhibitor. J. Med. Chem. 31:185–189.

Rockman, G. E., Amit, Z., Brown, Z. W. et al. (1982). An investigation of the mechanisms of action of 5-hydroxytryptamine in the suppression of ethanol intake. Neuropharmacol. 21:341–347.

Ross, S. B. and Renyi, A. L. (1967). Inhibition of the uptake of tritiated catecholamines by antidepressant and related agents. Eur. J. Pharmacol. 2:181–186.

Ross, S. B. and Renyi, A. L. (1969). Inhibition of the uptake of tritiated 5-hydroxy-tryptamine in brain tissue. Eur. J. Pharmacol. 7:270–277.

Roth, B. L., Meltzer, H. Y. and Craigo, S. (1992). Typical tricyclic antidepressants possess potent $5-HT_{1C}$ receptor activity. Abst. 22nd Ann. Meeting, Society for Neuroscience 18:522.

Rutter, J. J. and Auerbach, S. B. (1993). Acute uptake inhibition increases extracellular serotonin in the rat forebrain. J. Pharmacol. Exp. Ther. 265:1319–1324.

Sano, S. (1977). 5-Hydroxy-L-tryptophan; a fast-acting drug for endogenous depression. Drugs Exptl. Clin. Res. 1:239–242.

Scatton, B., Claustre, Y., Graham, D., et al. (1988). SL 81.0385: A novel selective and potent serotonin uptake inhibitor. Drug Dev. Res. 12:29–40.

Schmidt, M. J., Fuller, R. W. and Wong, D. T. (1988). Fluoxetine, a highly selective serotonin reuptake inhibitor: a review of preclinical studies. Brit. J. Psychiatr. 153 (Suppl 3):40–46.

Shaskan, E. G. and Snyder, S. H. (1970). Kinetics of serotonin accumulation into slices from rat brain: relationship to catecholamine uptake. J. Pharmacol. Exp. Ther. 175: 404–418.

Shaw, D. M., Camps, F. E. and Eccleston, E. G. (1967). 5-Hydroxytrytamine in the hind brain of depressive suicides. Brit. J. Psychiatry 113:1407–1411.

Slater, I. H., Jones, G. T. and Moore, R. A. (1978). Inhibition of REM sleep by fluoxetine, a specific inhibitor of serotonin uptake. Neuropharmacol. 17:383–389.

Slater, I. H., Rathbun, R. C. and Kattau, R. (1979). Role of 5-hydroxytryptamergic and adrenergic mechanism in antagonism of reserpine-induced hypothermia in mice. J. Pharm. Pharmacol. 31:108–110.

Snyder, S. H. and Coyle, J. T. (1969). Regional differences in 3H-norepinephrine and 3H-dopamine uptake into rat brain homogenates. J. Pharmacol. Exp. Ther., 165:78–86.

Snyder, S. H. and Yamamura, H. I. (1977). Antidepressants and muscarinic acetylcholine receptor. Arch. Gen. Psychiatry 34:236–239.

Sugrue, M. F. (1979). On the role of 5-hydroxytryptamine in drug-induced antinociception. Brit. J. Pharmacol. 65:677–681.

Sulser, F., Watts, J. and Brodie, B. B. (1962). On the mechanism of antidepressant action of imipramine-like drugs. Ann. N. Y. Acad. Sci. 96:279-

Thomas, D. R., Nelson, D. R. and Johnson, A. M. (1987). Biochemical effects of the antidepressant paroxetine, a specific 5-hydroxytryptamine uptake inhibitor. Psychopharmacol. 93:193–200.

U'Prichard, D. C., Greenberg, D. A., Sheehan, P. B. et al. (1978). Tricyclic anti-depressants; therapeutic properties affinity for a-noradrenergic receptor binding sites in the brain. Sci. 199:197–198.

Weil-Malherbe, H. and Szara, S. I. (1971). Brain amines and affective disorders, in "The Biochemistry of Functional and Experimental Psychosis," Charles C. Thomas Publisher; Springfield, IL, pp. 57–76.

Whitby, L.G., Axelrod, J. and Weil-Malherbe, H. (1961). The fate of H^3-norepinephrine in animals. J. Pharmacol. Exp. Ther. 132:193–201.

Wong, D. T. and Bymaster, F. P. (1976). The comparison of fluoxetine and nisoxetine with tricyclic antidepressants in blocking the neurotoxicity of p-chloroamphetamine and 6-hydroxydopamine in the rat brain. Res. Comm. Chem. Pathol. Pharmacol. 15: 221–231.

Wong, D. T., Bymaster, F. P., Horng, J. S. and Molloy, B. B. (1975a). A new selective inhibitor for uptake of serotonin into synaptosomes of rat brain: 3-(p-trifluoro-methylphenoxy)-N-methyl-3-phenylpropylamine. J. Pharmacol. Exp. Ther. 193:804–811.

Wong, D. T., Bymaster, F. P., Mayle, D.A. et al. (1993a). LY248686, a new inhibitor of serotonin and norepinephrine uptake. Neuropsychopharmacology 8:23–33.

Wong, D. T., Bymaster, F. P., Reid, L. R. and Threlkeld, P. G. (1983). Fluoxetine and two other serotonin uptake inhibitors without affinity for neuronal receptors. Biochem. Pharmacol. 32:1287–1293.

Wong, D. T., Bymaster, F. P., Reid et al. (1993b). Norfluoxetine enantiomers as inhibitors of serotonin uptake in rat brain. Neuropsychopharmacology 8:337–344.

Wong, D. T., Bymaster, F. P., Reid, L. R. et al. (1985a). Inhibition of serotonin uptake by optical isomers of fluoxetine. Drug Dev. Res. 6: 397–403.

Wong, D. T. and Fuller, R. W. (1987). Serotonergic mechanisms in feeding. Int. J. Obesity 11 (suppl 3):125–133.

Wong, D. T., Fuller, R. W. and Robertson, D.W. (1990a). Fluoxetine and its two enantiomers as selective serotonin uptake inhibitors. Acta Pharm. Nord. 2:171–180.

Wong, D. T., Horng, J.-S. and Bymaster, F. P. (1975b). dl-N-methyl-3-(o-methoxyphenoxy)-3-phenylpropylamine hydrochloride, Lilly 94939, a potent inhibitor for uptake of norepinephrine into rat brain synaptosomes and heart. Life Sci. 17:755–760.

Wong, D. T., Horng, J. S. and Fuller, R. W. (1973). Kinetics of serotonin accumulation into synaptosomes of rat brain: Effects of amphetamine and chloroamphetamine. Biochem. Pharmacol. 22:311–322.

Wong, D. T., Horng, J. S., Bymaster, F. P. et al. (1974). A selective inhibitor of serotonin uptake: Lilly 110140, 3-(p-trifluoromethylphenoxy)-N-methyl-3-phenylpropyl-amine. Life Sci. 15:471–479.

Wong, D. T. and Murphy, J. M. (1989). Serotonergic mechanisms in alcohol intake, in "Neurobiological and Metabolic Aspects of Alcohol," G. Y. Sun, Y. H. Wei, P. K. Rudeen et al. (eds.), Humana Press, pp. 133–146.

Wong, D. T., Reid, L. R., Bymaster, F. P. et al. (1985b). Chronic effect of fluoxetine, a selective inhibitor of serotonin uptake, on neurotransmitter receptors. J. Neural Transm. 64:251–269.

Wong, D. T., Reid. L. R., Thompson, D. C. and Robertson, D.W. (1990b). LY210448, a new selective inhibitor of serotonin (5-hydroxytryptamine, 5-HT) uptake and a potential antidepressant and antiobesity drug. Abst. 29th Ann. Meeting, Am. Coll. Neuropsychopharmacology, p. 133.

Wong, D. T., Robertson, D. W., Bymaster, F. P. et al. (1988). LY227942, an inhibitor of serotonin and norepinephrine uptake: biochemical pharmacology of a potential antidepressant drug. Life Sci. 43:2049–2057.

Wong, D. T. and Threlkeld, P .G. (1993). Tricyclic antidepressant drugs exhibit high affinity for serotonin (5-HT)1C receptors. FASEB J. 7:abst. 1530, p. A264.

Wong, D. T., Threlkeld, P. G., Best, K. L. and Bymaster, F. P. (1982). A new inhibitor of norepinephrine uptake devoid of affinity for receptor in rat brain. J. Pharmacol. Exp. Ther. 222:61–65.

Wong, D. T., Threlkeld, P. G. and Robertson, D. W. (1991). Affinity of fluoxetine, its enantiomers and other inhibitors of serotonin uptake for subtypes of serotonin receptors. Neuropsychopharmacology 5:43–47.

Yen, T. T., Wong, D. T., and Bemis, K. G. (1987). Reduction of food consumption and body weight of normal and obese mice by chronic treatment with fluoxetine: A serotonin reuptake inhibitor. Drug Dev. Res. 10:37–45.

THE IDENTIFICATION OF HETEROGENEITY OF 5-HT$_3$ RECEPTORS WITH [^{3}H]RS-42358-197

Erik H.F. Wong, Douglas W. Bonhaus, and Richard M. Eglen

Department of Neurosciences
Institute of Pharmacology
Syntex Discovery Research
3401 Hillview Avenue
California 94304

INTRODUCTION

The diverse pharmacological action of serotonin (5-HT) has been the subject of intense study since its identification in 1936.[1] The diversity of actions include activation or inhibition of smooth and cardiac muscle, exocrine and endocrine glands, cells of the hematopoietic and immune systems, as well as central and peripheral neurons.[2,3] The results of these investigations have led to the identification of at least 12 different 5-HT receptors based on operational (functional, antagonism, location), transductional (G-protein, ion channel), and structural (gene sequence, chromosomal location) criteria.[4] Based on these criteria, 5-HT receptors have been grouped into 5 types, namely 5-HT$_1$, 5-HT$_2$, 5-HT$_3$, 5-HT$_4$, and 5-HT$_5$ receptors.[5,6]

Out of the 5 classes of 5-HT receptors mentioned above the 5-HT$_3$ receptor has a unique place in pharmacological investigation not only because it is the only 5-HT receptor which is coupled to an ion channel[7], but also the localization in the CNS and PNS is distinct from other classes of 5-HT receptors. In addition, cloning work has also indicated similarity to the nicotinic receptor rather than other 5-HT receptors[8].

5-HT$_3$ receptors have been characterized in the peripheral nervous systems on the basis of measurements of depolarization or contraction in a variety of in-vitro and in-vivo preparations[9,10]. In the central nervous system, identification of these receptors has been facilitated by radioligand binding techniques using tritium labeled 5-HT$_3$ receptor antagonists such as [^{3}H]GR 65630[11,12], [^{3}H]ICS 205-930[13,14], [^{3}H]Quaternary ICS 205-930[15,16], [^{3}H]quipazine[12,17], [^{3}H]zacopride[18,19], [^{3}H]BRL 43694[20], [^{3}H]LY 278,584[21], [^{125}I]iodo-zacopride[22], and [^{3}H]RS-42358-197[23]. To date, both binding and functional studies indicate that 5-HT$_3$ receptors are located only on neurones[24,25].

A primary incentive for the study of 5-HT$_3$ receptor has been the promise of novel therapeutic agents in view of the ability of 5-HT$_3$ receptor antagonists to produce anti-emetic, anxiolytic, anti-psychotic and cognitive enhancing actions in a variety of animal models[26,27,28]. It seems logical to inquire whether these actions of 5-HT$_3$ receptor antagonists are mediated by one or more than one type of 5-HT_3 receptors. RS-42358-197 is a highly potent and selective 5-HT_3 receptor antagonist[29,30] with potent anxiolytic action but without anti-psychotic or cognitive enhancing actions[31]. Tritium labeling of this compound has allowed investigation of the mechanism of action of this compound as well as using this as a tool for studying 5-HT_3 receptor in peripheral and central tissues. The scope of the present study is to utilize the potency and selectivity of this radioligand to investigate the existence of heterogeneity of 5-HT_3 receptors. Studies in a variety of species and tissues have identified both inter- and intra-species heterogeneity of 5-HT_3 receptor, suggesting the existence of subtypes of 5-HT_3 receptors.

METHODS

Membrane Preparation

Tissues from CD-1 (Charles river) mice were obtained from animals killed by asphyxiation with CO_2. Sprague-Dawley rat brains were obtained from Pel-Freez. Methods for preparation of NG-108-15 cells, rabbit and guinea-pig ileum were reported by Wong et al.[29]. Excised tissues were rinsed and then homogenized in Tris-HCl 50 mM, Na_2EDTA 5mM buffer (pH 7.4 at 4° C) using a Polytron P-10 tissue disrupter (setting 5, 2 x 10 s bursts). The homogenate was centrifuged at 48,000 X g for 15 minutes. The pellets were then washed, by resuspension and centrifugation, once in homogenizing buffer and once in Tris-HCl 50 mM, EDTA 0.5 mM buffer (pH 7.4 at 4° C). Membranes were stored under liquid nitrogen until required.

Radioligand Binding Assays

For competition binding studies membranes were incubated with approximately 0.1 nM [^{3}H]RS 42358-197 in a Tris-Krebs buffer (NaCl 154 mM, KCl 5.4 mM, KH_2PO_4 1.2 mM, $CaCl_2$ 2.5 mM, $MgCl_2$ 1.0 mM, D-glucose 11 mM, Tris 25 mM; pH 7.4 at 25° C) for 45 minutes. Competing drugs were added in at least 10 concentrations (10 pM to 100 mM) prior to the addition of radioligand. For saturation binding studies membranes were incubated with 0.01 to 5 nM [^{3}H]RS 42358-197, 0.03 to 15 nM [^{3}H]GR 65630 or 0.1 to 30 nM [^{3}H]quipazine. Nonspecific binding was defined with 1 mM (S)zacopride. In all cases reactions were terminated by vacuum filtration through Whatman GF/B filters which had been pretreated with 0.3% polyethyleneimine.

Previous studies have demonstrated that under these conditions binding reaches steady-state and is linearly dependent upon protein concentration[29]. When 0.1 nM [^{3}H]RS 42358-197 was used the percent specific binding typically ranged from approximately 80% in NG-108-15 cells, 79% in rat cortex, 64% and 45% in rabbit and guinea-pig ileum to 32% in CD-1 mouse tissues. Specific binding ranged from 600 D.P.M. in rat cortex to 200 D.P.M. in the CD-1 mouse ileum.

Data Analysis

Competition binding data were analyzed by iterative curve fitting to a four parameter logistic equation. Hill coefficients and IC_{50} were obtained directly. pK_i (-log of K_i) of competing ligands were calculated from IC_{50} values using the Cheng-Prusoff equation[32]. Analysis of saturation binding data were performed with the program Ligand[33]. Comparison of binding affinities of a specific ligands in different tissues was made using analysis of variance with a post-hoc least significance difference test or independent t-tests as appropriate (CSS statistical package).

MATERIALS

RS-42358-197 ((S)-N(1-azabicyclo[2.2.2]oct-3-yl)-2,4,5,6-tetrahydro-1-H-benzo[de] isoquinolin-1-one hydrochloride), granisetron (endo-1-methyl-N-(9-methyl-9-methyl-9-aza-bicyclo [3.3.2]non-3-yl-1H-indazolecarboxamide), renzapride ($\pm$ -endo-4-amino-5-chloro-2-methoxy-N-(1-azabicyclo[3.3.1]non-4-yl) benzamide), ondansetron (1,2,3,9-tetra-hydro-9-methyl-3-[(2-methyl-1H-imi dazole-1-yl)methyl]-4H-carbazolone, $HCl.2H_2O$), tropisetron (ICS 205,930, endo-8-methyl-9-azabicyclo[3.2.1]oct-3-yl-1H-indole-3-carboxylate) and MDL 72222 (endo-8-methyl-8-azabicyclo[3.2.1.]oct-3-yl-3,5-dichloro-benzoate HCl) were synthesized at Syntex Research (Institute of Organic Chemistry) Palo Alto, California, U.S.A. [^{3}H]RS 42358-197 (55 Ci/mmole) was prepared by Dr. Howard Parnes in the Institute of Organic chemistry, Syntex Research. [^{3}H]quipazine (55-71Ci/ mmol) was purchased from Dupont NEN. [^{3}H]GR 65630 (80 Ci/mmol) was a gift from Dr. S. Hurt of Dupont NEN. (R) and (S)YM060 (5-[(1-methyl-3-indole)carbonyl]4,5,6,7-tetrahydro-1H-benzimidazole hydrochloride) were generously provided by Yamanouchi Pharmaceuticals, Japan. 1-(m-chlorophenyl)-biguanide was obtained from Cookson Chemicals. Other chemicals and reagents were purchased from Sigma Chemical Company or Research Biochemicals Incorporated.

RESULTS

Saturation Studies

In the different preparation studied, [^{3}H]RS-42358-197 bound to a single population of high affinity, saturable and heat-sensitive sites (Figure 1). Saturation binding studies in NG-108-15 cells, rat and mouse cerebral cortex, and rabbit ileum indicated similar affinity (K_d range of 0.10 to 0.20 nM). By contrast, the affinity of [^{3}H]RS-42358-197 in membranes of CD-1 mouse and guinea-pig ileum (K_d of 1.3 and 1.54 nM, respectively) were significantly less than that observed in other tissues (Table 1). The density of sites were different in the different preparations with the NG-108-15 cells showing the greatest density. In parallel studies, [^{3}H]RS-42358-197 and [3H]GR 65630 labeled similar number of sites in all the tissues tested. In contrast, [^{3}H] quipazine labeled 2-4 fold more sites than [^{3}H]RS-42358-197. [^{3}H]GR 65630 and [^{3}H]quipazine failed to label any specific sites in guinea-pig membranes.

Competition Studies

The pharmacological specificity of [^{3}H]RS-42358-197 was determined in the six tissues mentioned above (Table 2). All the 5-HT_3 receptor antagonists and agonists employed competed for [^{3}H]RS-42358-197 binding in a dose-dependent manner (Figure 2). The guinea-pig ileum exhibited affinities for 5-HT_3 receptor ligands consistently lower (up to 400 fold weaker for GR 65630) than that observed in other tissues. A comparison of the pharmacological profile of [^{3}H]RS-42358-197 in different tissues was carried out. In accordance with the notion that [^{3}H]RS-42358-197 selectively labels 5-HT_3 receptor binding sites, an excellent correlation ($r^2 = 0.96$) was observed between the pK_i of compounds in displacing [^{3}H]RS-42358-197 and [^{3}H]quipazine binding[12]. In contrast, the correlations were poorer when pK_i values of [^{3}H]RS-4235478-197 binding were compared between different tissues. The correlation coefficients for rat cortex vs guinea-pig ileum, rabbit ileum and NG-108-15 cells were 0.56, 0.67 and 0.72, respectively. The compounds which showed the greatest disparity in pK_i were the agonists phenylbiguanide and mCPBG.

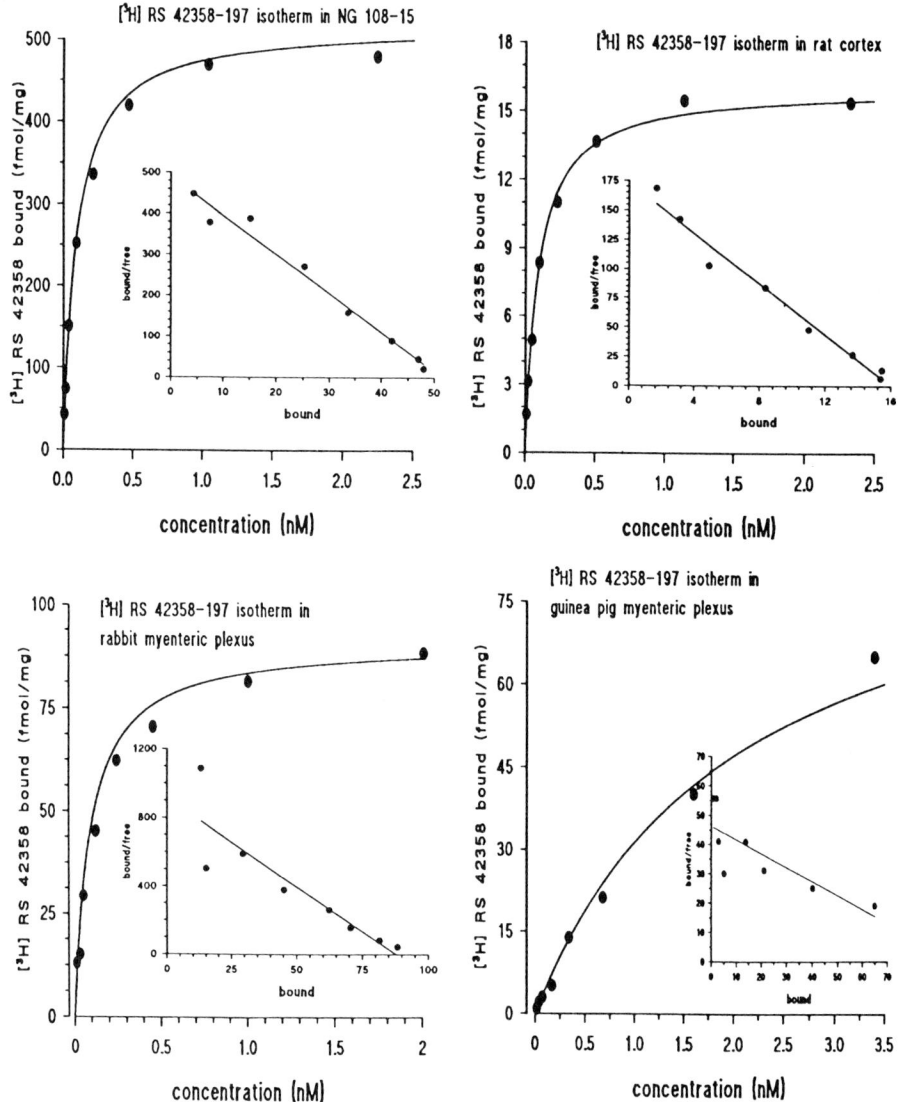

Figure 1. Saturation isotherm for [3H]RS-42358-197 binding in NG-108-15 cells, rat cortex, rabbit myenteric plexus and guinea-pig myenteric plexus. Curves are froma representative experiment. Insets show Scatchard transformation of the data.

In all tissues other than the mouse, all the antagonists tested inhibited binding with a Hill coefficient not significantly different from unity. In contrast, most of the agonists tested displaced [^{3}H]RS-42358-197 binding in rat cortical membranes with Hill coefficients significantly greater than unity[16]. Comparison of ligand affinity at 5-HT$_3$ receptors in rat cortex, CD-1 mouse cortex and CD-1 mouse ileum indicated clear differences (Table 3). Tropisetron and RS-42358-197 had higher affinity in the cerebral cortex than in the ileum, while mCPBG and (R)YM060 had higher affinity in the ileum than in the cortex (Figure 3).

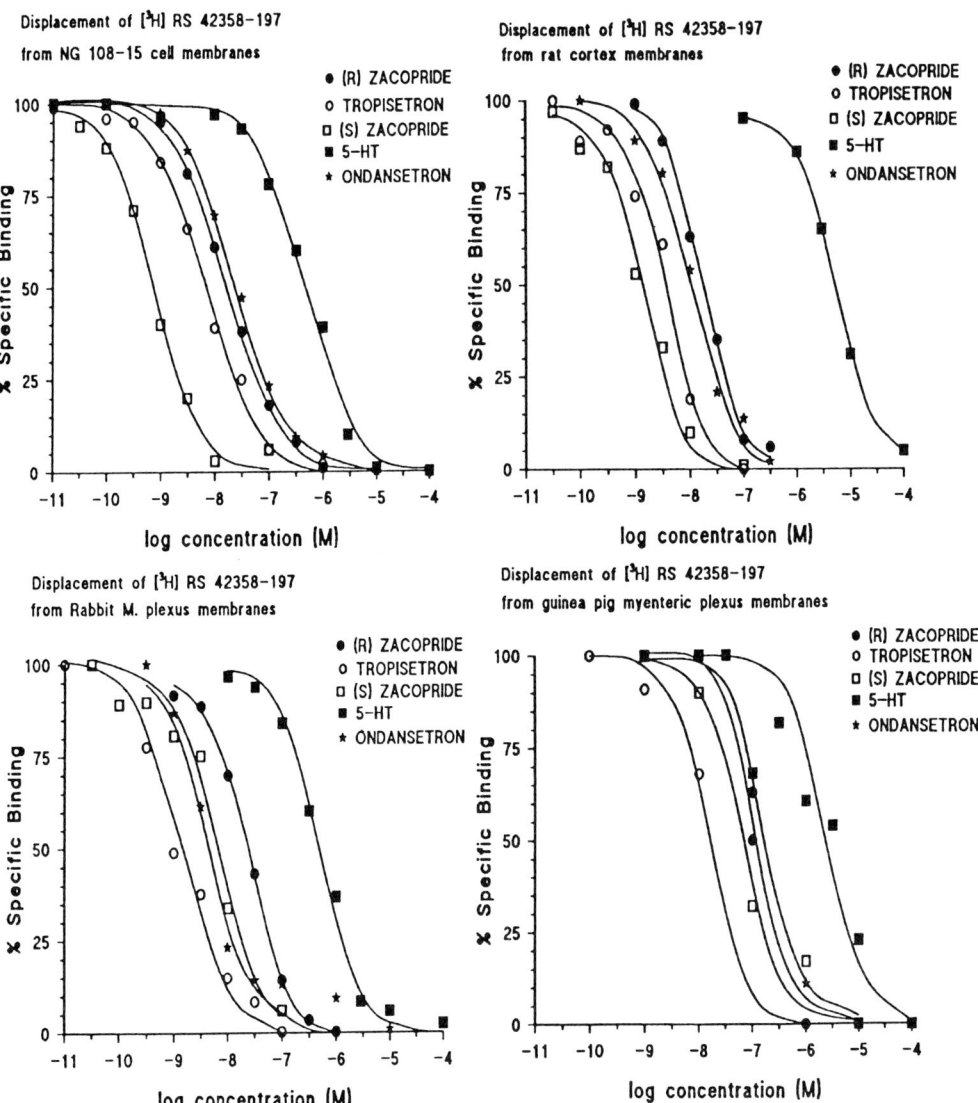

Figure 2. Displacement curves for standard 5-HT3 receptor agonist and antagonists against [3H]RS-42358-197 binding in NG-108-15 cells, rat cortex, rabbit myenteric plexus and guinea-pig myenteric plexus membranes.

DISCUSSION

[^{3}H]RS-42358-197 has been used to label high affinity binding sites in a variety of tissues. The pharmacological specificity of this molecule was indicated by the weak affinity (pK_i < 6.0) in 23 standard receptor binding assays[29]. The ability of 5-HT$_3$ receptor agonists and antagonists to inhibit [^{3}H]RS-42358-197 binding strongly suggests that [^{3}H]RS-42358-197 labeled 5-HT$_3$ receptors. This conclusion is further confirmed by the functional[30] and the behavioral[31] effects of this compound. Indeed, [^{3}H]RS-42358-197 appears to be the most potent (Table 1) and selective radioligand for the 5-HT$_3$ receptor (Table 1, 2 and 3).

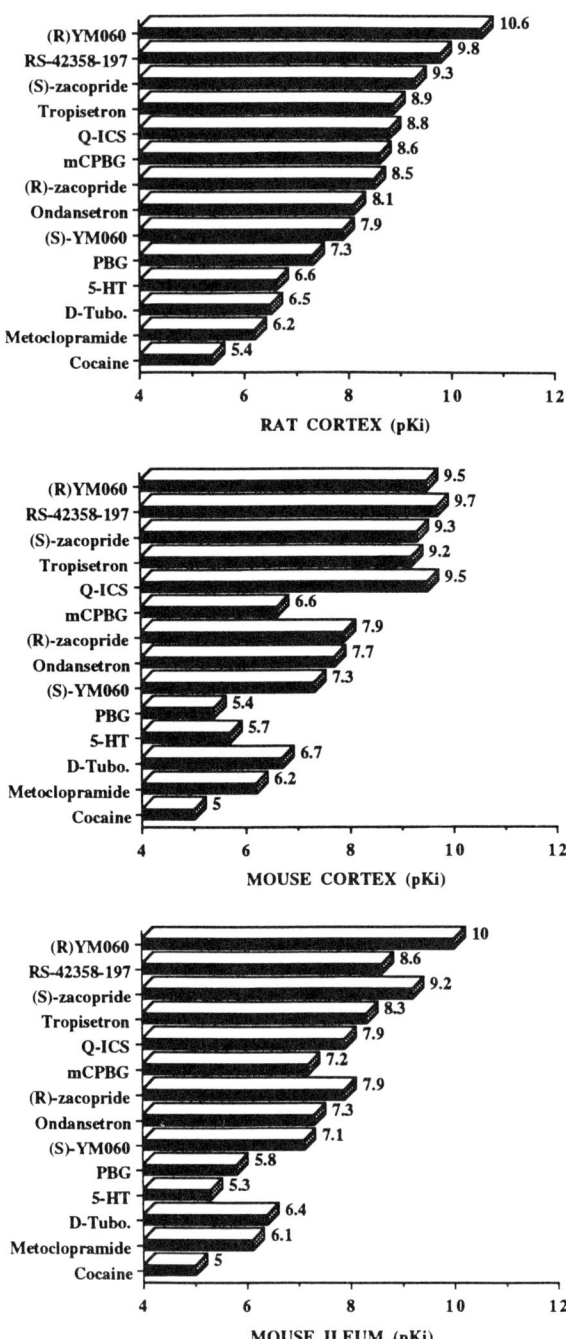

Figure 3. Comparison of the pharamcological profile of [3H]RS-42358-197 binding in rat cortex, CD-1 mouse cortex and CD-1 mouse ileum.

Table 1. Saturation studies of [3H]RS 42358-197, [3H]GR 65630, and [3H]quipazine

	[³H]RS 42358-197		[³H]GR 65630		[³H]quipazine	
	K_d (nM)	B_{max} (fmoles/ mg protein)	K_d (nM)	B_{max} (fmoles/ mg protein)	K_d (nM)	B_{max} (fmoles/ mg protein)
NG 108-15 cells	0.20±0.01	660±74	0.66±0.05	950±120	4.7±0.5	1280±120
Guinea-pig IMP	1.54±0.26	91±14	N.D.	N.D.	N.D.	N.D.
Rabbit IMP	0.10±0.01	88±12	0.19±0.01	69±6.0	4.9±1.1	126±19
Rat Cortex	0.16±0.11	45±13	0.55±0.06	15±1.5	4.3±0.8	64±8.2
CD-1 Mouse Cortex	0.20±0.07	30±6	-----	-----	-----	-----
CD-1 Mouse IMP	1.30±0.35	30±2	-----	-----	-----	-----

Values are means + S.E.M. of 3 - 14 separate determinations.
N.D. Not detectable up to 10 nM.

Table 2. Inhibition of [3H]RS-42358-197 binding for 5-HT3 receptor agonists and antagonists

	Rat Cortex	NG-108-15 cells	Rabbit Ileum	Guinea-pig Ileum
Antagonists				
RS-42358-197	9.8±0.1	10.2±0.1	9.9±0.1	8.4±0.2
(S)-zacopride	9.7±0.1	9.9±0.1	9.9±0.1	8.4±0.1
GR 65630	9.3±0.1	9.1±0.1	9.7±0.1	7.1±0.1
Tropisetron	9.0±0.1	9.2±0.1	9.6±0.1	7.0±0.1
Qua. ICS	8.8±0.1	9.0±0.1	9.7±0.1	7.0±0.2
Ondansetron	8.5±0.1	8.3±0.1	9.1±0.1	6.9±0.1
(R)-zacopride	8.3±0.1	8.6±0.1	8.6±0.2	7.1±0.1
MDL 72222	7.5±0.2	8.0±0.1	8.2±0.2	6.4±0.1
d-Tubocurarine	6.6±0.2	7.2±0.1	6.1±0.1	5.2±0.2
Metoclopramide	6.6±0.2	6.8±0.1	7.1±0.1	5.7±0.1
Agonists				
mCPBG	8.9±0.1	7.5±0.1	6.9±0.1*	5.6±0.1
Phenybiguanide	7.3±0.1	6.0±0.1	6.7±0.2*	5.5±0.2
2-methyl-5-HT	6.9±0.1	6.2±0.1	6.6±0.1	6.2±0.1
5-HT	6.2±0.2	6.8±0.1	5.9±0.1	4.9±0.5

Values are means ± SEM of at least three separate determinations.
* Statistically significant difference from values obtained using guinea-pig ileal membranes ($p < 0.05$, Student's t test).
Hill coefficients were reported in Wong et al.[29].

Inter-Species Heterogeneity

The first sign of heterogeneity in 5-HT$_3$ receptor binding sites was indicated by the recognition of significantly lower affinity of [³H]RS-42358-197 in CD-1 mouse ileum and the

Table 3. Binding parameters of 5-HT3 receptor ligands in C D-1 mouse ileum and cortex

	CD-1 mouse cortex		CD-1 mouse ileum	
	pK_i	nH	pK_i	nH
RS-42358-197	9.7 ± 0.3	0.70 ± 0.09	8.6 ± 0.2*	0.90 ± 0.12
(R)-YM060	9.5 ± 0.2	0.67 ± 0.10	10 ± 0.1*	0.94 ± 0.25
(S)-zacopride	9.3 ± 0.2	0.96 ± 0.08	9.2 ± 0.1	0.91 ± 0.14
Tropisetron	9.2 ± 0.3	0.88 ± 0.06	8.3 ± 0.2 *	0.89 ± 0.09
(R)-zacopride	7.9 ± 0.3	0.88 ± 0.10	7.9 ± 0.2	0.68 ± 0.12
Ondansetron	7.7 ± 0.2	0.81 ± 0.15	7.3 ± 0.1	0.78 ± 0.11
(S)-YM060	7.3 ± 0.1	0.96 ± 0.22	7.1 ± 0.3	0.66 ± 0.10
MDL 72222	7.2 ± 0.1	0.77 ± 0.10	6.7 ± 0.3	0.81 ± 0.11
d-Tubocurarine	6.7 ± 0.2	0.71 ± 0.08	6.4 ± 0.2	0.85 ± 0.07
Metoclopramide	6.2 ± 0.1	0.83 ± 0.06	6.1 ± 0.1	1.07 ± 0.08
Cocaine	5.0 ± 0.1	0.75 ± 0.16	5.0 ± 0.2	1.00 ± 0.20
mCPBG	6.6 ± 0.1	1.48 ± 0.20	7.2 ± 0.1 *	1.00 ± 0.08
5-HT	5.7 ± 0.2	1.15 ± 0.29	5.3 ± 0.1	1.00 ± 0.16
2-me-5HT	5.7 ± 0.1	0.96 ± 0.33	5.9 ± 0.2	0.95 ± 0.11
Phenylbiguanide	5.4 ± 0.2	1.02 ± 0.15	5.8 ± 0.3	0.84 ± 0.22

Values are mean ± SEM of 3-8 determinations. For each compound tested the upper value is the pKi (negative log of K_i) or K_d and the lower value is the Hill slope.
* Indicates a statistical ($p < 0.05$) difference from mouse cerebral cortex pK_i values.

guinea-pig ileum than in rat cortex (Table 1). Previous pharmacological[34,35] and electrophysiological[36,37] data have predicted the weak affinity of standard agonists and antagonists for the 5-HT$_3$ receptors in the guinea-pig. This might explain the inability of the currently available radioligands to label this site[11,38]. Unlike the other antagonists (Table 2), [^{3}H]RS-42358-197 apparently maintained a sufficiently high affinity for the guinea-pig 5-HT$_3$ receptors. The weak affinity of [^{3}H]RS-42358-197 had for the mouse ileum was somewhat unexpected. Nevertheless, it does re-enforce the idea that the guinea-pig is by no means unique in exhibiting evidence of 5-HT$_3$ receptor heterogeneity. While the mouse and guinea-pig ileum have lower affinity for [^{3}H]RS-42358-197 than the other tissues, their pharmacological specificity was different. The two ileal preparations can be distinguished on the basis of different affinities for d-tubocurarine and (S)-zacopride (16 to 6 fold higher affinity in the mouse ileum) and 2-methyl-5-HT (3 fold greater affinity in the guinea-pig ileum).

Analysis of the affinity of 5-HT$_3$ receptor agonists and antagonists for the [^{3}H]RS-42358-197 binding site gave further evidence of inter-species heterogeneity. (S)-zacopride, showed a species rank order of NG-108-15=rabbit ileum=rat cortex>mouse cortex=mouse ileum>guinea-pig ileum; ondansetron, on the other hand, gave a different profile: rabbit ileum>rat cortex=NG-108-15>mouse cortex>mouse ileum>guinea-pig ileum. Other antagonists that recognise differences (> 0.6 log unit) between species include: d-tubocurarine, GR 65630, tropisetron, qua. ICS 205930, MDL 72222. The most striking examples of species selectivity were found with the 5-HT$_3$ receptor agonists mCPBG[39] and PBG, with 2.3 and 1.9 log units difference in affinity between rat and mouse cortex. The apparent selectivity of d-tubocurarine and phenylbiguanide between the six tissues is consistent with reported functional data which has established the existence of species difference in the affinity of these antagonists[36,37,40]. However, it should be pointed out that the absolute affinity (K$_i$) of d-tubocurarine in NG-108-15 cells or rabbit ileum is weaker than that reported in the mouse and N1E-115 cells, or rabbit nodose ganglia by electrophysiological studies[37,41].

Intra-Species Heterogeneity

A close correspondence in the binding and functional[36] (superior cervical ganglia, SCG) potencies of d-tubocurarine for the rat (6.6 vs 7.1) and guinea-pig (5.2 vs 4.8 tissues was observed. In contrast, a discrepancy exist between binding and functional values for mouse cortex (6.4-6.7) and mouse superior cervical ganglia (8.1), arguing for the existence of subtypes of 5-HT$_3$ receptors in the mouse. This notion is suggested also by the detail analysis of the binding profiles of [^{3}H]RS-42358-197 in the CD-1 mouse cortex and ileum. These populations of receptors could be distinguished on the basis of RS-42358-197 and tropisetron (higher affinity in the cortex) and (R)-YM060[42] and mCPBG (higher affinity in the ileum) (Figure 3). The difference in affinity of ligands in this two tissues may reflect drug interactions at different proportions of heterogenous populations of receptors since the antagonist displacement curves were frequently shallower than those anticipated by assuming binding to a single, non-interating population of sites. Given the likelihood of allosteric interactions at the 5-HT$_3$ receptor as indicated by noncompetitive interaction between [^{3}H] antagonists and nonlabeled agonists[43,44] further work is needed to determine whether these differences in ligand affinity reflects different 5-HT$_3$ receptor subtypes or heterogeneity as a consequence of differences in the subunits comprising the heteromeric receptor. Nevertheless, the finding of tissue-specific differences in ligand affinity have important implications for the use of 5-HT$_3$ receptor antagonists in the treatment of central and peripheral disorders, e.g. central selective anxiolytic without gastrointestinal side-effects.

Intra-Tissue Heterogeneity

Comparison of the Bmax values of three radioligands in three tissue preparation (NG-108-15 cells, rabbit ileum and rat cortex has helped to further explore evidence of receptor heterogeniety (Table 1). Saturation binding isotherms for each ligand could be described by a single site model[45]. However, the density of binding sites labeled within a given tissue was dependent upon the ligand used. [^{3}H]quipazine labeled a greater density of sites in rat cortex than did [^{3}H]GR 655630 or [^{3}H]RS-42358-197. This "extra" site accounted for 75 % of the specific [^{3}H]quipazine binding sites. In rabbit myenteric plexus [^{3}H]quipazine also labeled a greater density of sites than did [^{3}H]GR 655630. However, in this case the "extra" site accounted for approximately half of the specific quipazine binding sites. By contrast in NG-108 cell membranes each of the ligands labeled an equivalent number of sites. This suggests that [^{3}H]quipazine was labeling sites in rat cortex and rabbit myenteric plexus which were not recognized by the other ligands in the range of concentrations tested and which are not present in high abundance in the NG-108-15 cells. Cicin-Sain and Jenner[46] have recently reported similar discrepancies in receptor density after surveying Bmax values obtained from rat cortical membranes in studies using [^{3}H]GR 65630 and other 5-HT$_3$ receptor radioligands.

The nature of the additional binding sites labeled by [^{3}H]quipazine remains to be fully determined. However, this site has pharmacological characteristics and localization[47] virtually identical to the site labeled by [^{3}H]GR 65630[48] (except for lower affinity for GR 65630 itself) and other 5-HT$_3$ receptors[49,12,13] and thus is likely a subtype of 5-HT$_3$ receptor. This site is not the 5-HT uptake site[49] since the inclusion of paroxetine would have prevented quipazine from binding to this site[50] and since the saturation binding curves of this ligand gave no indication of heterogeneity of binding sites. The idea that GR 65630 distinguishes among subtype of 5-HT$_3$ receptors is supported by the finding that, in the rabbit ileum and rat cortex, nonlabeled GR 65630 displaced [^{3}H]quipazine, but not [^{3}H]GR 65630 binding with high and low affinity components[52]. One straight-forward explanation of the data is that [^{3}H]quipazine non-discriminately labeled a heterogenous population of 5-HT$_3$ receptor binding sites and that, under the conditions used for the saturation binding isotherms, [^{3}H]RS-42358-197 and [^{3}H]GR 65630 labeled only a subpopulation of these sites.

In summary, the data obtained from these studies has confirmed the low affinity of 5-HT$_3$ receptor ligands for a receptor in guinea-pig. Further detailed analysis has indicated that such species differences was not restrictive to the guinea-pig. Binding data from the CD-1 mouse has further emphasized the existence of intra-species receptor heterogeniety. The fact that different radioligands for this receptor could recognise different population of sites within a given tissue such as the rat cortex suggested an interaction at this novel ion channel coupled receptor more complex than originally anticipated.

REFERENCES

1. S.J. Peroutka, 5-hydroxytryptamine receptor subtypes, *Ann. Rev. Pharmacol. Toxicol.* 67:373 (1990).

2. P.M. Whitaker-Azmitia and S.J.Peroutka The neuropharmacology of serotonin. *Ann. N.Y. Acad. Sci.* 600:233 (1990).

3. J.R. Fozard and P.R. Saxena, "Serotonin: Molecular Biology, Receptors and Functional Effects". Birkauser, Basel (1991).

4. P.P.A. Humprey, P. Hartig and D.Hoyer, A re-appraisel of 5-HT receptor classification. Proceedings of the 2nd International Symposium on Serotonin: From Cell Biology to Pharmacology and therapeutics, Kluwer, Houston (1992).

5. P.B. Bradley, G. Engel, W. Feniuk, J.R. Fozard,P.P.A. Humphrey, D.N. Middlemiss, E.J. Mylecharane, B.P. Richardson and P.R. Saxena, Proposals for the classification and nomenclature of functional receptors for 5-hydroxytrpyamine, *Neuropharmacol.* 25:563 (1986).

6. D. Clarke and J. Bockaert, 5-HT$_4$ receptor: Current Status, *Med. Res. Rev.*: in press (1993).

7. V. Derkach, A. Suprenant and R.A. North, 5-HT$_3$ receptors are membrane ion channels, *Nature* 339:706 (1989).

8. A.V. Maricq, A.S. Peterson, A.J. Brake, R.M. Myers and D. Julius, Primary structure and functional expression of the 5-HT$_3$ receptor, a serotonin-gated ion channel, *Science* 254:432 (1991).

9. J.R. Fozard, Neuronal 5-HT$_3$ receptors in the periphery, *Neuropharmacol.* 23:1473 (1984).

10. J.R. Fozard, "The peripheral actions of 5-hydroxytryptamine." Oxford University Press, Oxford (1989).

11. G.J. Kilpatrick, B.J. Jones and M.B. Tyers, Identification and distribution of 5-HT$_3$ receptor in rat brain using radioligand binding, *Nature* 330:746 (1987).

12. N.A. Sharif, E.H.F. Wong, D.N. Loury, E. Stefanich, A.D. Michel, R.M. Eglen and R.L. Whiting,Characterization of 5-HT$_3$ binding sites in NG108-15, NCB-20 neuroblastoma cells and rat cerebral cortex using [^{3}H]-quipazine and [^{3}H]-GR 65630 binding, *Br. J. Pharmacol.* 102:919 (1991).

13. D. Hoyer, D. and H.C. Neijt, Identification of serotonin 5-HT$_3$ recognition sites in membranes of N1E-115 neuroblastoma cells by radioligand binding, *Mol. Pharmacol.* 33:303 (1988).

14. H.C. Neijt, A. Karpf, P. Schoeffter, G. Engel and D. Hoyer,. Characterization of 5-HT$_3$ recognition sites in membranes of NG 108-15 neuroblastoma-glioma cells with [^{3}H] ICS 205,930, *Naunyn-Schmiedeberg's Arch. Pharmacol.* 33:493 (1988).

15. K.J. Watling, S. Aspley, S., Swain, C.J. and Saunders J. (1988) [^{3}H] Quaternised ICS 205-930 labels 5-HT$_3$ receptor binding sites in rat brain, *Eur. J. Pharmacol.* 149, 397-398.

16. R.M. McKernan, N.P. Gillard, K. Quirk, C.O. Kneen, G.I. Stevenson, C.J. Swain, and C.I. Ragan, Purification of the 5-hydroxytryptamine 5-HT$_3$ receptor from NCB20 cells, *J. Biol. Chem.* 265:13572 (1990).

17. D.M. Milburn and S.J. Peroutka, Characterization of [^{3}H]quipazine binding to 5-hydroxytryptamine$_3$ receptors in rat brain membranes, *J. Neurochem.* 52:1787 (1989).

18. L. Pinkus, N.S. Sarbin, D.S. Barefoot and J.C. Gordon, Association of [^{3}H]zacopride with 5-HT$_3$ binding sites, *Eur. J. Pharmacol.* 168:355 (1989).

19. J.M. Barnes, N.M. Barnes, B. Costall, J.W. Ironside and R.J. Naylor, Identification and characterization of 5-hydroxytryptamine$_3$ recognition sites in human brain tissue, *J. Neurochem.* 53:1787 (1989).

20. D.R. Nelson and D.R. Thomas, [^{3}H]-BRL43694 (Granisetron), a specific ligand for 5-HT$_3$ binding sites in rat brain cortical membranes, *Biochem. Pharmacol.* 38:1693 (1989).

21. D.T. Wong, D.W. Robertson and L.R. Reid, Specific [^{3}H]LY 278584 binding to 5-HT$_3$ recognition sites in rat cerebral cortex, *Eur. J. Pharmacol.* 166:107 (1989).

22. A.M. Laporte, T. Koscielniak, M. Ponchant, D. Verge, M. Hamon and H. Gozlan, Quantitative autoradiographic mapping of 5-HT$_3$ receptors in the rat CNS using [^{125}I]iodo-zacopride and [^{3}H]zacopride as radioligands, *Synapse* 10: 271 (1992).

23. E.H.F. Wong, I. Wu, R.M. Eglen and R.L. Whiting, Labeling of species variants of 5-hydroxytryptamine$_3$ (5-HT$_3$) receptors by a novel 5-HT$_3$ receptor ligand [^{3}H]RS-42358-197, *Br. J. Pharmacol.* 105:33P (1992).

24. B.P. Richardson and G. Engel G.The pharmacology and function of 5-HT$_3$ receptors, *Trends Neurosci.* 7:424 (1986).

25. J.A. Peters, J.J. Lambert and H.M. Malone, Physiological and pharmacological aspects of 5-HT$_3$ receptor function. In "Aspects of Synaptic Transmission: LTP, Galanin, Opioid, Autonomic, 5-HT" (ed. Stone T.W.) , pp. 283, Taylor & Francis, London (1991).

26. M.S. Aapro, 5-HT$_3$ receptor antagonists: An overview of their present status and future potential in cancer therapy-induced emesis, *Drug* 42:551 (1991).

27. B. Costall, R.J. Naylor and M.B. Tyers, The psychopharmacology of 5-HT$_3$ receptors, *Pharmacol. Ther.* 47:181 (1990).

28. M.D.Tricklebank, Interactions between dopamine and 5-HT$_3$ receptors suggest new treatments for psychosis and drug addiction, *Trends Pharmacol. Sci.* 10:127 (1989).

29. E.H.F. Wong, D.W. Bonhaus, E. Stefanich, and R.M. Eglen, Labeling of 5-hydroxytryptamine$_3$ receptors with a novel 5-HT$_3$ receptor ligand, [^{3}H]RS-42358-197, *J. Neurochem.* 60:921 (1993).

30. R.M. Eglen, C. Lee, W.L. Smith, L.G. Johnson, R.L. Whiting and S.S. Hedge, RS-42358-197, a novel and potent 5-HT3 receptor antagonist, in vitro and in vivo, *J. Pharmacol. Exp. Ther.*, in press (1993).

31. B. Costall, A.M. Domeney, M.E. Kelly, D.M. Tomkurs, R.J. Naylor, E.H.F.Wong, W.L. Smith, R.L. Whiting and R.M. Eglen, The effect of the 5-HT$_3$ receptor antagonist, RS-42358-197, in animal models of anxiety, *Eur. J. Pharmacol.* 234:91 (1993).

32. Y.C. Cheng and W.H. Prusoff, Relationship between inhibition constant (K_i) and the concentration of inhibitor which causes 50 percent inhibition (IC_{50}) of an enzymatic reaction, *Biochem. Pharmacol.* 92:881 (1973).

33. J.O. Marcusson, M Bergstrom, K. Eriksson and S.B. Ross, Characterization of [^{3}H]paroxetine binding in rat brain, *J. Neurochem.* 50:1783 (1988).

34. A. Butler, C.J. Elswood, J. Burridge, S.J. Ireland, K.T. Bunce, G.J. Kilpatrick and M.B. Tyers, The pharmacological characterization of 5-HT$_3$ receptors in three isolated preparations derived from guinea-pig tissues, *Br. J. Pharmacol.* 101:591 (1990).

35. R.M. Eglen, S.R. Swank, L.K.M. Walsh and R.L. Whiting, Characterization of 5-HT$_3$ and 'atypical' 5-HT receptors mediating guinea-pig ileal contractions *in vitro.*, *Br. J. Pharmacol.* 101:513 (1990).

36. N.R. Newberry, S.H. Cheshire and M.J. Gilbert, Evidence that 5-HT$_3$ receptors of the rat, mouse and guinea pig superior cervical ganglion may be different. *Brit. J. Pharmacol.* 102:615 (1991).

37. H.M. Malone, J.A. Peters.and J.J. Lambert, Physiological and pharmacological properties of 5-HT$_3$ receptors-a patch clamp-study, *Neuropeptide* 19:25 (1991).

38. J.R. Fozard, 5-HT$_3$ receptors in the context of the multiplicity of 5-HT receptors. In "Central and Peripheral 5-HT$_3$ receptors", ed. Hamon, M., Academic Press Ltd., London (1992).

39. G.J. Kilpatrick, B.J. Jones and M.B. Tyers, 1-(m-chlorophenyl)-biguanide, a potent high affinity 5-HT$_3$ receptor agonist, *Eur. J. Pharmacol.* 182:193 (1990).
40. G.J. Kilpatrick and M.B. Tyers, Inter-species variants of the 5-HT$_3$ receptor, *Biochem. Soc. Trans.* 20:118 (1992).
41. J.A. Peters, H.M. Malone and J.J. Lambert, Characterization of 5-HT$_3$ receptor mediated electrical responses in nodose ganglion neurones and clonal neuroblastoma cells maintained in culture. In "Serotonin: Molecular Biology, Receptors and Functional Effects" (ed. Fozard, J.R. and Saxena, P.R.), pp. 84-94. Birkhauser, Basel (1991).
42. K. Miyata, T. Kamato, A. Nishida, H. Ito, Y. Katsuyama, A. Iwai, H. Yuli, M. Yamano, R. Tsutsumi, M. Ohta, M. Takeda and K. Honda, Pharmacological profile of (R)-5-[(1-methyl-3-indolyl)carbonyl]-4,5,6,7-tetrahydro-1H-benzimida-zole hydrochloride (YM060), a potent and selective 5-hydroxytryptamine$_3$ receptor antagonist, and its enantiomer in the isolated tissue, *J. Pharmacol. Exp. Ther.* 259:15 (1991).
43. J.M. BarneS, N.M. Barnes, B. Costall, S.M. Jagger, R.J. Naylor R.J. and D.W. Robertson, Agonist interactions with 5-HT$_3$ receptor recognition sites in the rat entorhinal cortex labeled by structurally diverse radioligands, *Br.J.Pharmacol.* 105:500 (1992).
44. D.W. Bonhaus, R.M. Eglen and E.H.F. Wong, Allosteric interactions of agonists and antagonists at 5-hydroxytryptamine (5-HT$_3$) receptors, *Society for Neuroscience Abstracts* 18:1518 (1992).
45. E.H.F. Wong, J. Lee, D.N. Loury, R.M. Eglen and R.L. Whiting, Heterogeneity of 5-HT$_3$ receptors as labeled by [^{3}H]GR65630 and [^{3}H]quipazine, *J. Neurochem.* 57: S136 (1991).
46. L. Cicin-Sain and P. Jenner, Localization and characterization of 5-HT$_3$ receptors in the forebrain of the rat identified by the specific binding of [^{3}H]GR 65630, *Neurosci. Res. Comm.* 10:17 (1992).
47. N.A. Sharif, J.L. Nunes, Z.P. To, E.H.F. Wong, R.M. Eglen and R.L. Whiting, Quantitative autoradiographic distribution of 5-HT$_3$ receptors in rat, ferret and dog brain, *Br. J. Pharmacol.* 102:142P (1991).
48. G.J. Kilpatrick, B.J. Jonesand M.B. Tyers, Binding of the 5-HT$_3$ ligand, [^{3}H] GR65630, to rat area postrema, vagus nerve and the brains of several species, *Eur. J. Pharmacol.* 159:157 (1989).
49. S.C. Lummis, G.J. Kilpatrick and I.L. MArtin, Charaacterization of 5-HT$_3$ receptors in intact N1E-115 neuroblastoma cells, *Eur. J. Pharmacol.* 189:223.
50. A.W. Schmidt, S.D. Hurt and S.J. Peroutka, [3H]quipazine degradation products label 5-HT uptake sites, *Eur. J. Pharmacol.* 171:141 (1989).
51. J.O. Marcusson, M. Bergstrom, K. Eriksson and S.B. Ross, Characterization of [^{3}H]paroxetine binding in rat brain, *J. Neurochem.* 50:1783 (1988).
52. E.H.F. Wong, D.W. Bonhaus, J.A. Lee, I. Wu, D.N. Loury and R.M. Eglen, Different densities of 5-HT$_3$ receptors are labeled by [^{3}H]quipazine, [^{3}H]GR 65630 and [^{3}H]granisetron, *Neuropharmacol.* 32: in press (1993).

ADVANCES IN CLINICAL RESEARCH ON COMMON MENTAL DISORDERS WITH COMPUTER CONTROLLED ELECTRO-ACUPUNCTURE TREATMENT

Hechun Luo[1], Yunkai Jia[1] Xiugin Feng[1], Xueying Zhao[1] and Lily C. Tang[2]

[1]Institute of Mental Health
Beijing Medical University
Beijing, People's Republic of China
[2]SLCT Inc.
P. O. Box 1634
Bethesda, MD 20827, USA

INTRODUCTION

Depressive psychosis is a common psychological disorder. In the 1930s electric shock treatment has been generally used in the world, but was discarded due to undesired traumatic results. The last three decades, chemicals such as aminoketones, monoamine oxidase inhibitors, phenothiazines, serotonin uptake inhibitors, tetracyclics, tricyclics have been used with minimal success. These compounds all exhibit various side effects. Although the patients and their families accepted these treatments better than the electric shock but there exist many disadvantages: drugs toxicities and high recurrence rate as well as long term use of tachyphylaxis is inevitable. Recently, there is a great interest in seeking "alternative therapies". Among the "alternative therapies" acupuncture is one that has the most scientific support.

Acupuncture has been used in China for more than a thousand years not only for somatic disease but for mental disorders as well. In the last 50 years electric stimulation has been applied to acupuncture needle to act on the nerve cells around the acupoints. Many countries have adopted the use of acupuncture and electro-acupuncture (EA) for the relieve of localized pain with considerable success.[1-12] In Japan, since 1973 has been investigating the basic mechanism and clinical therapeutic effect on applying electron stimulation to induce sleep. That is by stimulating several acupoints, Yintang being one, for three minutes and 17 seconds, with 14 Hz to 0.1 Hz could induce sleep within 20 min.[13] Acupuncture has also been reported to be effective in treating mental disorders, depressive psychosis.[14] The(EA) appears to be much more effective than acupuncture in most of the treatments.[14,15] We have tried using EA to treat depressive patients as well as investigated its mechanism. EA has also been tried on male White Wister rats. It shows that 30-45 min after EA treatment at acupoints Baihui (Du20) and Yintang (EX-HN) also called (Glabella), there is a change in the animal behavior. The results obtained from the animal behavioral study indicated that the electrode itself does not affect the animal behavior but the stimulation of these acupoints presented sedation.[16]

In the last decade, China has promoted the use of combine therapies of western medicine and traditional Chinese medication. The early trials of EA treatment on depression, the choice of acupoints for EA stimulation of each medical facility and physician varied. The electric hertz applied was large and not regulated which produced unpredicted side effects. There were no standard procedure or uniform criteria of the patients' condition on the combination of EA and drugs treatments. The acupoints used or patients included in the treatments were not standardized. It is difficult to determine the efficacy of EA in treatment of depressive disorders.[17] There is a desperate need to review and evaluate all the clinical studies on treatment of depression by EA at the Institute of mental Health, Beijing Medical University and other hospitals in the nearby provinces of Beijing. In this paper we report the findings of our review on clinical data collected since 1981, on EA treatment of depression. Based on our review, we designed a strictly control clinical trial to verify our findings. These results were compared with those obtained on patients receiving antidepressant, amitriptyline (AM) treatment. In order to secure better efficacy of EA treatment we expanded our study by using computer controlled electro-acupuncture (CCEA) for combating depressive psychosis and schizophrenia.

MATERIAL AND METHODS

Equipment

The equipment used were Electro-acupuncture instrument, models G6805 and WQ-6F and CCEA instrument (Figure 1). The CCEA was designed and constructed with the collaboration of the Department of Pharmacology, Basic Medical Science, Beijing Medical University and the Electrical High Technology Research Institute, Academia Sinica.

Characteristics of Patients Included in the Clinical Trials:

Patients who conformed to the diagnosis with the criteria presented at the Huangshan Symposium on manic-depression in 1981[18] and the Handbook of Epidemiologic Survey of Mental Disease in China[19] were randomly assigned into various groups for our studies. Criteria for determining the patients included in the studies were according to the Present State Examination (PSE) and Schedule for A Standardized Assessment of patients with Depressive Disorders (SADD) recommended by World Health Organization (WHO). Patients were also assessed by the Hamilton Depressive Rating Scales (24 items).[20] The total Hamilton score of a patient has to be above 20 to be considered in the studies.

Methods

Trial Design We reviewed and analyzed the three clinical investigations carried on in the period from 1981 to 1988.

The first (1981-1983) pilot, study[21,22]: 47 patients from in-and out-patient Departments of the Institute of Mental Health, Beijing Medical university were divided into two groups:

1). The EA group: Twenty seven patients were treated by EA, one hour everyday except Sunday for five weeks, a total of 30 needling. 2). The AM group: Twenty patients were treated with AM for five weeks with daily dosage ranging form 100-200 mg with an average of 142 mg.

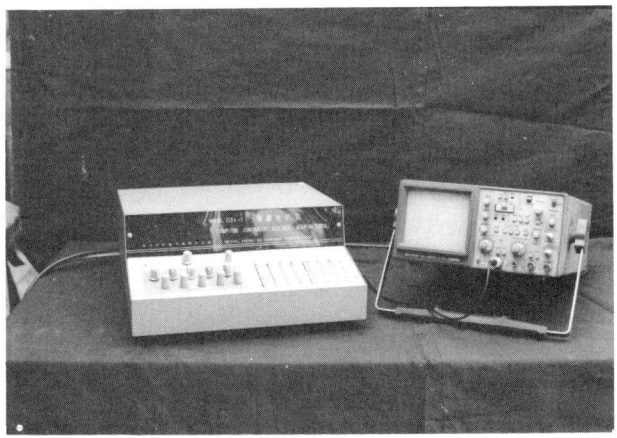

Figure 1.

The second study (1984-1985)[23], patients with a history of diagnosis, no longer than two years, of suffering from depressive disorders were hospitalized in the clinic of the Institute of Health, Beijing Medical University. The patients received no drug for at least one week and divided into three groups for this double blind program. Group I: Eight patients received EA treatment for six weeks, a total of 36 needling. These patients also received placebo capsules similar to those of AM in appearance during the period of EA treatment. Group II: Eleven patients were administered AM as described in the pilot study for six weeks, dosage adjusted to average 175 mg daily. Group III: The patients were treated by EA with the same protocol as group one in addition of receiving AM 25 mg t.i.d. for six weeks.

The third study (1985-1987)[24], is a collaborative study of ten hospitals to verify the efficacy of EA. All patients were hospitalized and remained drug free for at least a week and randomly divided into two groups. Group one consisted of 133 patients with an average history of 5.41 months was treated by EA and placebo tablets for five weeks. Group two has a total of 108 patients with an average history of 5.39 months receiving AM for five weeks with an average dose of 161 mg per day.

The fourth trial was carried out in 241 patients, 109 male and 132 female, by the Institute of Mental Health, Beijing Medical University in collaboration with the Psychiatric Hospitals in Xiang Fan, and Sha Shi, Hubei. The age of the patients ranges from 17 to 64, with an average of 36. Of the 241 cases, 193 were of the Depressive phase of manic-depression and 48 of reactive depression. Average duration of illness was 5.41 months. These patients were divided into two groups: the EA group and the AM group. 133 patients were assigned to the EA group, with 42 newly diagnosed and 91 recurrent cases. The AM group consisted of 108 patients, with 31 newly developed and 77 relapsed. Follow up study was conducted for 2-4 years on 101 of these patients, EA 54 and AM 47. 54 patients (32 male and 22 female) of the EA group were suffering from depressive phase of manic-depression, 10 with reactive depression 2 with depressive neurosis and 1 with atypical depressive neurosis. Among the AM group (19 male and 28 female), 37 patients was with depressive phase of manic-depression, 9 patients with reactive depression, and one patient with depressive neurosis.

Trial on CCEA treatment was designed with the purpose of comparing its efficacy with EA treatment. CCEA treatment was conducted on 98 patients with affective disorders, schizophrenia and neurosis, and 26 cases of manic-depression. Clinical diagnosis of the 98 cases, 40 cases were depressive psychosis, 13 male and 27 female, age 42.2 ± 14.0, average duration of illness being 68 - 97.9 months; 37 cases were neurosis, 13 male and 24 female, age 43.6 ± 13.7, with duration of illness 52.2 - 57.1 months; and 21 cases were schizophrenia, 7 male and 14 female, age 31.1 ± 8.2 and the duration of illness was 52.2 - 57.1 month. Of the 26 cases of manic depression, there were 8 cases of male and 18 female, age 41.8 ±13.4 and duration of illness was 63.4 - 100.2 months. 71 cases of affective disorders, schizophrenia and neurosis and 25 cases of manic-depression were treated with EA (WQ-6F). Among the 71 cases: there were 15 cases of depressive psychosis, 7 male and 8 female, age 42.9 ± 11.6 with duration of illness of 77.6 - 80.3 months; 40 cases were neurosis, 24 male and 16 female, age 34.1 ± 13.8 with 44.9 - 58.2 months of duration of illness; and 16 cases were schizophrenia, 5 male and 11 female, age 26.5 ± 6.8, total duration of illness was 56.5 - 63.8 months. In the 25 manic depression cases, there were 9 male and 16 female, age 44.7 ± 14.6, with total duration of illness being 55.4 - 88.2 months.

Clinical Treatments The acupoints applied were Baihui and Yintang. Acupoint Baihui was chosen for its capability in adjusting the action on the C-V system and power in Yang draining. The Yintang, an extra-channel acupoint, was chosen because it can activate and adjust the action with sedative and tranquilizing effects as well. Yantang can also relieve stasis and soothe the nervous system. Both Baihui and Yintang acupoints have been considered by the traditional Chinese medicine as an effective treatment for depressive psychotic conditions. The needle was inserted obliquely in the frontal direction beneath the scalp for 8 fens (1 fen = one tenth of a cun or inch) at Beihui. At the Yintang acupoint, the needle was inserted obliquely and upward, 8 fens beneath the skin. All needles were connected to an Electro-acupuncture Stimulator (G6805) or a Computer Controlled Electro-acupuncture apparatus. For the EA treatment, the current applied was 3-5 mA . Current was adjusted to optimum when slight twitching of the skin was visible around the needle and the patient remained comfortable. The frequency of the stimuli was 2 Hz. 3.0-5.0 V were used. Stimulation was 45-60 min for each treatment. With the CCEA study, four parameters were selected: the antidepressant parameter 1, AD1; antidepressant parameter 2, AD2; electro-acupuncture anesthetic parameter, EAA and parameter mimic the change of the EEG, MCE. The times of stimulation were ten minutes for AD1; 10 min for AD2; 10 min for EEA and 15 min for MCE. The sine of the wave was the used for standard adjustment. Wave frequencies for the 4 parameters were 12, 10, 8, 6 respectively. The first basic wave frequency was 250, while the second one was 750 Hz. Total time of the daily treatment was 45 min. Each course consisted of 36 treatments. In AM treatment, patients were administered amitriptyline 25 mg t.i.d. for the first week. The dosage was then adjusted according to the effects and side effects of the drug up to 50 mg t.i.d..

Clinical Assessment At least two physicians examined and evaluated independently the therapeutic efficacies before, after and weekly during treatments. The evaluation criteria were according to the Hamilton Depression Rating Scale, the Clinical Global Impression Chart (CGI)[25], and the Rating Scale for Side Effects (ASBERG)[26]. The Hamilton Rating Scale, with the score consists of 24 symptoms. Each of the symptom is rated as either 0-2 or 0-4. 0 being absent. The higher the number, the more severe the symptom is. CGI consists of three items: 1) severity of the disease; 2). general progress (the higher the readings the more severe the illness is and the poorer the therapeutic effect) and 3). the efficacy index which is derived from the scores of side-effects. The efficacy index is in inverse proportion to the side-effects. The higher the reading indicated the more effective the therapy. Grading system commonly employed in China for the assessment of therapeutic effects (GSC) consists of four categories: cured, markedly improved, improved and failed. The GSC was established as a general standard in China at the Nanking conference 1958. Patients were evaluated by at least two

Table 1. The Mean Scores and the Difference before and after Treatment

	N	Before	After	%	P
EA	27	28.4 ± 2.2	12.8 ± 2.0	-55.1	< 0.01
Amitriptyline	20	29.4 ± 1.4	14.2 ± 1.8	-51.7	< 0.01

Table 2. The OGI Sources and the Difference before and after Treatment

	EA (N = 133)				Amitriptyline (N = 108)			
	N	Before	After	P	N	Before	After	P
A	27	5.1 ± 0.7	2.9 ± 0.3	<0.01	20	5.2 ± 0.1	3.1 ± 0.4	<0.01
B	27	4.0 ± 0.0	2.0 ± 0.2	<0.01	20	4.0 ± 0.0	2.2 ± 0.2	<0.01
C	27	1.0 ± 0.0	3.0 ± 0.2	<0.01	20	1.0 ± 0.0	1.5 ± 0.1	<0.01

A : Severity of Disease; B: General Progress; C : Efficiency Index

physicians without knowing the treatment (double blind study). The agreements between the evaluating physicians were above 88%.

RESULTS

Trial 1, Pilot Study

The mean Hamilton Scores of the patients, 27 EA treated and 20 the AM group were displayed in Table 1. Both groups showed marked improvement after treatment. The p values are smaller than 0.01. Table 2 shows that after five weeks of treatment, both groups' CGI altered significantly as compared with the readings before treatment. There is no difference between the two groups in the categories of severity of disease and general progress but there is a mark difference in efficacy index. Using the GSC rating, after 5 weeks of treatment, in the EA group, there were twelve considered cured, seven markedly improved, five improved and three failed. The total successfully treated (cured and markedly improved) was 70%. In the AM group, there were six cured, seven markedly improved, seven improved and none failed. The total effective rate was 65%. It appears there is no significant difference between the EA and the AM therapies.

Trial 2, Double Blind Study

The Hamilton scores of all 239 patients of the three groups in this study reduced significantly after being treated for six weeks p < 0.01. Hamilton scores of the three groups are similar. No difference was observed among the three groups. (Table 3). The CGI scores of the groups studied were shown in Table 4. Again there is no difference between groups that

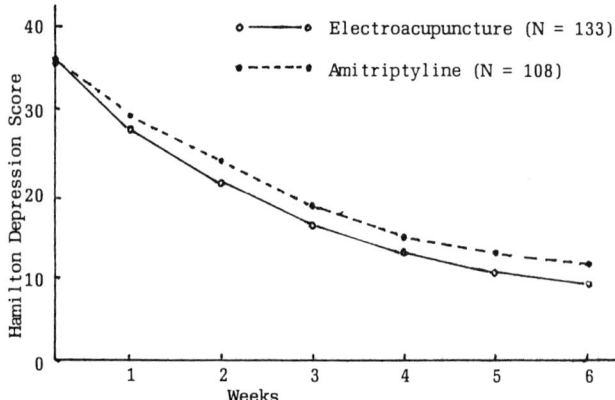

Figure 2. Hamilton Depression Scores in two groups of patients treated with electroacupuncture or amitriptyline for six weeks.

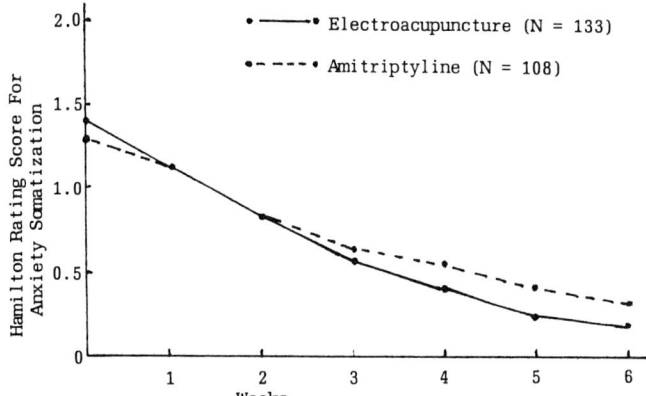

Figure 3. Hamilton Rating Scores for anxiety somatization in two groups of patients treated with electroacupuncture or amitriptyline.

indicates the efficacy of EA in treating depression in general is equal to the efficacy of administering antidepressant drug AM.

Trial 3, Collaborative Study of Institute of Mental Health, Beijing Medical University and Nine Other Psychiatric Hospitals

Results of this study are presented in Figure 2 and Table 5. Figure 2 showed the mean Hamilton scores of the 133 patients in the EA group and the 108 cases of the drug treated were both significantly reduced, six weeks after treatment as compared with the mean score before treatment (p value < 0.01). There exhibits no significant difference between groups (p > 0.05) which suggested that the therapeutic effect of EA is as efficient as AM treatment. This double blind study confirms the positive efficacy of EA on depressive disorders.

Table 3. The Hamilton's Depression Scores: The Difference before and after Treatment

	N	Before	After	P
EA + Placebo	8	24.6 ± 0.6	11.6 ± 0.9	< 0.01
Amitriptyline	11	28.9 ± 0.7	9.9 ± 0.7	< 0.01
EA + Amitriptyline	10	30.1 ± 0.6	13.0 ± 0.8	< 0.01

Table 4. OGI Scores and the Difference of Severity before and after Treatment

	N	Before	After	P
EA + Placebo	8	4.8 ± 0.10	2.5 ± 0.17	< 0.01
Amitriptyline	11	5.4 ± 0.07	2.6 ± 0.15	< 0.01
EA + Amitriptyline	10	5.4 ± 0.06	3.0 ± 0.18	< 0.01

Table 5. The effects of EA and Amitriptyline Treatments

	Total		Cured		Markedly Improved		Improved		Failed	
	N	%	N	%	N	%	N	%	N	%
EA	133	100	72	54.1	28	21.1	30	22.6	3	2.3
Amitryptyline	108	100	50	46.3	22	20.4	31	28.7	5	4.6

When the scores of individual syndrome of the Hamilton Rating Scale of the EA and AM treatments are examined carefully, the scores on the syndrome of anxiety somatization (Figure 3) and the disturbance of cognitive process were significantly lower in the EA than the AM treated group. (Table 5). These data imply that in these two categories EA treatment is more effective than tricyclics. The Hamilton' mean score of the EA group decreased form 36.46 ± 0.78 to 8.99 ± 1.00 whereas the AM treated group decreased from 36.36 ± 0.87 to 10.53 ± 1.33. $p < 0.05$. However, the EA treatment seems to be more effective in treating reactive depression than AM. The Hamilton's mean score of the EA group declined from 30.43 ± 1.74 to 5.24 ± 1.44, while the drug treated group, the score decreased from 32.43 ± 1.74 only to 10.04 ± 1.98 after treatment. ($p=0.05$)

Similar to the Hamilton scores of the patients treated with EA and AM in this trial, when examined using the General Grading System commonly used in China for assessment of therapeutic efficacy; there is no significant difference between these treatments. The total cured and markedly improved rate was 75.2% for EA group and 66.7 % for AM group.

Table 6. Rating Scale for Side Effects (ASBERG)

	EA (n = 133)		Amitriptyline (n = 108)	
	Slight	Severe	Slight	Severe
Physical tireness	26	6	38	8
Headache	14	1	25	5
Dizziness	14	0	28	7
"Orthostatic" syndrome	2	0	18	5
Palpitations	16	0	43	11
Tremor	5	0	36	10
Perspiration	9	0	30	5
Dryness of month	16	0	42	38
Constipation	8	0	30	24
Micturition disturbance	0	0	12	20
Drowsiness	0	0	0	0
Sleep disturbance	18	10	27	7
Interference with sexual function	10	0	13	2

Table 7. Factors in Hamilton's Depression Rating Scale for Patients Treated with EA and Amitriptyline

Factor	EA (M = 133)		Amitriptyline (N = 108)	
	Before	After	Before	After
1. Anxiety somatization	1.35 ± 0.05	0.17 ± 0.03	1.24 ± 0.06	0.32 ± 0.05*
2. Weight change	0.98 ± 0.07	0.08 ± 0.03	0.95 ± 0.09	0.11 ± 0.03
3. Cognitive process disturbance	1.02 ± 0.05	0.14 ± 0.03	1.09 ± 0.09	0.27 ± 0.06*
4. Diurnal variation	1.02 ± 0.07	0.19 ± 0.04	1.06 ± 0.08	0.19 ± 0.04
5. Retardation	2.23 ± 0.05	0.57 ± 0.07	2.33 ± 0.06	0.74 ± 0.08
6. Insonia	1.12 ± 0.06	0.21 ± 0.04	1.09 ± 0.06	0.30 ± 0.05
7. Helplessness	2.45 ± 0.08	2.43 ± 0.09	2.43 ± 0.09	0.75 ± 0.09

* $P < 0.05$ compared with the corresponding value in EA group

All others $P < 0.01$ compared with corresponding value in EA group

The ASBERG rating scale of this trial were shown in Table 6. Side effects are more severe in the AM group than in the EA ones. These side effects are mostly of the undesired symptoms on the cardiovascular, extrapyramidal and cholinergic systems.

Trial 4, Short and long term recurrence study of EA and AM treatments of depressive diseases

Data obtained from this trial, among the 133 patients with depressive psychosis treated by EA reduced the Hamilton mean score of 35.3 ± 0.7 to 8.3 ± 0.7; $p < 0.01$, while the 108 AM treated patients' Hamilton score reduced from 35.3 ± 0.8 to 10.4 ± 1.1; $p < 0.05$. Again this trial showed that EA and AM treatment both benefit the depressive patients.

When examined the follow-up data of the 101 patients symptoms recurred in 35 of the 53 patients of the EA group, a recurrence rate of 64%; while 26 of the 45 patients of the AM group suffered relapse, a recurrence rate of 53%. The long term curative effect of the two groups as assessed by chi square ($p > 0.05$) are not significantly different. Similar results were obtained when evaluating the long-term therapeutic efficacy by examining the patients' data using the GSC system.

Further examined the seven critical symptoms of the Hamilton Scale revealed that the scores of both EA and AM groups reduced significantly after treatments ($p < 0.01$). However, there is no difference between the two groups on five of the symptoms $p > 0.05$; except the scores on the anxiety somatization and cognitive process disturbance, $p < 0.05$ (Table 7). These results verified the findings of our collaborative trial with nine other hospitals. Data from this study suggests that EA may be superior in treating patients with anxiety somatic and cognitive disorder syndromes.

Therapeutic efficacies of both EA and AM treatments on the Depressive Phase of manic-depression assessed by the Hamilton Scale, are similar with no significant difference. The p values of the scores of the EA group obtained before and after treatments, is < 0.01 and the AM group is < 0.05. Both EA and AM treatments reduced the Hamilton score in regard to Reactive Depression. However, EA was found more effective ($p < 0.01$) than AM ($p < 0.5$) in respect to clinical observation.

Analysis of the data obtained from the follow-up study, the recurrence rates in the various syndromes: 26 of 40 patients with depressive phase of manic-depression developed recurrence and 7 out of 10 patients with reactive depression did not benefit from EA treatment. In the AM group, 18 out of 35 patients with depressive phase of manic depression relapsed and 7 out of 9 patients with reactive depression showed negative effect. The two groups as a whole do not exhibit significant difference ($p > 0.05$).

Among the EA group, 36 (68%) patients received supplementary drug therapy after they were discharged from the hospital. The supplementary drug is one of the following: AM, doxepin, imipramine, lithium carbonate or other antidepressants. 24 (68.8%) patients of the AM group were given similar supplements as the EA group. In spite of the supplementary treatment, there were still high incidence of recurrence in both groups. There was no difference in the recurrence rates in the EA and AM groups.

We have carefully examined the effects of the family members' concern on the patients' conditions. In the EA group, 29 (67%) of 43 patients who received close concern and attention from their relatives developed recurrence, whereas 6 (78%) out of 8 patients whose families have little or no concern of them relapsed. Within the AM group, 27 (64%) of 42 patients who have close concern from their family members developed recurrence, while 1 (33%) of 3 patients whose relatives showed little or no concern had relapsed. These data suggested that the concern of the family of the condition of the patient have little bearing on the recurrence of the illness.

Table 8. The Effects of OCEA and EA Treatments

	Total		Cured		Markedly Improved		Improved		Failed	
	N	%	N	%	N	%	N	%	N	%
CCEA	98	100	44	44.9	32	32.7	15	15.3	7	7.1
EA (WQ-6F)	71	100	19	26.7	24	33.8	17	24.0	11	15.5

Table 9. The Effects of OCEA and EA Treatments on Patients with Depressive Disorders

	Total		Cured		Markedly Improved		Improved		Failed	
	N	%	N	%	N	%	N	%	N	%
CCEA	40	100	22	55.0	11	27.5	4	10.0	3	7.5
EA (WQ-6F)	15	100	5	33.3	3	20.0	5	33.3	2	13.4

Another important factor had also been analyzed. That is the relationship between recurrence and family history of mental disorders. The analysis of the data showed that 9 (60%) of the EA group and 7 (54%) of the AM group that have family history of mental illness developed recurrence.

In the EA group two patients attempted suicide; one succeeded. One patient in this group died of somatic disease. In the AM group four had tried to commit suicide but none succeeded.

CCEA and EA studies

Data obtained from 98 patients treated with CCEA and 71 cases with EA (WQ-6F) are listed in Table 8. They clearly showed that CCEA therapy yielded significant advantage over EA treatment by the evaluation of GSC rating. The statistical analysis presented significant difference between these two groups of patients (Chi Square = 5.70, $p < 0.05$).

Results obtained by analyzing the GSC grading system of the patients with depressive disorders are presented in Table 9. From this table, one can see the therapeutic efficacy of CCEA is 82.5% and EA (WQ-6F) is 53.3%. The chi square being - 4.89 ($p < 0.05$). There is significant difference between the effects of these two treatments. These data reflect the superiority of CCEA over EA.

The GSC ratings of the CCEA and EA treated neurosis patients are tabulated in Table 10. Therapeutic efficacy of the CCEA group is 75.6% whereas the EA group is 60%. There is no significant difference between the two groups, chi square = 2.15, $p > 0.05$. This indicates that CCEA and EA treatments are similar in treating neurosis.

Table 11 shows the data obtained from the treatment of CCEA and EA on schizophrenia. The statistical analysis of these data exhibited no significant difference in this two types of treatment, chi square = 0.03, $p > 0.05$. Success in CCEA therapy is 71.4% while the success of EA treatment is 68.8%. The physician may choose either CCEA or EA treatment since the therapeutic efficacy on schizophrenia of the two is similar.

Table 10. The GSC Ratings of the OCEA and EA Treated Neurosis Patients

	Total		Cured		Markedly Improved		Improved		Failed	
	N	%	N	%	N	%	N	%	N	%
OCEA	37	100	17	45.9	11	29.7	7	19.0	2	5.4
EA (WQ-6F)	40	100	10	25.0	14	35.0	9	22.5	7	17.5

Table 11. The GSC Ratings of the OCEA and EA Treated with Schizophrenic Patients

	Total		Cured		Markedly Improved		Improved		Failed	
	N	%	N	%	N	%	N	%	N	%
OCEA	21	100	5	23.8	10	47.6	4	19.1	2	9.5
EA (WQ-6F)	16	100	4	25.0	7	43.8	3	18.7	2	12.5

The therapeutic efficacies of CCEA and EA on manic-depression are presented in Table 12. In the CCEA group 80.1% showed significant improvement, whereas, the improvement rate of the EA group is 68%. The chi square = 1.09, $p > 0.05$, indicated no difference between the two treatments.

DISCUSSION

For the past 30 decades, chemicals such as tricyclic and monoamine oxidase inhibitors have been considered the drugs of choice for antidepression. Yet the use of these chemicals was far from satisfactory. These chemicals produce various side effects, develop tachyphylaxis and with high recurrence rate. Since 1981 we have conducted series of studies with electro-acupuncture. For hundreds of years, acupuncture has been reported to be an efficient traditional Chinese medical treatment for combating depressive disorders. Our pilot study on 47 patients in 1981-1983, double blind study of 239 patients in 1984-1985 and our extended study of 241 patients in collaborations with other psychiatric hospitals on electro-acupuncture, all appear to have positive results. While in comparison with antidepressant, amitriptyline, EA is as efficient as amitriptyline if not better toward treatment of newly developed depressive disorders. It is believed that the metabolism of monoamine neuro-transmitters are jeopardized in patients suffering from depression.[27,28] The basic principle of using tricyclic or monoamine oxidase inhibitors for treatment of depressive psychosis are their actions on potentiation of the monoamines' effects on the central nervous system (CNS).[29] EA has been shown to release monoamines in the CNS.[30] Both EA and AM treatments have shown to have the effect of enhancing the effects of monoamines in CNS.

Others have reported that AM and other antidepressants develop various degrees of side effects.[31,32] In our short and long term recurrence follow up study in 1988-89, as assessed by Asberg's Table for side reactions, AM seems to have effects on the cardiovascular (C-V) system and anticholinergic reaction. 22.7% of the patients had abnormal ECG and 54.8 % had

Table 12. The GSC Ratings of the OCEA and EA Treated Manic-Depressive Patients

	Total		Cured		Markedly		Improved		Failed	
	N	%	N	%	N	%	N	%	N	%
OCEA	26	100	12	46.1	9	34.6	4	15.4	1	3.9
EA (WQ-6F)	25	100	8	32.0	9	36.0	6	24.0	2	8.0

elevated SGPT. In contrast, EA elicited none of these effects. It is advisable that patients with C-V or liver diseases, should be considered receiving EA treatment instead of AM.

When the Hamilton scores obtained from our study in 1989-1990 were analyzed in detail, especially on the most prominent factors, the EA appears to have better therapeutic efficacy. EA is more effective in treating patients with reactive depression, relieving anxiety somatization and correcting the disturbance of cognitive process. These results also suggest that EA is similar to AM in affecting the monoamines actions in CNS, but have addition different mechanistic actions for treatment of depression. Biochemical study of our patients with depressive psychosis, the $MHPG.SO_4$, metabolite of Norepinephrine (NE) in 24 hours' urine is significantly lower than those of the normal people. This suggests inhibition of the metabolism of NE in the CNS of the depressive patients. Patients on EA treatment, the 24 hours' urine $MHPG.SO_4$ increase while the AM treated patients' 24 hours' urine $MHPG.SO_4$ remain low. These findings suggested that the mechanism of EA differs from the AM. Its therapeutic efficacy possibly exhibits through acting on the metabolic mechanism of NE in CNS. Also routine electro-encephalography showed that EA mainly present slow waves, reducing α wave but increasing $alpha_2$ and Beta waves. Our EA patients conformed well with the changes in the induced potential cerebral perception.

Results of our follow-up study have proven neither EA nor drugs therapy can prevent recurrence of manic depression. Family history or genetic does have some relation with development of depression but family concerns do not play an important role in recurrence. Most cases relapse in spite of close concern by family members. It is imperative that future research should emphasize on seeking therapy that reduces the tendency of recurrence. Another important issue is to give more attention to the patients' safety issue since two of the EA patients and four of the AM group attempted suicide.

Our preliminary result of the trial in using CCEA treatment on affective disorders, neurosis, and schizophrenia is clearly more effective than EA. CCEA is better than EA treatment possibly due to better control of the stimulating frequency. The greatest advantage of CCEA is that one can choose different parameters that have not been able to accomplish earlier with EA. We have based on the changes in the electro-encephalogram of our previous successful electric shock treated patients to develop the different stimulating parameters. We use these parameters to stimulate the Beihui and Yintang acupoints in the experiments conducted on rats that induce changes in their behavior. We convey our laboratory findings and applied to clinical therapy. The four different parameters determined in the experiments on rats were employed in the CCEA procedure in treating the depressive patients. CCEA uses the sine function of the wave to adjust its frequency and amplitude for obtaining maximum efficacy instead of the constant frequency uses in the EA would definitely be more advantageous. This clearly express the efficacy of CCEA is greater than EA. CCEA treatment provides a new efficient and safe therapeutic avenue for controlling affective disorders, schizophrenia, neurosis and manic depression. Since there are significantly fewer and less severe side effects encountered in patients receiving CCEA and EA (CCEA is more well controlled than EA) treatments than the general antidepressants, CCEA should be given priority on treating depressive and schizophrenic patients. Our studies presented here have been mainly on newly diagnosed patients. With our positive therapeutic efficacy on CCEA treatment, we plan to extend our investigation on treating chronic depressive psychosis and schizophrenia. Further research attempting to seek more stimulating parameters and understanding the mechanisms of other acupoints and their characteristics and functions are needed. Basic mechanisms of CCEA in treatment of depressive and schizophrenia with complicated classification and pathogenesis demand further study.

REFERENCES

1. Zhang,ST: The action of acupuncture and moxibustion on pain. (1973) Chinese Science 1:28-52.
2. Kim,YS: (1975) Letters to the editor: Some comments on acupuncture and cancer, Am. J. Chin. Med. 3:302-3.
3. Sjolund,B., Terenius,L. and Eriksson,M.: (1977) Increased cerebrospinal fluid levels of endorphins after electro-acupuncture. Acta Physiol Scand 100:382-384.
4. Firsova,PP and Zubova, ND: (1979) Electropuncture electroanalgesia in oncology. Vopr Onkol 25:69-71.
5. Mao,W., Ghia,JN, Scott,DS, Duncan,GH and Gregg,JM: (1980) High versus low intensity acupuncture analgesia for treatment of chronic pain: effects on platelet serotonin. Pain 8:331-342.
6. Clement-Jones,V, McLoughlin,L, Tomlin,S, Besser,GM,R Rees,LH and Wen,HL: (1980) Increased beta-endorphin but not met-enkephalin levels in human cerebrospinal fluid after acupuncture for recurrent pain. Lancet 2:946-949.
7. Zubova,ND, Bradychev,MS and Guseva,LI: (1981) Use of needle reflex therapy in the treatment of the pain syndrome of oncology patients. Vestn Akad Med Nauk SSSR 8:87-90.
8. Kaiser,RS, Khatami,MJ, Gatchel, RJ, Huang,XY, Bhatia,K and Altshuler,KZ: (1983) Acupuncture relief of chronic pain syndrome correlates with increased plasma met-enkephalin concentration. Lancet 2:1394-1396.
9. Hans,JS, Fei,H and Zhou,ZF: (1984) Met-enkephalin-Arg6-Phe7-like immunoreactive substances mediate electroacupuncture analgesia in the periaqueductal gray of the rabbit. Brain Res 322:289-96.
10. Szczudlik,A and Kwasucki,J.: (1984) Beta endorphin-like immunoreactivity in the blood of patients with chorinic pain treated by pinpoint receptor stimulation (acupuncture. Neurol Neurochir Pol 18:415-420.
11. Iguchi,Y, Ozaki,M, Kishioka,S, Tamura,S and yamamoto,H: (1985) The role of the pituitary in the devopment of electroacupuncture analgesia in rats. Nippon Yakurigaku Xasshi 85:453-465.
12. Jurni,J: (1986) The nociceptive system is modified. Analgesics and "non-analfesics". Med Klin Suppl 1:6-8.
13. Itil,TM: (1978) the progress of EEG in psychiatry. Presented at twentieth annual meeting. Group-without-A-name International Psychiatric Research society p m45-56.
14. Acupuncture and Moxibustion. (1976) ed. by Traditional chinese medical research acedemia. p. 336-344.
15. Tany,M.: (1977) Electrical stimulation acupuncture therapy. Am. J. Acup. 6:315-322.
16. Han,JS st al: (1979) Regeneration of
17. Luo,HC, Jia,YK and Zhou,DF: (1992) Perspective of treating mental disorders with conbination of Western and traditional chinese medicine: Review on research on electro-acupuncture on treatment of depression. Electro-acupuncture treatment on depression and related new techniques on conbining the western and chinese medicine. p. 7-8.
18. Clinical Diagnosis standards for depressive disorders: (1985) Chinese J. of psychiatry 18:317
19. Handbook of examination of psychitry: (1985) first edition Beijing People Public Health Publisher.
20. Kaplan,HI et al: Comprehensive Textbook of Psychiatry V, Vol.1,2, Williams & Wilkins Press, Baltimore, 1989.
21. Luo,HC, Jia,YK, Zhan,L (1985) electro-acupuncture verses Amitriptyline in the treatment of depressive states. J. of traditional Chinese medicine 5:3-8.

22. Luo,HC, Jia,YK, Zhan,L and Wang,B: (1984) the observation of the therapeutic effects on the treatment of depressive state of affective psychosis by electro-acupuncture. chinese J. Acupuncute 4:1-4.
23. Luo,HC, Zhou,D, Shu,N and Jia,YK: (1985) A double blind control study on the treatment of depression with electro-acupuncture and amitriptyline. Chinese J of Neurological and Psychiatry 18: 273.
24. Luo,HC, shen,YC, Jia,YK and Zhou,D: (1988) Clinical observations of electro-acupuncture on 133 patients with depression in comparison with amitriptyline. Clin J Inter. Med 8:45-47.
25. Electro-acupuncture treatment of depression and related new techniques on conbination of western and chinese medicine. Luo,HC and Jia,YK ed. (1992) p.77.
26. Luo,HC, Jia,YK, Wu,XH and Dai,WM: (1990) Electro-acupuncture in the treatment of depressive psychosis. International J of Clin Acupuncture 1:7-13.
27. Morris,JB and Beck,AT: (1974) The efficacy of antidepressant drugs, a review of research (1958-1972). Arch Gen Psychiatry 30:179-182.
28. Zhou,DF et al: (1987) Dexamethasone suppression test and unrinetry $MHPG.SO_4$ excretion in depressive patients. Psychiatry 22:883.
29. Carlsson,A et al: (1969) Effects of some antidepressant drugs on the depletion of intraneuronal brain catercholamine stores caused by 4-dimethyl-meta-tyramine. Eur. J Pharmacol 5:367-73.
30. Riederer,P et al: (1975) Manipulation of neurotransmiters by acupuncture (a preliminary communication). J Neural Transm 37:81-4
31. Richelson,E and Nelson,A: (1984) Antagonism by antidepressants of neurotransmitter receptors of normal human brain in vitro. J Pharmacol Exp Ther 230:94-102.
32. Beasley,CM, et al : (1990) Fluosetine: relationship among dose, response, adverse events, and plasma concentration in the treatment of depression. Psychopharmacol 87:253-9.

THE IMPORTANCE OF GLUTAMATE RECEPTORS IN BRAIN ISCHEMIA

Anker Jon Hansen

Pharmaceuticals Division
Department of Neuropharmacology
Novo Nordisk A/S
Novo Nordisk Park
DK-2760 Målov
Denmark

INTRODUCTION

The chapter reviews some aspects of the hypothesis that excitatory amino acids under pathological conditions stimulate neuronal glutamate receptors and cause the cells to die. A number of neurological conditions have been related to this excitotoxic hypothesis, but we limit the description to conditions in which acute and substantial changes are at hand, such as brain ischemia. Here excitatory amino acids are released from brain cells and major changes of the interstitial ion composition are observed. In the following a description of the chain of events in the central nervous system leading to cell necrosis will be given. We shall concentrate on the events in global ischemia because these are the conditions that have been most extensively described. First, some details of the glutamate transmitter system are given.

CHARACTERISTICS OF THE GLUTAMATE SYSTEM

L-glutamic acid is the most important excitatory neurotransmitter in the mammalian brain. The transmitter system comprises several receptor types which are named after the most potent agonist: N-methyl-D-aspartate (NMDA), alpha-amino-3-hydroxy-5-methyl-4-isoxazolepropionic acid (AMPA) and kainate receptors (kainic acid). These three receptor subtypes are ionotropic because they are coupled to ion channels that mediate movements of cations across the cell membrane. Two other glutamate receptors have been described. The metabotropic receptor (or 1-amino-cyclopentane-trans-dicarboxylate (transACPD) receptor) which via a GTP-binding (G) protein is coupled to the phosphotidyl inositol second messenger system (Sugiyama et al, 1989), and the L-2-amino-4-phosphonobutanoate (L-AP-4)receptor, which is presynaptically located and inhibits transmitter release, probably via an effect of G-proteins on calcium channels (Trombley and Westbrook, 1992). This receptor robably belongs to the metabotropic family (mGluR4) (Thomsen et al.). Although many other classes of drugs have been described that are effective as neuroprotectants in cerebral ischemia, only drugs that act directly on glutamate receptors will be discussed.

BRAIN ISCHEMIA

There are in principle two forms of brain ischemia: global and focal ischemia. In global ischemia the blood supply to the whole brain is transiently arrested; in focal ischemia the blood supply to a part of the brain is permanently arrested. When the blood or oxygen supply to brain tissue is compromised, a number of reactions take place. Within seconds the electrical activity is arrested, oxidative phosphorylation stops and energy metabolism is now supported by anaerobic glycolysis. After a few minutes the energy-rich phosphates are consumed and the cells unable to support ion homeostasis resulting in an extinction of ionic gradient across the cell membrane (e.g., Hansen, 1985). In case of global ischemia, consciousness is lost and respiratory movements cease; in focal ischemia the symptoms are less dramatic and relate to the affected brain region. If energy supply to the brain is reinstigated there is a normalization of most of the parameters, but some nerve cells are destined to die depending upon the duration of the energy failure. This so-called *delayed neuronal death* entails the phenomenon that certain nerve cells, after an interval in which they appear normal, succomb, while neighbouring neurons are left unharmed, e.g., pyramidal cells of the CA1 region in the hippocampus vs. the resistant interneurons in the same region (Johanson et al, 1983). After global ischemia, delayed neuronal death affects different brain regions with different time scales. Some neurons in the hilar region in hippocampus succomb after a few hours, while striatal and thalamic neurons die after 24 hours, while still others, like the CA1 nerve cells of the hippocampus and the cerebral cortex pyramidal cells of layer 3,5,6 die after several days (Pulsinelli et al., 1982a; Smith et al., 1984)(Fig.1). The pyramidal neurons of the CA1 region are particularly vulnerable and most often used to assess the effects of drugs or treatment.

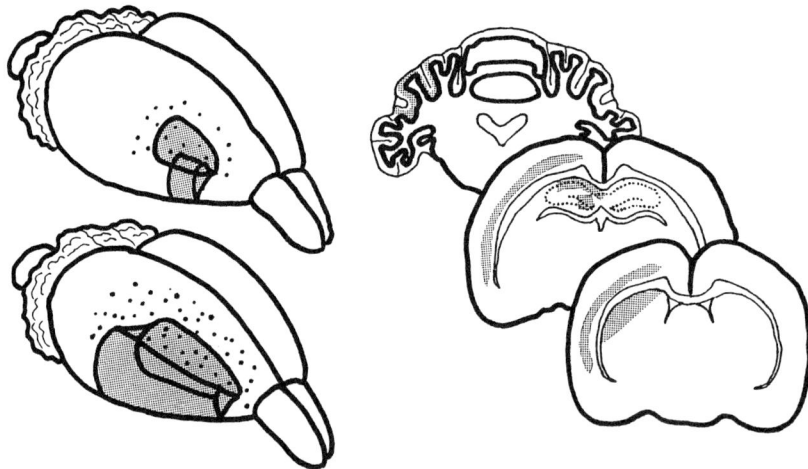

Figure 1. Schematic illustrations of localization of cell damage in the rat brain after focal and global cerebral ischemia. *Left* part shows the development of the right middle cerebral artery (upper part) and following a permanent occlusion (lower part) when evaluated several days after the insult. The dots outside the infarct denote selective neuronal necrosis. *Right* part illustrates the occurence of selective neuronal necrosis following an episode of global, cerebral ischemia. For further details please see text.

THE RISE OF EXCITOTOXIC HYPOTHESIS

The series of events leading to death of neurons as well as other types of cells is largely unknown, but a pivotal role has been given to calcium. The contention is that an overload of calcium of the cells is the basic mechanism responsible for cell death (Schanne et al., 1979); Siesjo and Bengtsson, 1989). Besides the direct role of calcium for activation of certain potassium and chloride ion channels and its well-known role for transmitter release, it also activates a number of intracellular enzymes, such as proteases, phosphatases, lipases and protein kinases, which can lead to cell degeneration (Siesjp and Bengtsson, 1989).

It is known that glutamate and its analogs are toxic to the brain and are able to induce damage which resembles that of ischemia (Olney et al., 1971). And when cultured cells are exposed to glutamate or NMDA agonists for only 5 min they degenerate over the next day (Choi et al., 1987). The toxic effect is blocked by glutamate receptor antagonists suggesting the damage to be a receptor-mediated phenomenon (Frandsen et al., 1989). Hence the damage is, as a rule, located to regions where the glutamate receptors are abundant. Apart from its known important role as an excitatory transmitter, glutamate is now thought to be important for the neurodegeneration in a number of pathological situations such as brain ischemia, e.g., stroke or cardiac arrest, trauma, epilepsia, Hungtintons disease, Alzheimers disease, amyotrophic lateral sclerosis, or the Wernicke/Korsakoff syndrome (Choi, 1989; Olney, 1989).

The link between glutamate and cellular calcium influx was established when it was shown that the NMDA receptor was coupled to an ion channel with a high calcium conductance (Mayer and Westbrook, 1987a); the AMPA\kainate receptor, on the other hand was considered only to conduct monovalent cations, e.g., Na^+ and K^+ (Mayer and Westbrook, 1987a).

In 1984, Benveniste et al. showed that glutamate was released from brain cells *in vivo* during ischemia, and the necessary connection between ischemia and glutamate was established. Furthermore, in ischemia most of the calcium located in the interstitial space moves into the cellular compartment (Hansen and Zeuthen, 1981). The importance of glutamatergic transmission was further substantiated by the finding that ablation of the glutamatergic input to the hippocampal CA1 region before ischemia (e.g., cutting the Schaffer collaterals) was able almost to prevent the neuronal loss following an episode of global ischemia (Wieloch et al., 1985; Benveniste et al., 1988). The Schafferotomy also prevented the ischemia-related rise in interstitial glutamate concentration (Benveniste et al., 1989) as well as the accumulation of calcium (Benveniste et al., 1988). Accordingly, the excitotoxic hypothesis could be formulated as follows: During ischemia glutamate reaches abnormally high levels in the synaptic cleft, which triggers excessive calcium influx in the postsynaptic neuron. A chain of events is hereby initiated that leads to cell death.

The hypothesis finally came on solid ground when it was shown that NMDA antagonists were able to attenuate the extent of neuronal damage in the CA1 region following an episode of global ischemia (Simon et al., 1984; Gill et al., 1987).

THE TRANSIENT FALL OF THE EXCITOTOXIC HYPOTHESIS

However, the effect that everybody expected of these compounds could unfortunately not be repeated consistently; different experimental techniques and animals were thought to be the explanation.

It is now established that there are at least two physiological variables that are especially important for the outcome after an episode of global ischemia: temperature and plasma glucose level. When the pre-and/or post-ischemia temperature of the *brain* is lowered the neuronal damage is significantly diminished (Busto et al., 1989). And hyperglycemia in the pre-and/or post-ischemia period aggrevates the ischemia damage (Pulsinelli et al., 1982b). It turned out that these NMDA antagonists were able to reduce brain temperature. When the temperature of the brain was controlled, no beneficial effect of the NMDA antagonists on severe global ischemia could be found (Buchan and Pulsinelli, 1990; Buchan et al., 1991).

The inability of NMDA antagonists to give protection after global ischemia prompted a new surge for other types of drugs. At present there are many types of drugs that have been claimed to counteract the cell damage encountered after brain ischemia. These include calcium channel antagonists, sodium channel antagonists, adrenergic agonists, adenosine agonists, serotonin antagonists, NO agonists, NO antagonists, NMDA antagonists, AMPA antagonists, free radical scavengers, as well as others. If there is a common denominator for the actions of these different classes of drugs it still remains to be identified. Only a few of these drugs have passed the final test, which means that their effects have been reproduced in several, independent laboratories. One of these drugs turned out to be another kind of glutamate receptor antagonists.

THE EXCITOTOXIC HYPOTHESIS REVIVED BY AMPA RECEPTOR BLOCKERS

This compound interferes with the AMPA/kain receptor and was efficacious against cell damage after global ischemia (Sheardown et al., 1990b). This receptor mediates fast synaptic transmission within the glutamate system (Mayer and Westbrook, 1987b). It has been shown that certain quinoxalinediones are potent and selective AMPA antagonists (Honore et al, 1988). The compound NBQX (2,3-dihydroxy-6-nitro-7-sulfamoyl-benzo(f)quinoxaline) has a 30-fold higher selectivity for AMPA receptors than for the high-affinity kainate receptor (Sheardown et al., 1990b). The fact that these compounds are very effective in ameliorating ischemic cell damage (Sheardown et al., 1990b; Diemer et al., 1992; Nellgard and Wieloch, 1992a; Pulsinelli et al., 1992) has promoted renewed interest in AMPA receptors and their antagonists for clinical drug testing. This opens up not only for a therapeutic role of these compounds, but also for a new insight into mechnisms leading to ischemic cell death. Earlier it was believed that the brain damage took place during the ischemic episode, but it is now clear that much of the damage is generated in the hours following the ischemic episode. The recent introduction of another class of AMPA receptor antagonists, amino-methyl-2,3,-benzodiazepine, GYKI 52466, which has shown beneficial effects on ischemic cell demage, has confirmed the importance of this receptor type for ischemic cell damage (Le Peillet et al., 1992).

HOW DOES AMPA RECEPTOR BLOCKADE AMELIORATE ISCHEMIC CELL DAMAGE?

There is no simple answer to this question. It is known that NBQX is very effective in blocking synaptic transmission, e.g., in hippocampus (Sheardown et al., 1990a), and that neuronal activity is increased, and excitatory transmission in the CA1 region of the hippocampus is enhanced for several hours after an ischemic episode (Urban et al., 1989). Calcium movements, induced by high frequency stimulation of the perforant pathway, to the cells of the CA1 region were increased 250% following ischemia, compared to the control situation, whereas the evoked field potentials were not changed (Andine et al., 1992). Although Andine et al, 1992 did not measure the calcium flux but only changes in $[Ca^{++}]_e$, it is probably safe to concluded that ischemia augmented the entry of calcium into pyramidal cells of the CA1 region for several hours after the ischemic episode. They also found that NBQX completely blocked the induced calcium movements. The NMDA antagonist, MK-801, on the other hand, only blocked 50% of the calcium movements. Since NBQX does not block neuronal voltage-dependent calcium channels (Birn and Hansen, unpublished), the observation by Andine et al., suggests that calcium movements evoked by the orthodromic stimulation in the post-ischemic phase are mainly due to activation of AMPA receptors and to a lesser degree NMDA receptors. If calcium movements are important for the development of cell necrosis, the beneficial effect of AMPA/kainate blockade could have at least two reasons: 1) An inhibition of Na^+ and K^+ movements across the cell membrane thereby attenuating depolarization, and hence opening of voltage-dependent calcium channels or NMDA receptor-dependent calcium channels. The evidence against this possibility rests on the use of pharmacological compounds with actions on these receptors. The inability of NMDA antagonists, as described above, or calcium channel blockers to protect hippocampus pyramidal neurons after global ischemia seems to dismiss this explanation. It should, however, be kept in mind that most studies so far have used calcium channel blockers which are not specific for the neuronal types of calcium channels. 2) Another explanation lies in the observation that the AMPA/kainate gated channels are in fact able to conduct calcium ions (Ozawa et al., 1991). The receptor complex probably contains five subunits of which four different kinds exist, gluR1...glu4. When combinations of subunits are expressed in oocytes or cell lines, it was shown that the proteins encoded by gluR2 are responsible for the lack of calcium conductance (Hollmann et al., 1991; Verdoom et al., 1991). This property has been ascribed to the position of a single amino acid residue (Hume et al., 1991; Verdoom et al., 1991). In gluR2 it is taken by an arginine, but in the other subunits by a glutamine. Sommer et al., (1991) showed that this important arginine residue in gluR2 is present only because of editing of a single nucleotide of the mRNA. In the genom, the codon for gluR2 encodes a glutamine residue like the other glutamate subunits. Thus there are several points for the regulation of the properties of non-NMDA glutamate receptors. An intriguing possibility is that the observation by Pellegrini-Giampietro et al. (1992) that AMPA complex in CA1 region selectively loses gluR2 subunits following ischemia, suggests a way by which these cells become more calcium permeable and that the fast-transmission conduction by this system could lead to augmented calcium influx and cell necrosis.

This explanation deals with delayed neuronal death and may as such not apply to other types of brain injury in which damage readily is established.

FOCAL ISCHEMIA

Since cerebral arteries are end-arteries and only few anastomoses to other vessels are present (i.e. the other cerebral arteries or the dura vessels) any occlusion leads inevitably to an infarct. The interstilial potassium concentration, $[K^+]_e$, in brain cortex increases markedly during ischemia, and $[Ca^{++}]_e$ decreases coccomittantly indicating Ca^{++} flux into the cells (Hansen, 1985). When interstitial ions are measured in brain cortex following occlusion of middle cerebral artery (MCA) in the rat there are characteristic changes. The center of the infarct *in spe* has such a severe blood flow reduction that the energy demands for ion homeostasis are not met, resulting in a permanent elevation of $[K^+]_e$ (and lowering of $[Ca^{++}]_e$) (Hansen and Nedergaard, 1988), and of $[glutamate]_e$ (Hillered et al., 1989; Butcher et al., 1990). On the periphery of the infarct zone the reduction of blood flow interferes with nerve cell function but the ion homeostasis remains intact, albeit somehow incapacitated. This constitutes the so-called penumbral zone. There are transient, intermittent elevations of $[K^+]_e$/decreases of $[Ca^{++}]_e$ (Hansen and Nedergaard, 1988; Gill et al., 1992a). These events could be episodes of spreading depression induced from the center of the infarct zone by the high $[K^+]_e$ or $[glutamate]_e$. In order to handle these episodes, energy metabolism must increase to restore the ionic gradients. If the energy demands are not met, ion homeostasis is compromised, resulutng in increased intracellular [Ca^{++}], cell swelling, and local edema. It is proposed that the entry of calcium, in consonant with the decreased local blood supply, starts deleterious processes which lead to cell necrosis. A vicious circle is established which forms the basis for an expansion of the infarct (Fig.1).

If a drug is able to reduce the infarct volume permanently, the following effects are required: 1) The drug increases the blood flow to the infarct from the sourroundings (opening of patent vessels or angiogenesis). 2) The drug counteracts processes in the "young" infarct which makes the infarcted volume bigger with time. 3) A combination of 1) and 2).

A number of drugs have been shown to decrease the size of the infarct following experimental focal ischemia. Blockers of NMDA receptors seem to be the best studied and have repeatedly been shown efficacious (McCulloch, 1992). The AMPA blocker NBQX is now also established as an effective drug for these conditions (Gill et al., 1992b; Smith and Meldrum, 1993) as well as the non-competetive AMPA blocker, GYKI 52466 (Smith and Meldrum, 1992). The mechanisms by which these drugs act are still desputed. It has been shown that MK-801 is able to arrest the transient changes in [Ca^{++}]$_e$ at the infarct zone (Gill et al., 1992a), which is in line with the suggestions made before, Whether the other classes of effective drugs have similar actions are at present unknown.

CONCLUSION

It is now well established that glutamate receptor antagonists are efficacious in conditions with acute brain injury. This chapter describes the conditions in brain ischemia but the antagonisrs are also able to ameliorate the neyronal damage encountered following a period of severe hypoglycemia (Nellgard and Wieloch, 1992b) and to improve neurological function following a brain trauma (Faden et al., 1989).

NMDA-receptor antagonists are not effective in "severe" ischemia (global ischemia) in which the energy failure is pronounced in contrast to AMPA receptor antagonists, whereas they are of beneficial value when the energy failure is moderate in, e.g., hypoglycemia and in focal ischemia. Why AMPA antagonists are omnipotent, while NMDA antagonists only show effects in focal ischemia is at present unknown but gives a hint of the different nature of these two conditions.

In *global ischemia* the grave pertubations of ions and transmitters incl. glutamate are readily reversed but the damage encurred only after several hours/days. It is thought that these acute changes somehow trigger other processes which, after a free interval, manifest themselves and destroy the host cell. The intimate nature of these processes are not disclosed but they can obviously be interfered with by blocking the AMPA receptors. The recent observation by Pulsinelli et al., (1992) that hippocampal damage, induced by global ischemia, can be antagonized by ingestion of NBQX even 12 h after the ischemic insult gives great promises to the treatment of this condition.

In *focal ischemia* cell damage occurs within the first hours due to the permanent occlussion of the blood supply. Some of the brain cells are exposed to an ionic milieu which resembles that of permanent ischemia while in other regions, the microenvironment is constantly changing. The conditions are distinctly different from global ischemia. The ability of the glutamate antagonist to function may be due to two effects, which may be interrrelated. That they are able to decrease the passive leak of ions across the cell membrane and hereby lessen the ATP turn-over needed for the ion pumps in the cells already suffering from limited substrate supply. And perhaps they are also able to increase the local blood flow. It should be mentioned that the only rational therapy is, of course, to remove the obstacle of the blood supply.

This chapter has mainly dealt with two classes of glutamate antagonists. The possibility, however, exists, that the other glutamate receptors somehow are involved in the pathogenesis of brain damage. Some reports, albeit conflicting, are available on the metabotropic receptors (Chiamulera et al., 1992; Sucaan an Schoepp, 1992), whereas no one has tried agonists of the L-AP4 receptor even though this receptor is able to control the release of glutamate.

REFERENCES

Andine, P., Jacobson, I., and Hagberg, H. (1992) Enhanced calcium uptake by CA1 pyramidal cell dendrites in the postischemic phase despite subnormal evoked field potentials:Excitatory amino acid receptor dependency and relationship to neuronal damage. J. Cereb. Blood Flow Metab. 12:773–783

Benveniste, H., Drejer, J., Schousboe, A., and Diemer, N. (1984)Elevation of extracellular concentrations of glutamate and aspartate in rat hippocampus during transient cerebral ischemia monitored by intracerebral microdialysis. J. Neurochem. 43:1369–1374.

Benveniste, H., Jogensen, M.B., Diemer, NH, and Hansen, A.J. (1988) Calcium accumulation by glutamate activation is involved in hippocampal cell damage after ischemia. Acta Neurol. Scand. 78:529–536.

Benveniste, H., Jorgensen, M.B., Sandberg, M., Hagberg, H., and Diemer, N.H. (1989)Ischemia-induced damage in the hippocampal CA1 region depends on glutamate and intact innervation from CA3. J. Cereb. Blood Flow Metab. 9:629–639.

Buchan, A., Li, H., and Pulsinelli, W. (1991)The N-methyl-D-asparte antagonist, MK-801, fails to protect against neuronal damage caused by transient, severe forebrain ischemia in adult rats. J. Neurosci. 11(4):1049–1056.

Buchan, A., and Pulsinelli, W.A. (1990)Hypothermia but not the N-methyl-D-asparte antagonist, MK-801, attenuates neuronal damage in gerbils subjected to transient global ischemia. J. Neurosci. 10(1):311–316.

Busto, R., Dietrich, W.D., Globus, M.Y.-T., and Ginsberg, M.D. (1989) Postischemic moderate hypothermia inhibits CA1 hippocampal ischemic neuronal injury. Neurosci. Lett. 101:299–304.

Butcher, S.P., Bullock, R., Graham, D.I., and McCulloch, J. (1990) correlation between amio acid release and neuropathologic outcome in rat brain following middle cerebral artery occlusion. Stroke 21:1727–1733.

Chiamulera, C., Albertini, P., Valerio, E., and Reggiani, A. (1992) Activation of metabotropic receptors has a neuroprotective effect in a rodent model of focal ischemia. Eur. J. Pharmacol. 216:335–336.

Choi, D.W., Maulucci-Gedde, M.A., and Kriegstein, A.R. (1987)Glutamate neurotoxicity in cortical cell culture. J. Neurosci. 7:357–368.

Choi, D.W. (1988)Glutamate neurotoxicity and diseases of the nervous system. Neuron 1:623–634.

Diemer, N.H., Jorgensen, M.B., Johansen, F.F., Sheardown, M.J., and Honore, T. (1992)Protection against ischemic hippocampal CA1 damage in rat with a new non-NMDA antagonist, NBQX. Acta Neuro. Scand. 86:45–49.

Faden, A.I., Demediuk, P., Panter, S.S., and Vink, R. (1989)The role of excitatory amino acids and NMDA receptors in traumatic brain injury. Science 244:798–800.

Frandsen, A.A., and Schousboe, A. (1987)Time and concentration dependency of the toxicity of excitatory amino acids on cerebral neurones in primary culture. Neurochem. Int. 10:583–591.

Gill, R., Nordholm, L., and Lodge, D. (1992b)The neuroprotective actions of 2,3-dihydroxy-6-nitro-7-sulfamoyl-benzo(F)quinoxaline (NBOX) in a rat focal ischemia model. Brain Res. 580:35–43.

Gill, R., Foster, A.C., and Woodruff, G.N. (1987)Systemic administration of MK-801 protects against ischemia-induced hippocampal neurodegeneration in gerbil. J. Neurosci. 7:3343–3349.

Gill, R., Andine, P., Hillered, L., Persson, L., and Hagberg, H. (1992a)The effect of MK-801 on cortical spreading depression in the penumbra zone following focal ischemia in the rat. J. Cereb. Blood Flow Metab. 12:371–379.

Hansen, A.J. (1985)Effects of anoxia on ion distribution in the brain. Physiol. Rev. 65:101–148.

Hansen, A.J., and Zeuthen, T. (1981)Extracellular ion concentrations during spreading depression and ischemia in the rat brain cortex. Acta Physiol. Scand. 113:437–445.

Hansen, A.J., and Nedergaard, M. (1988)Brain ion homeostasis in cerebral ischemia. Neurochem. Pathol. 9:195–209.

Hillered, L, Hallstrom, A., Segersvard, S., Persson, L., and Ungerstedt, U. (1989)Dynamics of extracellular metabolites in the striatum after middle cerebral artery occlusion in the rat monitored by intracerebral microdialysis. J. Cereb. Blood Flow Metab. 9:607–666.

Hollmann, M., Hartley, M., and Heinemann, S. (1991)Calcium permeability of KA-AMPA-gated glutamate channels depends on subunit composition. Science 252:851–853.

Honore, T., Davis, S.N., Drejer, J., Fretcher, E.J., Jacobson, P.Lodge, D., and Nielsen, F.E. (1988) Quinoxalinediones: Potent comparative non-NMDA glutamate receptor antagonists. Science 241:701–703.
Hume, R.I., Dingledine, R., and Heinemann, S.F. (1991)Identification of a site in glutamate receptor subunits that controls calcium permeability. Science 253:1028–1031.
Johansen, F.F., Jorgensen, M.B., and Diemer, N.H. (1983)Resistance of hippocampal CA1 interneurones to 20 min of transient ischemia in the rat. Acta Neuropath. 61:135–140.
Le Peillet, E., Arvin, B., Moncada, C., and Meldrum, B. (1992)The non-NMDA antagonists, NBQX and GYKI 52466, protect against cortical and striatal cell loss following transient global ischemia in the rat. Brain Res. 571:115–120.
Mayer, M.L., and Westbrook, G.L. (1987a)Permeation and block of N-methyl-D-aspartic acid receptor channels by divalent cations in mouse culture central neurons. J. Physiol. 394:501–527.
Mayer, M.L., and Westbrook, G.L. (1987b)The physiology of excitatory amino acids in the vertebrate central nervous system. Prog. Neurobiol. 28:197–276.
McCulloch, J. (1992)Excitatory amino acid antagonists and their potential for the treatment of ischemia brain damage in man. Br. J. Clin. Pharac. 34:106–114.
Nellgard, B., and Wieloch, T. (*1992a)Postischemic blockade of AMPA but not NMDA receptors mitigates neuronal damage in the rat brain following tranient severe cerebral ischemia. J. Cereb. Blood Flow Metab. 12:2–11.
Nellgard, B., and Wieloch, T. (1992b)Cerebral protection by AMPA- and NMDA-receptor antagonists administration after severe insulin-induced glycemia. Exp. Brain Res. 92:259–266.
Olney, J.W. (1989) Excitatory amino acids and neuropsychiatric disorders. Biol. Psychiatry 26:505–525.
Olney, J.W., Ho, O.L., and Rhee, V. (1971)Cytotoxic effects of acidic and sulphur containing amino acids on the infant mouse central nervous system. Exp. Brain Res. 14:61–76.
Ozawa, S., Iino, M., and Tsuzuki, K. (1991)Two types of kainate responses in cultured rat hippocampal neurons. J. Neurophysiol. 66:2–11.
Pellegrini-Giampietro, D.E., Zukin, R.S., Bennett, M.V.L., Cho, S., and Pulsinelli, W.A. (1992)Switch in glutamate receptor subunit gene expression in CA1 subfield of hippocampus following global ischemia in rats. Proc. Natl. Acad. Sci. 89:10499–10503.
Pulsinelli, W., Dimagl, U, Jacewicz, M., and Buchan, A. (1992)Antagonists of excitatory amino acid neurotransmitters: A comparison of their effects on global versus focal ischemia. In: Drug Research Related to Neuroactive Amino Acids (Schousboe, A., Diemer, N., Kofok, H., eds.), Munksgaard, Copenhagen, pp. 225–238.
Pulsinelli, W.A., Waldeman, S., Rawlinson, S., and Plum, F. (1982b)Moderate hyperglycemia augments ischemic brain damage: a neuropathological study in the rat. Neurology 32:1239.
Pulsinelli, W.A., Brierely, J.B., and Plum, F. (1982a)Temporal profile of neuronal damage in model of transient forebrain ischemia. Ann. Neurol. 11:491.
Sacaan, A.I., and Schoepp, D.D. (1992)Activation of hippocampal metabotropic excitatory amino acid receptors leads to seizures and neuronal damage. Neurosci. Lett. 139:77–82.
Schanne, F.A.X., Kane, A.B., Young, E.E., and Farber, J.L. (1979)Calcium dependence of toxic cell death: A final common pathway. Science 206:700–702.
Sheardown, M.J., Hansen, A.J., Eskesen, K., Suzdak, P., Diemer, N.H., and Honore, T. (1990a)Blockade of AMPA receptors in the CA1 region of the hippocampus prevents ischemia induced cell death. In: Pharmacology of Cerebral Ischemia (Krieglstein, J., Oberpichler, H., eds), Stuttgart, Wissenschaftliche Verlagsgesellschaft mbH, pp. 245–253.
Sheardown, M.J., Nielsen, E.O., Hansen, A.J., Jacobson, P., and Honore, T. (1990b) 2,3-dihydroxy-6-nitro-7-sulfamoyl-benzo(F)quinoxaline: A neuroprotectant for cerebral ischemia. Science 247:571–574.
Siesjo, B.K., and Bengtsson, F. (1989)Calcium fluxes, calcium antagonists, and calcium-related pathology in brain ischemia, hypoglycemia, and spreading depression: A unifying hypothesis. J. Cereb. Blood Flow and Metab. 9:127–140.
Simon, R.P., Swan, J.H., Griffiths, T., and Meldrum, B.S. (1984)Blockade of N-methyl-D-aspartate receptors may protect against ischemic damage in the brain. Science 226:850–885.
Smith, S.E., and Meldrum, B. (1992)Cerebroprotective effect of a non-N-methyl-D-aspartate antagonist, GYKI 52466, after focal ischemia in the rat. Stroke 23(6):861–864.
Smith, M.-L., Auer, R.N., and Siesjo, B.K. (1984)The density and distribution of ischemic brain injury in the rat following two to ten minutes of forebrain ischemia. Acta Neuropathol. 64:319.
Smith, S.E., and Meldrum, B.S. (1993)Cerebroprotective effect of a non-methyl-D-aspartant antagonist, NBQX, after focal ischemia in the rat. Functional Neurol., in press.

Sommer, B., Kohler, M., Sprengel, R., and Seeburg, O. (1991)RNA editing in brain controls a determinant of ion flow in glutamate-gated channels. Cell 67:11–19.

Sugiyama, H., Ito, I., and Watanabe, M. (1989)Glutamate receptor subtypes may be classfied into two major categories: a study on Xenopus oocytes injected with rat brain mRNA. Neuron 3:129–132.

Thomsen, C., Kristensen, P., Mulvihil, E., Haldeman, Band Suzdak, P. (1992)L-2-amino-4-phosphonobutyrate(L-AP4) is an agonist at the type-iv metabotropic glutamate receptor which is negatively coupled to adenylate cyclase. Eur. J. Pharmacol. 227:361–362.

Trombley, P.Q., and Westbrook, G.L. (1992)L-AP4 inhibits calcium currents and synaptic transmission via a G-protein-coupled glutamate receptor. J. Neurosci. 12(6):2043–2050.

Urban, L., Neill, K.H., Crain, B.J., Nadler, J.V., and Somjen, G.G. (1989)Postischemic synaptic physiology in area CA1 of the gerbil hippocampus studied in vitro. J.Neurosci. 9(11): 3966–3975.

Verdoorn, T.A., Burnashev, N., Monyer, H., Seeburg, P.H., and Sakmann, B. (1991)Structural determinants of ion flow through recombinant glutamate receptor channels. Science 252:1715–1718.

Wieloch, T., Lindvall, O., Blomqvist, P.and Gage, F.H. (1984)Evidence for amelioration of ischemic neuronal damage in the hippocampal formation by lesions of the perforant path. Neurol. Res. 7:24–26.

BIOCHEMICAL STUDY OF THE POSTISCHEMIC NEURONAL DAMAGE

Dehua Jiang, Xue Rong, Qingyou Li and Zhongyou Wei

Lab. of Neurochemistry, Tianjin Neurological Institute
Tianjin, China 300052

In order to elucidate the role of biochemical perturbation in the evolution of primary neuronal damage at the initial stage of cerebral ischemia, a rat model of irreversible, middle-cerebral artery occlusion was chosen and serial experiments were designed to verify whether some putative, deleterious neurotransmitters bear the same role as they play in the development of secondary neuronal damage which emerges during the phase of reperfusion. Our findings were presented as follows:

PART A.

An animal model of irreversible focal cerebral ischemia was performed with microsurgical techniques, according to the methods described by Tamura et al, (1981)[1] and Bederson et al, (1986)[2] with slight modification. We prefer to ligate the right middle cerebral artery at two sites. One locates nearby the olfactory tract and another nearby the inferior cerebral vein. 24 hours following the microsurgical operation, the rats were sacrificed. The area dimension of infarct was identified with Evans blue, stain and 2,3,5-triphenyl tetrazolium chloride (TTC) stain, as well as the histopathological findings. A well-demarcated, with slight variation in the dimension of infarct, involving the outer two-third os striatum and front-temporal cortical region, could be found in all the animals operated (n=23).[3] Accordingly, we considered the parieto-occipit al cortical area as the peri-ischemic area.

PART B. MONOAMINE LEVELS IN THE ISCHEMIC TISSUE AS A FUNCTION OF OCCLUSION TIME

The contents of monoamines, dopamine (DA), norepinephrine (NE) and 5-hydroxytryptophan (5HT) and their metabolites 3,4-dihydroxyphenylacetic acid (DOPAC), homovanillic acid (HVA) and 5-hydroxyindoleacetic acid (5HIAA), at the ischemic (outer two-third of striatum and fronto-temporal cortical region) and peri-ischemic (posterior parieto-occipital

Table 1. The change of DA concentration in the striatal dialysate as a function of time

	10	30	50	70	90	110	130	150	170	(min)
NOR (n=6)	ND	ND	ND	ND	ND	ND	ND	ND	ND	
SHAMO (n=6)	ND	ND	ND	ND	ND	ND	ND	ND	ND	
MCAO (n=6)	144.3 ± 87.6	145.1 ± 118.8	91.9 ± 65.9	78.5 ± 54.1	78.1 ± 50.6	69.5 ± 27.4	69.2 ± 29.9	56.1 ± 21.5	46.1 ± 31.9	

Data are shown Mean ± SD (pmole/ml). perfusion rate: 3.7 μl/min. ND denotes undetectable.

region) area were assessed as a function of occlusion time (up to 60 minutes). Our data showed that, in general, NE, DA and 5HT levels in both the ischemic and peri-ischemic regions were significantly reduced at occlusion time from 5 to 60 minutes, with the exception that in the ischemic straitum, an initial temporary enhancement of NE was found. The DOPAC level was increased and 5HIAA reduced concomitantly with the significant reduction of DA and 5HT at the ischemic tissues during the occlusion time observed. Meanwhile, the HVA level remained unchanged.[4] It denotes that even at the initial time, as early as 20 min., following arterial occlusion, the intracellular oxidative deamination of DA is accelerated significantly at the ischemic tissue. Since DA is converted to DOPAC by the catalization of monoamine oxidase (MAO).

$$DA + H_2O_2 + O_2 \xrightarrow{MAO} DOPAC + H_2O_2 + NH_3$$

Thus, an intracellular over-production of H_2O_2 might be a reasonable deduction. Obviously it does not deal with any reperfusion of ischemic area.

Table 2. The effect of intrastriatal injection of Muscimol on the SP concentration of striatum and substantia nigra

	striatal sp	nigral sp
Normal control	1.43 ± 0.29 (7)	8.79 ± 2.32 (8)
Normal saline i.s.	1.47 ± 0.30 (8)	7.97 ± 1.99 (6)
Mus. 3 μg/μl i.s.	1.17 ± 0.15* (7)	7.60 ± 3.30 (7)

Data are expressed as M ± SD. (pmol/mg protein). Number in the parenthesis is the number of experiment. Volume of injection: 1 μl. Rats were sacrified at 40 min. following injection. * $p<0.05$.

Biochemical Study of the Postischemic Neuronal Damage

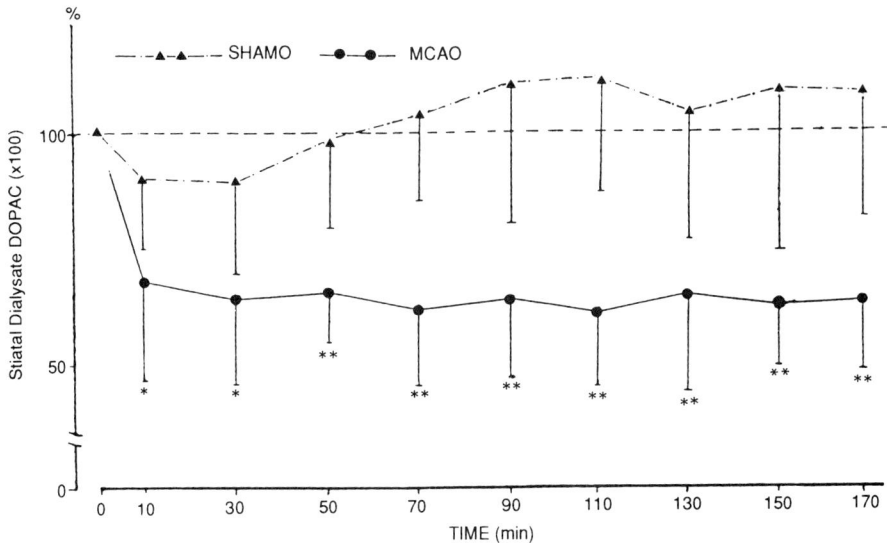

Figure 1. Data shown were calculated individually and taken the average concentration of DOPAC at 4 consecutive samples just before the craniotomy, as 100% control. The concentration of DOPAC in each of group control was: 324.08 13.84 pmole/ml (SHAMO), 293.87 27.36 pmole/ml (MCAO). Mean SD * $p < 0.05$, ** $p < 0.01$.

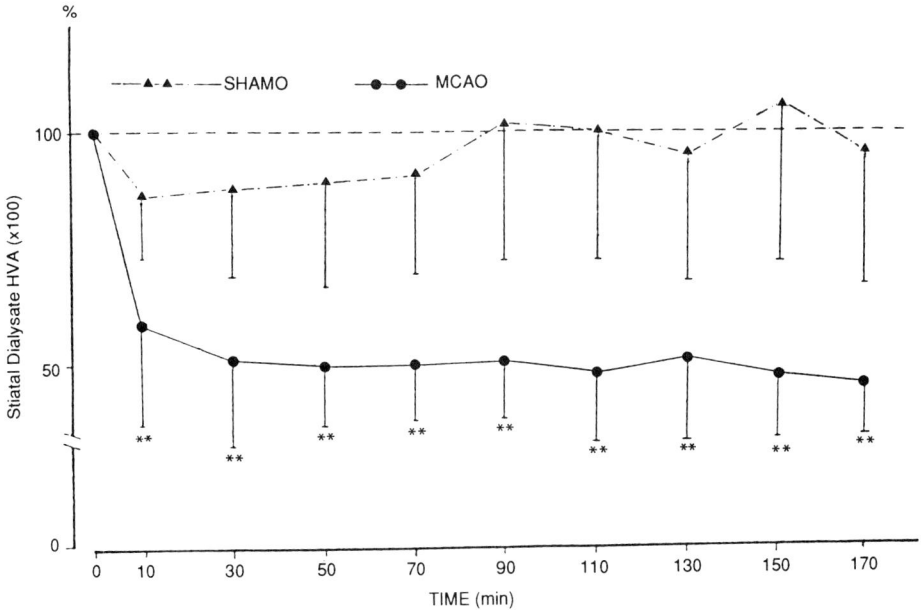

Figure 2. Data shown were calculated individually and taken the average concentration of HVA at 4 consecutive samples just before the craniotomy, as 100% control. The concentration of HVA in each of group control was: 283.46 25.16 pmole/ml (SHAMO), 271.20 8.99 pmole/ml (MCAO). Mean SD, n = 6, * $p < 0.05$, ** $p < 0.01$.

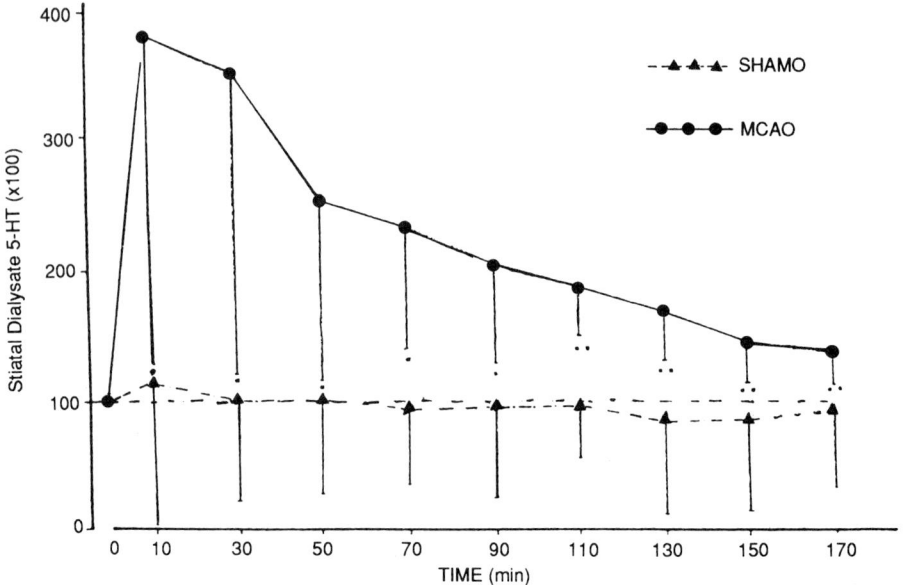

Figure 3. Data shown were calculated individually and taken the average concentration of 5-HT at 4 consecutive samples just before the craniotomy, as 100% control. The concentration of 5-HT in each group control was: 2.49 0.07 pmole/ml (SHAMO), 4.13 0.58 pmole/ml (MCAO). Mean SD, n = 6, * p < 0.06, ** p < 0.01.

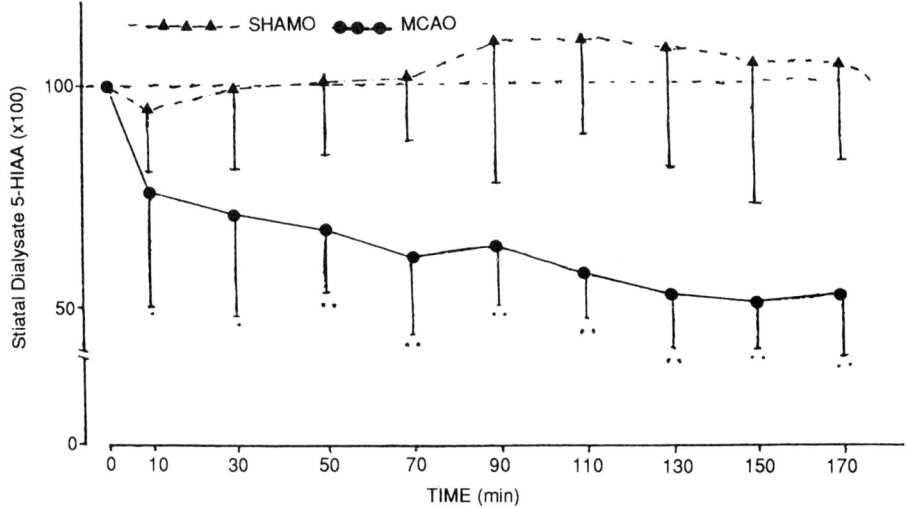

Figure 4. Data shown were calculated individually and taken the average concentration of 5-HIAA at 4 ocnsecutive samples just before the craniotomy, as 100% control. The concentration of 5-HIAA in each group control was: 101.96 8.16 pmole/ml (SHAMO), 92.62 1.92 pmole/ml (MCAO). Mean SD, n = 6, * p < 0.05, ** p < 0.01.

Biochemical Study of the Postischemic Neuronal Damage

PART C. FURTHER SUPPORT FOR THE ACCELERATION OF INTRACELLULAR OXIDATIVE DEAMINATION OF DA AT THE EARLY STAGE OF STRAITAL ISCHEMIA, AND THE MASSIVE LEAKAGE OF DA AND 5HT TO EXTRANEURONAL SPACE.

Using the intracerebral microdialysis technique and HPLC-EC method, the contents of extracellular endogenous striatal DA, 5HT and their metabolites DOPAC, HVA and 5HIAA were determined consecutively for 3 hours following MCAO. As compared with their steady level of DA, DOPAC, HVA, 5HT and 5HIAA. A steep rise of extracellular DA and 5HT, accompanying with significant reduction of DOPAC, HVA and 5HIAA was found 10 min. following occlusion. From then on, the enhanced level of DA and 5HT decreased step by step. However, by 3 hours of occlusion, DA and 5HT levels were still higher as compared to their basal levels, meanwhile their metabolites were steadily kept at low levels (Tab. 1 and Figs. 1-4). In conclusion, the massive leakage of DA and 5HT occurs early during the arterial occlusion, and thus both the neurotoxic effect of dopamine and intraneuronal overproduction of H_2O_2 might be considered as one of the initial pathophysiological mechanisms triggering the primary neuronal damage.

PART D. THE REGULATION OF DA METABOLISM IN THE ISCHEMIC STRIATAL TISSUE

Using the RIA and HPLC-EC methods, the effects of the intrastriatal administration of bicuculline (Bic), a GABA antagonist, muscimol (Mus), a GABA agonist, as well as the intranigral injection of [D-Arg^1, D-Phe^5, D-Trp^{7,9}, Leu^{11}]-substance P, (Spant), a Substance P antagonist on the striatal DA metabolism were studied following MCAO of rats.

Mus was injected intrastriatally, instead of its solute, normal saline, the striatal but not nigral SP was significantly reduced (Table 2). Such and effect of Mus, however, could be found in MCAO rats (data not shown).

When 80 nmol Bic was injected intrastriatally, a concomitant increase of the striatal DOPAC emerged, but not DA and HVA. The effect of Bic on striatal DOPAC was even more significant in MCAO rats. However, the effect of Bic was completely reversed with the intranigral injection of Spant (5 nmol). (Table 3). It implies that the effect of striatal Bic on striatal DOPAC is mediated exclusively by the nigral SP receptors.

Table 3. The effect of Bicuculline and Apant on the striatal DA DOPAC and HVA levels

	DA nmol/mg protein	DOPAC nmol/mg protein	HVA nmol/mg protein
Normal Saline i. s. (A)	24.2 ± 5.2 (5)	2.8 ± 0.9 (5)	2.6 ± 1.0 (5)
Bic. 80nmol i. s. (B)	25.5 ± 3.2 (8)	5.6 ± 2.2* (8)	3.8 ± 1.0 (8)
Bic. 80nmol i. s. +MCAO (C)	27.3 ± 4.1 (7)	7.0 ± 2.7** (7)	3.4 ± 1.2 (7)
SPant 5nmol i. n. +Bic. 80nmol i. s. +MCAO (D)	17.3 ± 1.9*** (5)	2.8 ± 0.9*** (5)	2.2 ± 0.4 (5)

Data are expressed as M ± SD. Number in parenthesis denotes the number of experiments. i. s. : intrastriatal injection , i. n. : intranigral injection . A: B, A: C. * P<0.05, ** P<0.01, C: D, *** P<0.001.

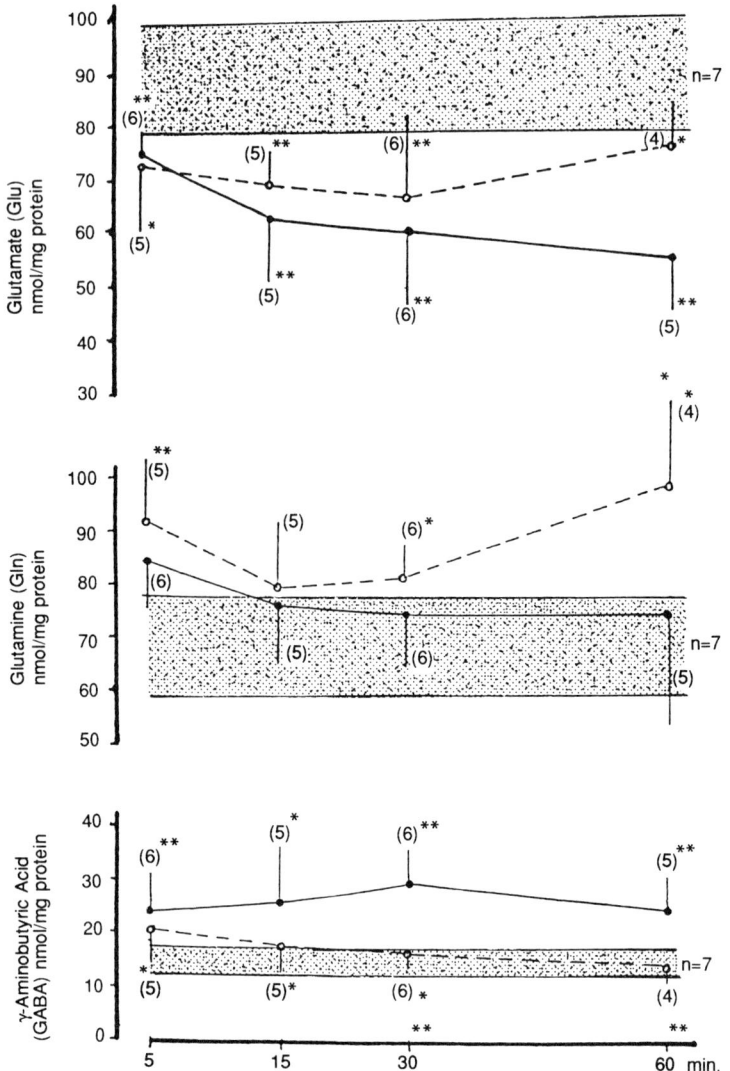

Figure 5. Striatal free Glu, Gln, and GABA level as a function of occlusion time. Values are expressed as M + SD. Numbers in parentheses denote the number of experiments. Shaded bar: normal control (M + SD), broken line indicates sham-operated control. Solid line indicates ischemia.

In conclusion, on the occasion of striatal ischemia, the regulatory effect of GABA and SP on the striatal DA metabolism still works. Providing the oxidative deamination of DA might be one of the main deleterious factors, as shown in our experiments, in the evolution of primary neuronal damage of striatal ischemia, SP antagonist should be beneficial theoretically in the attenuation of H_2O_2 production in DA neurons and could be used as a candidate in the free radical scavenging strategy.

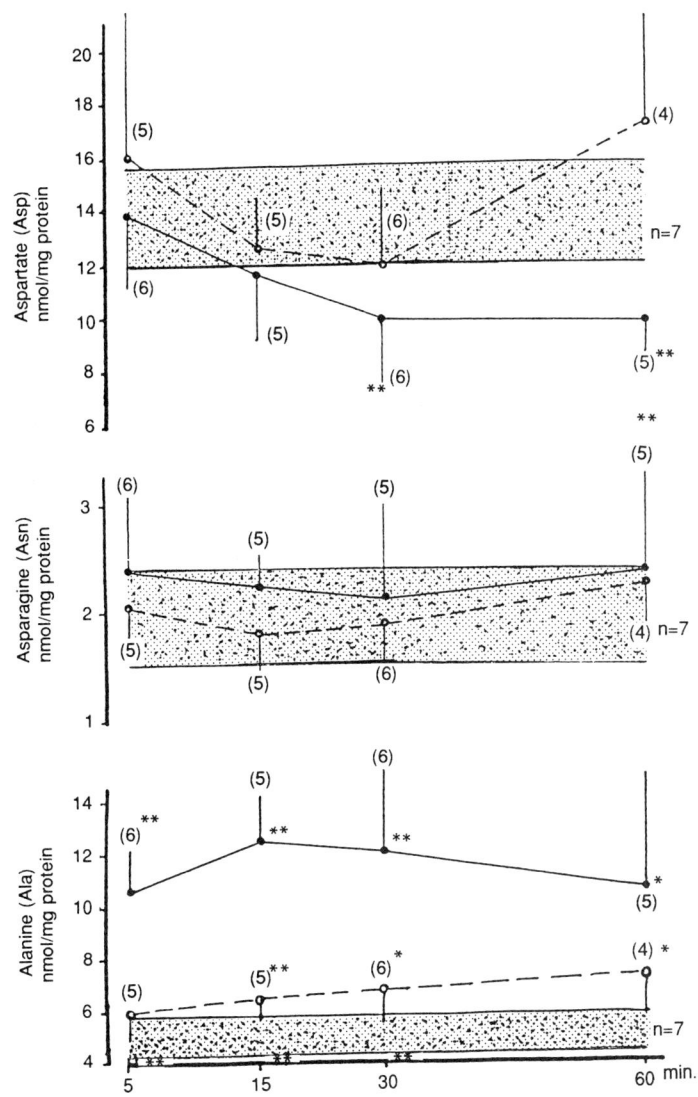

Figure 6. Strital free Asp, Asn, and Ala level as a function of occlusion time. Same notes as shown in Figure 5.

PART E. THE EXCITATORY AND INHIBITORY AMINO ACID NEUROTRANSMITTERS IN THE ISCHEMIC TISSUE AS A FUNCTION OF OCCLUSION TIME.

14 free amino acids (FAAs) were assessed as a function of time following the MCAO of rats, using HPLC method coupled with a UV detector and Pico-Tag^{TM} amino acid analysis system (Waters, USA).

Our data demonstrated were show as follows: (Fig. 5–7).

1. GLU decreased a 5 min. and approached to its maximum at occlusion time of 60 min.

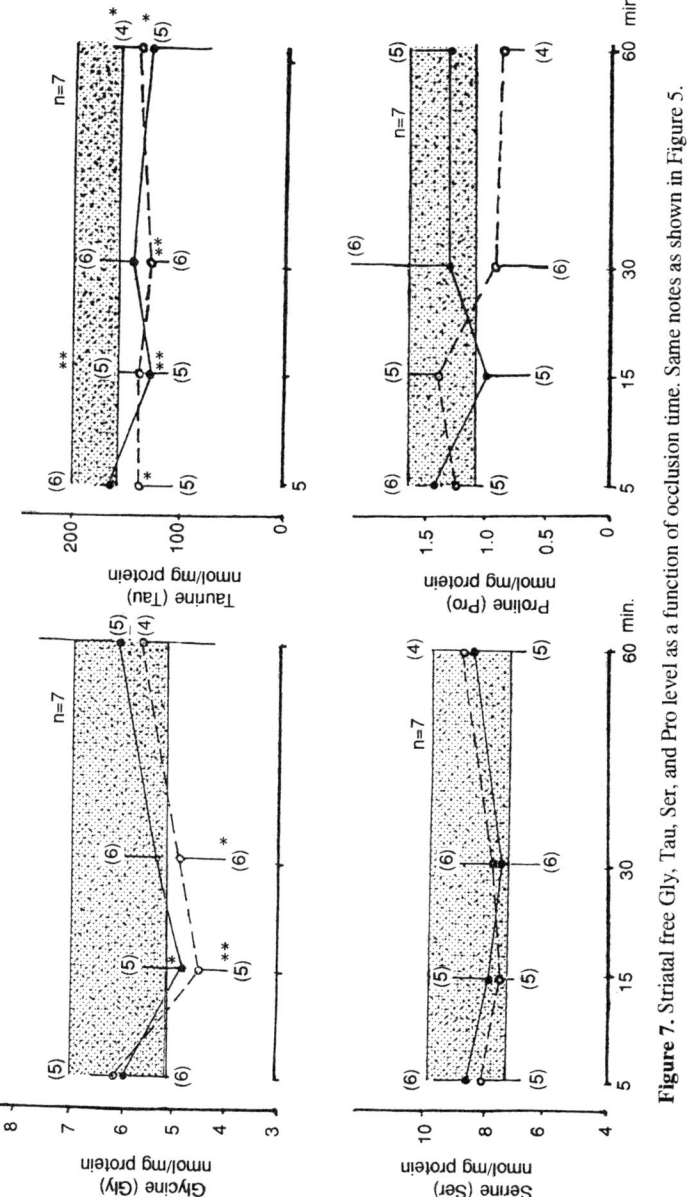

Figure 7. Striatal free Gly, Tau, Ser, and Pro level as a function of occlusion time. Same notes as shown in Figure 5.

2. GABA was significantly increased at 30 min. and 60. min. of occlusion.
3. The levels of GLN and PRO, however, didn't change concomitantly with the decrease of GLU, indicating that the alternative metabolic pathways of GLU to GLN and PRO to GLU were not involved. The conversion of GLU to GABA and the leakage of GLU were accelerated at the initial phase of cerebral ischemia would be a reasonable explanation.
4. ASP was reduced too at 5 min. and more significant at 60 min. of occlusion. ASN had no change and ALA increased profoundly at all the time observed.

It is well known that the content of lactate and pyruvate was increased at the ischemic tissue, thus, the engymatic reactions induced by ALA aminotransferase and ASP aminotransferase would shift to the synthesis of ALA and oxaloacetic acid. As a consequence, the ALA level increased and ASP decreased in the ischemic tissue. No change of GLY, TAU, and SER could be found. Furthermore, the effect of ischemia on the levels of TYR, ARG, HIS, PRO, THR were not verified, since the stress response of operation could not be ruled out.

In conclusion, the perturbation of FAAs included by ischemia manifested itself exclusively in the metabolism of GLU, ASP, GABA and ALA. Our data provide further support to the proposal that either excitatory or inhibitory amino acid neurotransmitters might be involved in the creation of ischemic neuronal damage, so far as the neurotoxic effect of the GLU be concerned.

PART F. ANTIOXIDANT ENZYME ACTIVITIES IN THE ISCHEMIC TISSUE.

The activities of striatal Cu-Zn superoxide dismutase (Cu-Zn-SOD) catalase (Cat) and Glutathione peroxidase (GSH-Px), which constitute the enzymatic scavenging system of toxic oxygen species, were determined according to the methods described by Sun and Zigman (1978)[5], Aebi (1983)[6], and Flohe and Gunzler (1984)[7] respectively. Taking the sham-operated rats as the controls, the Cu-Zn-SOD and GSH-Px, but not Cat activities were significantly reduced at an occlusion time of 30 min.[8] It denotes that the scavenging capacity was prominently reduced and provides further support to the hypothesis that oxygen stress might be one of the trigger factors of neuronal damage at the initial stage of ischemia.

Taken together, a unifying pathophysiological mechanism of primary ischemic neuronal damage that encompasses a diversity of responses must consider the intervention of toxic oxygen species and various antioxidants could potentially be used at the window of opportunity during treatment in order to interrupt the free radical reaction, and thus ameliorate the outcome of ischemic neurons.

REFERENCES

1. Tamura, A. et al, (1981) Focal cerebral ischemia in the rat: 1. description of techniques and early neuropathological consequences following middle cerebral artery occlusion, J. Cereb Blood Flow Metab, 1:53–60.
2. Bederson, J.B.: (1986) Rat middle cerebral artery occlusion: evaluation of the model and development of a neurologic examination. Stroke 17: 472–476.
3. Gu, W.G. et al, (1990) The ischemic model of rat middle cerebral artery. Chinese J. Neurosurg, 6(supl):85.
4. Lin, W. et al, (1990)The time course of changes of tissue monoamines in the acute experimental cerebral ischemia of rat. Chinese J. Neurosurg. 6(suppl.) 46–49.
5. Sun, M., Zigman S. (1978) An improved spectrophotometric assay for superoxide dismutase based on epinephrine autoxidation. Anal. Biochem 90:81.

6. Aebi, H. (1983) Catalase, In 'Methods of enzymatic analysis' (Bergmeyer, H.V., edit.) vol. 3, New York Academic Press, p. 273–286.
7. Flohe, L., Gunzler, W.A. (1984) Assay of glutathione peroxidase. In Methods in Enzymology, Packer, L. edit. vol. 105, New York Academic Press, p. 114–121.
8. Jiang, D.H. et al, (1992) The mechanism of neuronal damage caused by cerebral ischemia. Natl. Med. J. China 72:487–490.

EFFECTS OF ILEXONIN A ON CIRCULATORY NEUROREGULATION

Luo, Rong Jing, Chen, Jie Wen, Zhou le Quan, Chen Li, Jia Ke Liang, Chen Chao Feng, Hu Wei An, Luo Zhuo Ling, Yang Bei Xin

Guangzhou College of Traditional Chinese Medicine
Quang Zhou
People's Republic of China

ABSTRACT

Ilexonin A (IA), a pentacyclic triterpene, has been semisynthesized in china for the first time. It is extracted from the root of Ilicis pubescentis, a commonly used herbal medicine in Guangdong for the treatment of cardiovascular, cerebrovascular and peripheral vascular diseases and heart failure with satisfactory effects. The pharmacokinetic studies indicated that the elimination half-life after oral and i.v. dosing were 17.7 ± 2.4 h and 22.5 ± 2.9 h respectively. The total clearance was 4.6 ± 0.5 l/h. The bioavailability of IA capsules was 0.39 ± 0.14 and LD_{50} was 234 mg/Kg.

We have adopted modern techniques, including cellular electrophysiology, isotope tracing methods, molecular biology, electromicroscopy, etc., to probe into the pharmacologic mechanisms of the effects of IA on cardiovascular system. The results indicated that IA can increase the contractility of isolated guinea pig auricular myocardium, attenuate vascular smooth muscle tension induced by noradrenaline in the rabbit aorta. IA can exert a biphasic regulatory effect on arterial blood pressure. IA also can prolong A-V duration of Hiss bundle electrograph (HBE) in rabbits and prolong the action potential duration and the effective refractory period (ERP) of myocardial cells in guinea pigs. The results showed that IA can increase the cAMP content in the smooth muscle of aorta and exert a calcium-blockade effect. Therefore, the peripheral resistance vessels are relaxed and the cardiac afterload is lowered. IA-blocked calcium channels are correlated with both the potential-dependent channel and the receptor operated channel in vascular smooth muscles. IA also increases the cAMP content of myocardium and accelerates the cellular calcium influx and efflux, and this may be responsible for the direct mechanism of the positive inotropic action of IA. Under electron microscopy, it is observed that IA can alleviate the defect of succinate dehydrogenase in the myocardial mitochrondria of rabbit chronic congestive heart failure (CF) model and reduce the microstructural damage of the failed myodardium, therefore the anoxic tolerance of myocardium is increased. the effect of IA on the platelet stretching activity and microstructure in the patients with CF is also studied. It is found that IA can reduce the hypercoagulability of blood, decrease

Table 1. IA increased the contractility of isolated guinea pigs' auricular myocarcium connected with sinotrial node preparation X SD. The increased contractility rate of IA on myocardium compare with that of baseline and normal saline perfusion. * $p < 0.01$ ** $p < 0.001$

IA	control	5min.	25min.
10^{-6} M(n=6)	100%	113.2 ± 4.0 *	160.7 ± 15.1 **
5×10^{-6} M(n=6)	100%	126.5 ± 4.4 **	167.3 ± 11.5 **
NS (n=6)	100%	101.5 ± 0.9	102.3 ± 1.7

the severity of blood stagnation and improve the status of microcirculation. Effects of IA introventricular and cardiovascular central microinjection (nucleus tractus solitarius, paraventricular nucleus) on arterial blood pressure and heart rate were studied. It demonstrated that IA possess circulatory neuroregular effects by the medium of α-receptor and β-receptor of cardivascular motoneurons.

INFLUENCE OF IA ON MYOCARDIAL CONTRACTILITY OF GUINEA PIG AND ON VASCULAR SMOOTH MUSCLE TENSION OF RABBIT AORTA

Twenty-six isolated guinea pigs' right auricular myocardium connect with sinoatrial node preparations were used in this investigation. The preparations were perfused with Kreb-Hanseleit solution (37°+ 0.5° C, pH 7.3, 95% O_2 + 5% CO_2). Myocardial contractilities were measured with tension transducer recorded by RM-6000 polygraph system (Nihon Kohden). After the auto-rhythmic contraction of preparation has stabilized 30–40 minutes, change to perfuse with IA + Krebs-Hanseleit solution. Myocarial contractility were recorded 35 minutes. The results showed that different doses of IA gradually enhance myocardial contractility and the dose-effect relationship presented and S-shape curve. (Table 1)

Effect of IA on vascular smooth muscle tension of Rabbit aorta were studied. Twenty aortic strips (3X20mm) of rabbit were used. The preparations were perfused with Krebs solution (37° + 0.5 C, pH 7.3, 95% O_2 + 5% CO_2). After the preparation stabilized 1.5 hours and the aorta strips completely relaxed, vascular smooth muscle tension were measured as above. the results showed that IA can not change normal aorta strips tension but IA (10^{-5}M/L) can inhibit the attenuated effect of noradrenaline induced aorta tension (Figure 1).

THE INFLUENCE OF IA ON ARTERIAL BLOOD PRESSURE AND CAROTID SINUS REFLEX IN RABBITS

Twenty-three rabbits (2.5–3.5 Kg) were anesthetized with pentobarbital sodium, 30 mg/Kg iv. Supplemental doses of anesthetic were given as necessary throughout the experiment to maintain the anesthetic state of the experimental animals. The animals were incubated with a cuffed endotracheal tube and mechanically ventilated by use of respirator. Systemic arterial pressure was measured through a catheter placed in the femoral artery and connected to a

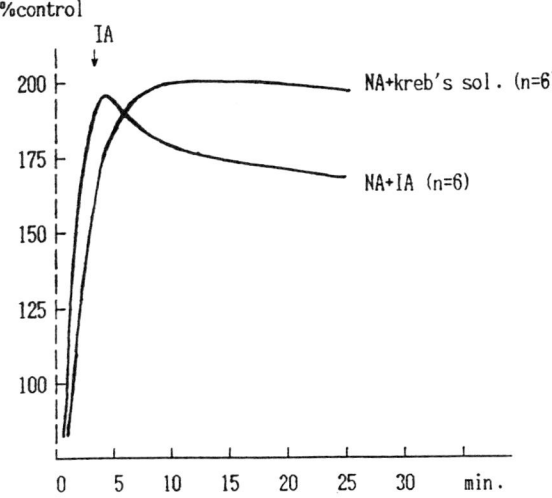

Figure 1. NA (2 10–7 mol/L) attenuate aortic tension and IA (10–5 mol/L) can inhibit the attenuate effect of NA-induced aortic tension in rabbits.

Stathan pressure transducer. Body temperature was maintained during the experiment by a heating pad.

The left carotid sinus was isolated from the rest of the circulation system. The internal and external carotid artery any small branches originating from the carotid bifurcation were completely ligated. The common carotid artery was then cannulated in a rostral direction, where as the external carotid artery was cannulated in a caudal direction, effectively isolating the carotid sinus between the two cannular. The right carotid sinus was denervated by pinching and local immersion in alcohol. Aortic depressor nerves were sectioned bilaterally. Carotid sinus pressure was measured by means of catheter connected with the outflow cannula and a screw clamp resistor was placed on the outflow cannula to permit rapid changes in carotid sinus pressure during constant flow perfusion.

In six rabbits, artery blood pressure, carotid sinus pressure and EGC were recorded simultaneously on a RM-6000 polygraph. The influence of IA (10^{-5} mol/L) on carotid sinus reflex sensitivity in rabbits were studied. The result indicated that IA can increase sensitivity of carotid sinus reflex in rabbits (Table 2).

In order to investigate the effect of IA on arterial blood pressure, hypotensive animal model and hypertensive animal model were used in this study. Seven bypotensive rabbits were induced by sodium nitroprusside iv 0.02 g/L/min., MBP constantly decreased 5.33–6.67 Kpa than normal. Five hypertensive rabbits were induced by noradrenaline iv 0.44-.088 mg/min. MBP constantly increased 4.5 Kpa than normal. The model animals were treated with IA iv (1 g/L, 20mg/Kg, 1–2 mg/min.). The results showed that arterial blood pressure increased in hypotensive animal especially increase systolic pressure and that arterial blood pressure decreased in hypertensive animal especially decrease diastolic pressure respectively (Table 3).

The result indicated that IA exert a biphasic regulatory effect on arterial blood pressure. It suggests that IA can mainly increase myocardium contractility in hypotensive animal,; on the other hand, IA relaxes peripheral blood vessels in the hypertensive animal.

Table 2. The influence of IA on the carotid sinus reflex sensitivity in six rabbits. Increasing CSP caused a decrease in MAP. It showed that IA can increase the carotid sinus reflex sensitivity.

		N.S	IA	P.value
$\triangle$CSP=6.67kpa	$\triangle$MAP(%)	-1.17±2.14	-3.43±2.69	> 0.05
	$\triangle$HR(%)	-1.04±1.38	-0.39±0.95	> 0.05
$\triangle$CSP=13.3kpa	$\triangle$MAP(%)	-20.08±6.99	-24.84±5.02	< 0.05
	$\triangle$HR(%)	-2.63±0.75	-3.34±1.01	< 0.05
$\triangle$CSP=20.0kpa	$\triangle$MAP(%)	-33.13±4.44	-38.89±3.02	< 0.01
	$\triangle$HR(%)	-4.28±0.49	-4.91±1.26	< 0.05

Table 3. IA exert a biphasic regulatory effect on arterial blood pressure in rabbit left: **SP IA iv. after compare with IA iv. before, P < 0.001. It indicates that IA can increase myocardial contractility in hyposensitive animals. Right: **DP IA iv. after compare with IA iv. before, P < 0.001. *HP iv. after compare with IA iv. before, P < 0.05. It indicates that IA can relax peripheral blood vessels.

	hypotensive model (n=7)		hypertesive model (n=5)	
	control	IA	control	IA
S.P.	10.73±2.26	12.29±1.26 **	21.20±1.20	19.41±5.05
D.P	6.21±1.28	6.74±0.97	16.36±2.38	14.63±3.27 **
HR.	186.9±39.8	173.22±36.5	216.40±48.1	187.7±52.3 *

EFFECT OF IA ON HISS BUNDLE ELECTROCARDIOGRAM IN RABBITS

Experiments were carried out on twenty rabbits (2–2.5 Kg) to observe the effect of IA on cardiac conducting system. 4f electrode was inserted from left common carotid artery into left ventricle to display HBE with a dual beam memory oscilloscope, VC-10 (Nihon Konden) and to record with RM-6000 polygraph system. The results indicated that after IA (30 mg/Kg) iv 5–10 min, A-H duration significantly prolonged (16.2 ± 6.0 ms) than normal saline iv or than IA iv before. H duration slightly prolonged IA iv after compared than before, but H-V duration did not change. It suggests IA can delay the A-V conduction time and reduce the excitability of cardiac conducting system. (Table 4).

Table 4. Effects of IA on A-V duration of HBE in rabbits' values are means SD.
*Significantly different from corresponding NS iv. control (p < 0.001). Significantly different from corresponding IA iv. before (P < 0.001).

	A-H duration		H duration		H-V duration	
	before	after	before	after	before	after
NS (n=10)	49.4 ± 6.64	51.5 ± 10.11	14.2 ± 1.90	14.8 ± 5.37	24.1 ± 1.90	24.1 ± 7.90
IA (n=10)	47.3 ± 8.22	63.5 ± 6.00* ▲	16.8 ± 4.42	17.4 ± 3.48	26.0 ± 4.11	26.4 ± 4.74

Table 5. Effect of IA(10–5M) on action potential of right papillary musclea of guinea pigs. Compare with IA perfused before *P < 0.05. Compare with Tyrode's solution control, p < 0.05 n = 10).

	APD_{50} (ms)		APD_{90} (ms)		ERP (ms)	
	before	after	before	after	before	after
IA(10 mol/L)	173.0 ± 18.6	188.6 ± 14.4*△	200.0 ± 17.7	220.6 ± 10.2*△	208.6 ± 14.7	221.0 ± 7.1*△
Tyrode's sol.	182.4 ± 5.3	182.0 ± 5.1	210.2 ± 10.3	209.0 ± 11.4	215.2 ± 9.5	215.8 ± 10.9

EFFECTS OF IA ON ACTION POTENTIAL OF PAPILLARY MUSCLE CELLS IN GUINEA PIGS

Guinea pigs' pepillary muscle cells were perfused with Tyrode's solution (pH 7.25–7.3, T. 35º + 0.5ºC, 95% O_2 + 5% CO_2). Electrical stimulate strength is two multiply threshold. Action potentials were amplified by (MEE-8201) and were observed with a dual-bean memory oscilloscope, VC-10 (Nihon Konden) and recorded with RM-6000 polygraph system. The result showed : 1). IA can prolong action potential duration of 50% depolarization (ADP_{50}), ADP_{90}, and effective refractory period (ERP) (Table 5). Different doses of IA gradually prolong ADP and ERP. The dose-effect relationship presented an S-shaped curve. 2). Nifedipine block the effect of IA in prolonging action potential. ADP_{50}, ADP_{90} and ERP significantly shorten by IA (2.8μmol/L), compare with IA perfused before (p<0.001). After nifedipine (2.8×10^{-6}μmol) perfused 10 min., papillary muscle cells were perfused with IA + nifedipine. It was found that effect of IA in prolonging ADP and ERP disappeared (Fig. 2). It suggested nifedipine can completely block the effect of IA in prolonging action potential may be mediated by blocking calcium channel. In other word, IA can promote Ca^{++} inflow mainly by activating "calcium channel".

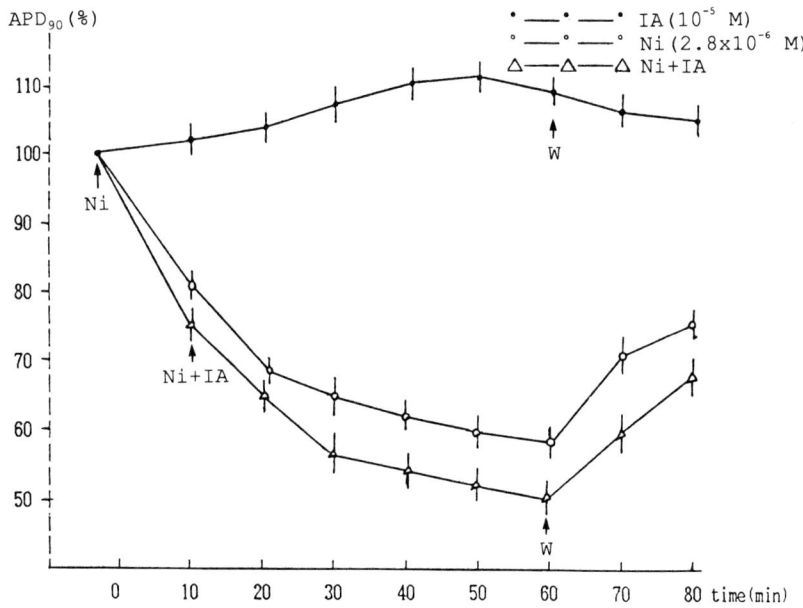

Figure 2. Influence of Nifedipine on the effect of Ilexonin A on action potentials of right papillary miscles of guinea pigs. (ADP90)

EFFECTS OF IA ON CALCIUM FLUXES AND CYCLIC NUCLEOTIDES IN CARDIOVASCULAR MUSCLES

In order to investigate the mechanisms by which IA produced the vasodilative and cardiotonic effects the effects of IA on calcium ions fluxes and cyclic nucleotides in the aortic smooth muscle and myocardium were determined. New Zealand rabbits (2–2.5 Kg, both sexes, radioactive calcium (Ca^{45}) and liquid scintillation counter were used.

EFFECTS OF IA ON CALCIUM FLUXES IN AORTIC SMOOTH MUSCLE

The results indicated that either Ca^{++} influx of efflux of rabbit aorta induced by IA itself in the normal medium was insignificantly different from the corresponding control, which indicated that IA did not affect the calcium ions permeability of the aorta smooth muscle cells in the normal medium.

Effects of IA (100 micromole/L) and verapamil (0.2 micromole/L) on high -K^+ or NE-induced Ca^{++} influx in rabbit aorta were studied. The Ca^{++} influx was measured at 10 min., after exposure to the contractile agents. The data were expressed as increase in Ca^{++} influx over control (Fig. 3).

It illustrated that IA inhibited Ca^{++} influx stimulated either by adrenoceptor activation of norepinephrine (10 micromole/L) or by high K^+ (140 micromole/L) depolarization. It suggested that IA blocked the Ca^{++} influx was correlated with both the potential dependent channel and the receptor operated channel in the vascular smooth muscle.

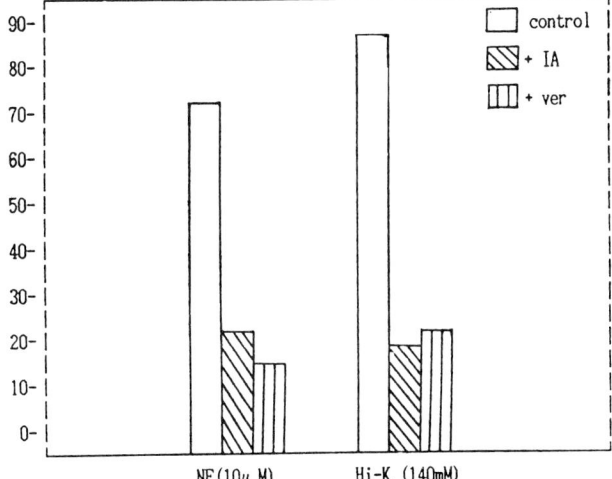

Figure 3. Effects of IA 100 micromole/L and verapamil 0.02 micromole/L for high-K+, 1 micromole/L for NE) on high K+ or NE-induced Ca++ influx in rabbit aorta. The Ca++ influx was measured at 10 min after exposure to the contractile agents. The data were expressed as decrease in Ca++ influx over control. An asterisk * indicated significant difference from corresponding control. The data were given as means +/ S.E.M. n = 12. Ordinate: Ca++ influx nanomole/g tissue.

Table 6. Effects of IA (10 micromole/L on Ca++ influx in rabbit myocardium. The Ca++ influx was measured at 3 min after exposure to radioactive calcium. The data were given as means +/S.E.M. n = 12. * $p < 0.05$ (Compared to control value).

	Ca++ influx (nanomole/g)
control	1034 +/− 18
IA	1372 +/− 22 *
isoproterenol	1388 +/− 24 *

EFFECTS OF IA ON Ca^{++} FLUXES IN MYOCARDIUM

Effect of IA (10 micromole/L) on Ca^{++} influx in rabbit myocardium were determined. The Ca^{++} influx was measured at 3 min., after myocardium exposed to radioactive calcium. The results showed that IA increased Ca^{++} influx significantly in the rabbit myocardium (Table 6).

It showed that IA increased Ca^{++} influx significantly in the rabbit myocardium. This experiment also demonstrated the effect of isoproterenol (1 micromole/L) on Ca^{++} influx in the heart.

The effects of proprenolol (10 micromole/L), phentolamine (10 micromole/L) and nifedipine (1.4 micromole/L) on IA (10 micromole/L)-induced Ca^{++} influx in rabbit myocardium were also studied too. The Ca^{++} influx was measured at 3 Min. after exposure to IA. Before exposure to IA, tissue were pretreated with propranolol, phentolamine or nifedipine respectively for 30 min.. It illustrated that the effect of IA on Ca^{++} influx was inhibited by

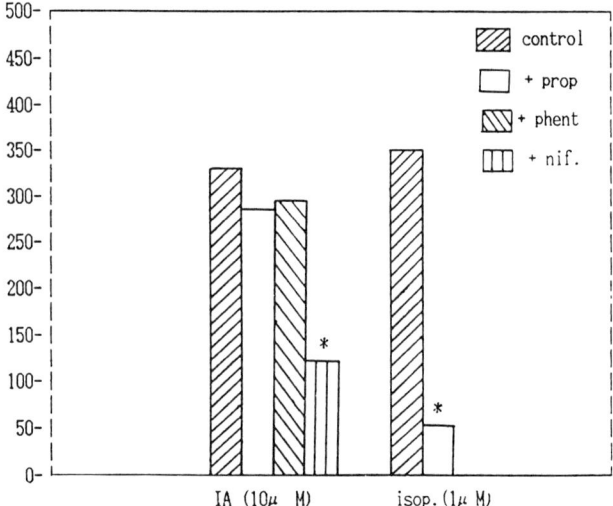

Figure 4. Effects of propranolol (10 micromole/L) phentolamine (10 micromole/L) and nifedipine (1.4 micromole/L) on IA (10 micromole/L)-induced Ca++ influx in rabbit myocardium. The Ca++ influx was measured at 3 min after exposure to IA. Before exposure to IA, tissues were pretreated with propranolol, phentolamine or nifedipine respectively for 30 min. The data were expressed as increase in Ca++ influx over control. An asterisk * indicated significantly different than corresponding control at $p < 0.01$. The data were given as mean +/ S.E.M. n = 12. Ordinate: Ca++ influx nanomole/g tissue.

calcium antagonist nifedipine, but not by phentolamine or propranolol. However, the inhibitory effect of propranolol for isoproternol was observed in this experiment. The result suggested that on one hand, IA indeed possessed the action of Ca^{++} influx in the myocardium, and on the other hand, the action of IA on Ca^{++} influx was not likely associated with myocardial alpha- and beta- adrenergic receptor (Fig.4).

EFFECTS OF IA ON CYCLIC NUCLEOTIDES IN RABBIT MYOCARDIUM AND AORTIC SMOOTH MUSCLE

Table 7 showed that IA enhanced the cyclic AMP levels 10 min. after exposure to IA in the smooth muscle significantly, but IA did not affect the cyclic GMP levels after either 5 min. or 10 min. of exposure to IA in the aortic smooth muscle.

Table 8 showed that IA increased markedly the content of cyclic AMP and also enhanced significantly cyclic GMP levels in the myocardium.

EFFECT OF INTROVENTRICULAR INJECTION OF ILEXONIN A ON ARTERIAL BLOOD PRESSURE

Effects of introventricular injection of Ilexonin A (IA) on arterial blood pressure and rate were studied. The result showed that different doses (0.001–29g/L) of IA gradually decreased

Table 7. Effects of IA (10 miscromole/L) on cyclic AMP and cyclic GMP levels in rabbit aortic smooth muscle. The data were given as means +/-S.E.M. The figures in brackets represented the numbers of rabbit. * $p < 0.05$ (compared to control value).

	cyclic AMP	cyclic GMP
	(picomole/g tissue)	
control	126.8 +/- 13.43 (n=8)	26.1 +/- 5.18 (n=5)
IA, 5min	134.8 +/- 7.79 (n=6)	27.3 +/- 1.67 (n=7)
IA, 10min	187.7 +/- 8.64 * (n=5)	25.4 +/- 4.22 (n=6)

arterial blood pressure and the dose-effect relationship presented an S-shaped curve. Systolic pressure and diastolic pressure decreased significantly especially the latter, but heart rate was insignificantly different from the normal control. It is suggested that IA can affect neural regulation of cardiovascular system and blood pressure decreases mainly by inflating blood vessel and by decreasing peripheral resistance. The result also showed : 1). MBP decreased $23.14 \pm 9.5\%$ that induced by IA (1g/L, 19 µl) intoventricular injection. MBP decreased $3.23 \pm 3.28\%$ that induced by phentolamine (50 µg) + IA (1g/L, 10µL) introventricular injection. It indicated that depressive effect of IA introventricular injection might be blocked by α-receptor blockade phentolamine. 2). MBP increased $11.01 \pm 2.00\%$ that induced by isoprenaline (10 µg) introventricular injection; isoprenalime + IA (1g/L, 10µL) introvatricular injection increased MBP $0.23 \pm 4.23\%$; isoprenalime (10µg) + propranololi (50µg) introventricular injection increased MBP $1.39 + 3.81\%$. It suggested that IA possess similar effect of B-receptor blockade propranololi which can block pressor effect of isoprenalime;

3). MBP increased $21.75 \pm 4.46\%$ induced by neostigmine (20µg) introventricular injection, IA + neostigmine introventricular injection increased MBP $25.41 \pm 7.76\%$. It was significantly different between the two groups. It suggested that IA does not affect pressor

Table 8. Effects of IA (10 micromole/L) on cyclic AMP and cyclic GMP levels in rabbit myocardium. The data were shown as means +/ S.E.M. * and ** indicated respectively $P < 0.05$ and $p < 0.001$ (compared corresponding to control values

	cyclic AMP	cyclic GMP
	(picomole/g tissue)	
control	150 +/- 7.7 (n=8)	5.8 +/- 0.34 (n=10)
IA, 2min	305 +/- 7.2 ** (n=4)	8.8 +/- 0.78 * (n=5)
IA, 5min	228 +/- 22.9 ** (n=5)	7.7 +/- 0.82 (n=5)
isop,2min	361 +/- 11.7 ** (n=4)	6.8 +/- 0.83 (n=5)
isop,5min	279 +/- 17.5 ** (n=7)	---

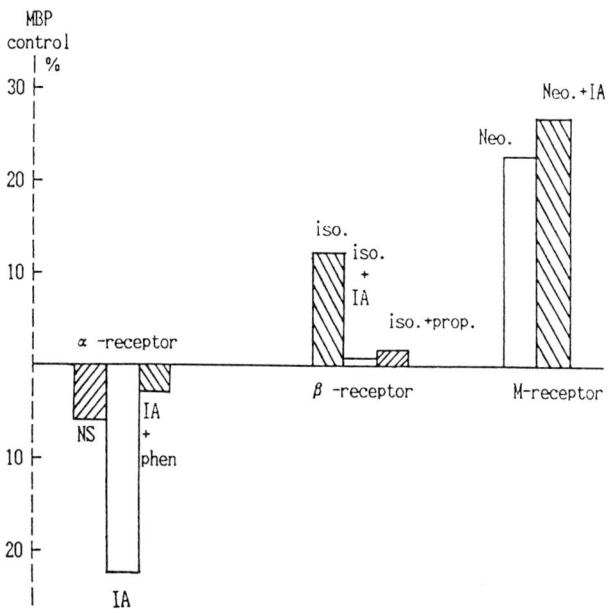

Figure 5. Effects of IA introventricular injection on cardiovascular motoneurons - , ß - , M-receptor in rat. It illustrated that depressive effect of IA introventricular injection (20 g. 10 1) can be blocked by -receptor blockade phentolamine (50 g) ; IA possess similar effect of ß -receptor blockade propranololi (50 g) which block the pressor effect of isoprenalime (10 g); IA does not affect the pressor effect of neostigmine (20 g). It suggested IA possess circulatory neuro-regular effects by the medium of -receptor and ß -receptor of cardiovascular motoneurons.

effect of neostigmine. The results indicated IA possesses circulatory neuroregular effects by the medium of α-receptor and β-receptor of cardiovascular motoneurons. (Fig. 5).

DEPRESSIVE EFFECT OF SOLITARY TRACT NUCLEUS AND PARAVENTRICULAR NUCLEUS PRODUCED BY IA MICROINJECTION

It has been demonstrated that solitary tract nucleus (STN) and paraventricular nucleus (PVN) are important cardiovascular center. Circulatory responses to IA microinjection into STN or PVN were investigated in 50 SD rats. Animals were anesthetized with pentobarbital sodium (30mg/Kg iv) and incubated with a cuffed endotracheal tube and placed on positive-pressure ventilation by use of respirator. Systemic blood pressure was monitored from a catheter in the femoral artery via a transducer (Statham) and recorded continuously on a RM-6000 polygraph, simultaneously heart rate and the electrocardiogram were monitored with disk electrodes in a lead II configuration.

The head of animal was fixed in a stereotoxic frame. STN was located according to stereotoxic atlas of rat brain (Paxinos and Watson) from 3.4–3.6mm to interaural line, from 1.4–1.6mm lateral to the medline, and from the cranial surface to 7.0–7.2mm deep. PVN was located following stereotaxic atlas of rat brain (Konig and Klippel) from 6.2–6.4 mm to interaural line, from 0.1–0.2 mm to the midline, and from the cranial surface to 7.8–8.2 mm microinjections of IA (1g/L, 1.0 µl. pH 7.4) through glass microelectrodes with a tip diameter of 30–40 µm, which were injected by microsyrings. For a control, similar microinjections of

Table 9. Effects of IA microinjection into PVN on Blood pressure in Rates. X n: number of microinjection sites Blood pressure decrease percentage of IA microinjection compared with CSF microinjection P. It showed IA can decrease blood pressure significantly especially decrease DP.

	CSF (N=41) 10μ l			IA (n=50) 10μ l. g/l		
	before	after	change percentage	before	after	change percentage
SP(Kpa)	18.10±1.11	17.42±1.12	-3.74±1.125	17.70±1.84	14.70±1.91	-16.93±4.77*
DP(Kpa)	14.24±1.50	13.70±1.49	-4.02±1.76	14.22±1.39	11.45±1.48	-19.71±6.13*
MBP(Kpa)	16.17±1.25	15.53±1.28	-3.78±1.18	15.96±1.50	13.08±1.58	-18.17±4.44*

Table 10. Effects of IA microinjection into NTS on blood pressure in Rats XSD n: number of microinjection sites. *Blood pressure decrease percentage of IA

	CSF (N=48) 10μ l			IA(n=70) 1g/l 10μ l		
	before	after	change percentage	before	after	change percentage
SP(Kpa)	18.68±1.54	18.05±1.50	-3.38±1.14	18.29±2.01	15.28±1.78	-16.35±3.89*
DP(Kpa)	15.26±1.74	14.75±1.73	-3.36±1.44	14.69±1.81	11.89±1.59	-18.81±4.76*
MBP(Kpa)	16.96±1.59	16.40±1.57	-3.34±1.00	16.49±1.84	13.59±1.58	-17.52±3.16*

artificial cerebrospinal fluid (CSF), 1% pontamine sky blue, were made. At the end of the experiment the medulla was removed and frozen. Subsequently, frozen transverse sections (40 μm) were cut serially. The position and depth of the nucleus were determined by microscopic examination.

The results indicated that IA microinjection into STN or PVN can decrease blood pressure significantly and especially diastolic pressure. (Tables 9 and 10).

CONCLUSION

IA is a pentacyclic triterpen compound which is semisynthesized from an extract of the original plant. In recent years, it has been applied for the treatment of cardiovascular, cerebrovascular and peripheral vascular diseases with satisfactory therapeutic effect. A series of experimental studies to probe into the pharmacologic mechanism of the effects of IA on cardiovascular system have been carried out in our laboratory. The research results indicated that IA not only directly affect circulatory effectors, but also possess circulatory neuroregular effects by the medium of cardiovascular motoneurons and arterial baroreceptor reflex.

STUDY ON CEREBROVASCULAR DISEASE OF THE ELDERLY IN CHINA

Wang Xin-de

Department of Neurology, Beijing Hospital
Beijing, People's Republic of China

ABSTRACT

The average incidence, prevalence and mortality rates of cerebrovascular disease (CVD) in China were markedly increased with increase of age and were much higher in senile stage and males than those in presenile stage and females. The constituent ratio of CVD consisted cerebral infarction for 67.5% and cerebral hemorrhage about 24.8%. There was no difference between the characters of lesions confirmed by CT scan in senile and presenile groups. The majority of CT lesions in the two groups was lacunar infarction, being 76.3% and 85.9% respectively. There were more cases of lobar hemorrhage in the senile group. The most important risk factor for CVD was hypertension (65.8%). Heart disease and diabetes mellitus take second place, accounting 19.0% and 10.7% respectively. The incidence of mixed type of hypertension was high in CVD especially the isolated systolic hypertension. The incidence of cerebral stroke was obviously higher than myocardial infarction in China. The percentage of positive findings of atherosclerosis in extracranial portion of carotid artery system in elderly patients with thrombosis and transient ischemic attacks was 60–100% and 55–100%.

INTRODUCTION

Cerebrovascular disease (CVD), coronary heart disease and cancer are the three major causes in the death of the world population. China is of no exception. Among this three leading high mortality diseases, in China the incidence, prevalence, mortality, fatality and recurrence rate of CVD is significantly higher than the other two. We, thererore, did a epidemiology studies of CVD in various enviromental settings and conditions in China. The findings are as follows:

Table 1. Incidence, Prevalence, and Mortality of CVD in Urban and Rural Area in China (/100,000/year)

	Incidence				Prevalence				Mortality			
	Cases	Gross rate	NAR	IAR	Cases	Gross rate	NAR	IAR	Cases	Gross rate	NAR	IAR
Six Cities Study in 1985 63 195 populat.	115	182	–	219	392	620	–	719	56	89	–	116
Beijing Study in 1985 459 602 populat.	861	187	177	232						69.9		
Rural Area Study in 1985 246 6812	200	113	111	185	625	253	251	393	204	83	79	141

NAR= National adjusted rate
IAR= International adjusted rate

1. THE INCIDENCE, PREVALENCE AND MORTALITY RATE OF CVD IN THE URBAN AND RURAL AREAS IN CHINA

Based on the records of several epidemiological studies of CVD in the urban and rural area in China, there are 113–187 incidences of CVD, 253–620 prevalence cases and 69.9–89 mortality per 100,000 population per year respectively.[1,2] Table 1 showed a significant difference of the incidence and prevalence rate of CVD in the urban and rural areas in China. It was lower in rural area than in the urban area. Of all the epidemiological studies, the prevalence rate of CVD in six cities from all studies and Beijing city were very similar. There was no significant difference of mortality of CVD in the urban and rural areas.

2. THE SEQUENCE OF THE PREVALENCE RATE OF CVD AND OTHER NEUROLOGICAL DISEASES

According to the epidemiological survey in rural area in 1985, the different aged group had a different sequence of the prevalence rate of neurological disease, particularly CVD.[1] All the age groups were divided into three: 0–44 years aged group, 15–53 years age group (presenile stage) and 60 years and over aged group (senile stage). The sequence of prevalence rate of CVD in all neurological disease were as follow: in 0–44 years aged group migraine was the first, the prevalence rate was 430.0. However, CVD was the 11th, the prevalence rate was 33.89. Interestingly, in 45–59 years aged group CVD went up to the second, the prevalence rate was greatly increased from 33.9 to 781.9. In 60 years and over aged group CVD took the first place. The prevalence rate was further increased up to 2210.3 (Table 2–4).

Table 2. The Sequence of Prevalence of Nervous Disease (0–44 years)

Sequence	Disease	Prevalence(100 000)	%
1	Migraine	487.9	21.0
2	Epilepsy	351.3	15.1
3	Brain injury	345.2	14.9
4	Meningitis	259.1	11.2
5	Bell's palsy	217.5	9.4
6	Mental retardation	140.5	6.1
7	Encephalitis	130.7	5.6
8	Poliomyelitis	119.4	5.1
9	Neuralgia sciatica	41.6	1.8
10	Peripheral neuropathy	38.5	1.7
11	CVD	33.9	1.4

Table 3. The Sequence of Prevalence of Nervous Disease (45–99 years)

Sequence	Diseases	Prevalence(100 000)	%
1	Migraine	1214.6	24.9
2	CVD	781.9	16.0
3	Bell's palsy	743.1	15.2
4	Brain injury	698.8	14.3
5	Epilepsy	393.7	8.1
6	Meningitis		
	Peripheral neuropathy		
	Neuralgia Sciatica	127.6	2.6
7	Extrapyramidal diseases Trigeminal neuralgia	77.6	1.6

Table 4. The Sequence of Prevalence of Nervous Disease (60– years)

Sequence	Diseases	Prevalence(100 000)	%
1	CVD	2210.3	41.3
2	Bell's palsy	600.2	11.2
3	Brain injury	451.9	8.4
4	Migraine	409.6	7.7
5	Epilepsy	402.5	7.5
6	Parkinson's disease	395.5	7.4
7	Extrapyramidal disease Trigeminal neuralgia	162.4	3.0
8	Neuralgia sciatica	112.9	2.1

Table 5. Incidence of CVD in the Presenile and Senile Stages (/100,000/year)

	Male	Female
Presenile Stage (45–59 years)	271.3	182.3
Senile Stage (60 years)	1186.0	788.0

3. THE COMPARISON OF THE INCIDENCE AND MORTALITY RATE BETWEEN THE SENILE STAGE AND PRESENILE STAGE

Based on the epidemiological study in the rural area in 1985, the incidence and mortality rate of CVD in the senile stage were increased 3.4 times in men and 3.3 times in women than in the presenile stage. The mortality rate of CVD in the senile stage increased 4.2 times in men and 4.8 times in women in the presenile stage. Overall, the prevalence and mortality rate of CVD were higher in men than in women in both senile and presenile stages (Table 5,6).

4. THE CONSTITUENT RATIO OF CVD LESIONS

Four hundred cases of CVD lesions verified by CT scan were studied prospectively. There were 281 males, 70.3% and 119 females, 29.7%. In the 204 cases of the senile group (71.6%) the age ranged from 60 to 87 with average of 66.7 years and 196 cases of the non-senile group (49.0%), the age ranged from 24 to 59 years with average of 51.9 years. The following changes were noticed: cerebral infarction 270 cases, cerebral hemorrhage 99 cases, subarachnoid hemorrhage 3 cases and normal CT scan 28 cases. Eight among 400 cases in whom the final diagnosis was subarachnoid hemorrhage by a test of cerebral spinal fluid (CSF). Cerebral infarction and hemorrhage in the non-senile group accounted for 68.9% and 21.9%, while in the senile group accounted for 66.2% and 27.5% respectively. There was no difference in constituent ratio of cerebral infarction and hemorrhage between the non-senile and senile groups. However, the total constituent ratio of CVD was 67.5% of cerebral infarction and 24.8% of cerebral hemorrhage (Table 7). It indicated that cerebral ischemic lesions were apparently greater than that of cerebral hemorrhage. In addition, the constituent ratio of cerebral hemorrhage (24.8%) in China was significantly higher than that reported by American and European scientists.

Table 6. Mortality of CVD in the Presenile and Senile Stages (/1000,000/year)

	Male	Female
Presenile Stage	77.7	63.0
Senile Stage	401.3	363.3

Table 7. Constituent Ratio of Lesions of CVD Verified by CT

	Cerebral infarction		Cerebral hemorrhage		Subarachnoid hemorrhage		Normal CT scan	
	Cases	%	Cases	%	Cases	%	Cases	%
Non Senile Group	135	68.9	43	21.9	2	1	16	8.2
Senile Group	135	66.2	56	27.5	1	0.5	12	5.8
Total	270	67.5	99	24.8	3	0.7	28	7.0

The constituent ratio of CT scan lesions of 400 cases with CVD in the non-senile group and senile group was also analysed in detail. The lesions were divided into ischemic and hemorrhagic ones. The former one consisted of the lacunar infarct and wedge-shaped infarction. The latter one comprised of the hemorrhage of basal ganglion, cerebral lobe, pons, cerebellum and cerebral ventricles. No difference in lacunar and wedge-shaped infarction between the non-senile group and senile group was found. There was also no difference of hemorrhage in basal ganglion between senile and non-senile groups. The constituent ratio of lobar hemorrhage in the non-senile and senile groups was 2.3% and 10.7%, respectively. It showed that the occurrence of lobar hemorrhage was slightly higher in the senile group than in the non-senile group (Table 8). It was considered to be related to the amyloid degeneration of cerebral arteries in the senile stage. Of all 400 cases of CVD there were 7 cases (3.6%) with leucoaraiosis in the non-senile group and 32 cases (15.9%) with leucoaraiosis in the senile group. It was apparently greater in the senile group than the non-senile group.

Table 8. Constituent Ratio of Lesions of CVD between the Non-senile and Senile Group in Detail

	Non-senile group		Senile group	
	Cases	%	Cases	%
Ischemic lesions:				
Lacunar infarct	116	85.9	103	76.3
Wedge-shaped infarct	19	14.1	32	23.7
Hemorrhagic lesions:				
Basal ganglion	39	90.7	46	82.1
Cerebral lobe	1	2.3	6	10.7
Pons	3	7.0	0	0
Cerebellum	0	0	3	5.4
Ventricles	0	0	1	1.8

Table 9. Comparison of Risk Factors between the Non-senile and Senile Groups

	Non-senile group n=196		Senile group n=204	
	Cases	%	Cases	%
Hypertension	126	64.3	137	67.2
Heart disease	27	13.8	49	24.0
Diabetes mellitus	17	8.7	26	12.7

5. THE ANALYSIS OF RISK FACTORS OF CVD BETWEEN THE NON-SENILE GROUP AND SENILE GROUP

The risk factors such as hypertension, heart disease and diabetes mellitus seemed to play an important role in the cause of CVD. In the study of above risk factors of 400 cases with CVD confirmed by CT scan, the results revealed that the first one was hypertension, 263 of 400 (65.8%), the second one was heart disease, 76 of 400 (19%) and the third one was diabetes mellitus, 43 of 400 (10.7%). Heart disease included coronary heart disease 71 cases, rheumatic heart disease 4 cases and congenital heart disease one case. There is no difference of percentage of hypertension between the senile and non-senile groups. The percentage of heart disease is significantly higher in the senile group than in the non-senile group (Table 9). Six cases with acute myocardiac infarction were present in 1.5% of 400 cases with CVD. It was called cerebrocardiac stroke. In addition, the number of cases and their clinical course of diabetes mellitus and hypertension between the non-senile group and senile group were compared. The results concluded that there was no obvious difference of the number of cases of hypertension between the non-senile group and the senile group, and a much higher number of cases in the senile group with diabetes mellitus than in the non-senile group. From a clinical point of view, cases of diabetes mellitus within 19 years would have cerebral stroke in only 33.5%. Diabetes mellitus was found in 12.7% of 400 cases in senile groups (Table 10).

The increase of fasting blood sugar level in 53 of 102 cases with CVD in the senile group (51.9%) was found. Thirty three cases with stress hyperglycemia (32.4%) were confirmed by glycosylated hemoglobin test. In 30 cases the diabetes mellitus was highly suspected. Except

Table 10. Incidence of Cerebral Stroke in Clinical Course of Diabetes Mellitus and Hypertension

Years	Diabetes M.		Hypertension	
	Senile G.	Non-senile G.	Senile G.	Non-senile G.
Uncertain			1	
1–9	20	14	34	36
10–19	5	3	61	52
20–29	1	0	30	28
30–39	0	0	11	10
	26	17	137	126

Table 11. Relationship between the Cerebral Infarction and the Lesions of Myocardial Infarction

Location of lesions	Cases
Anterior wall	7
Inferior wall	4
Posterior wall	1
Subendomyocardial	1

two fatal cases, the rest of those cases in the stable stage and finally were diagnosed as diabetes mellitus by means of the oral glucose tolerance test.

The cerebral infarction caused by the myocardiac infarction was also studied, called cardiocerebral stroke.[3] The result showed that in 13 of 297 (4.4%) cases with acute myocardiac infarction had cerebral stroke. In seven of them (2.4%) cases developed cerebral stroke within two weeks after myocardiac infarction during hospitalization. The fatality rate of myocardiac infraction cases with or without cerebral stroke in the hospitalization accounted for 28.6% and 11.9%, respectively. The cerebral stroke occured more frequently in anterior myocardiac infarction (53.8%) (Table 11). 199 of 297 cases were examined by sonocardiography. The mural thrombus was found in 10 myocardiac infraction cases (5.0%) and two of the 10 cases had cerebral stroke. The etiology of myocardiac infarction with cerebral stroke was considered to be related to the lesions of myocardiac infarction, particularly anterior myocardiac infarction, cardiogenic shock and severe heart failure.

The blood pressure of selected 268 cases with a confirmed diagnosis of CVD recorded pre, post and onset of stroke. The correlation of various types of hypertension with cerebral stroke and age was studied. The criteria of classification of hypertension was as follows: the mixed type of hypertension systolic blood pressure (SBP) 160mmHg; diatolic blood pressure (DBP) 95mmHg; the isolated systolic hypertension SBP 160mmHg, DBP 95mmHg; diatolic hypertension SBP 160mmHg, DBP 95mmHg. The results in our study suggested that CVD with the mixed type of hypertension were very common. CVD with the isolated systolic hypertension was predominantly higher in the senile stage than in the presenile stage (Table 12). In olden days, it was considered that the isolated systolic hypertension was not a risk factor for the elderly, but now it has been confirmed to be a predominant risk factor for myocardiac

Table 12. Constituent Ratio of Hypertension Types of CVD in the Senile and Pre-senile Stages

	Presenile stage		Senile stage	
	Cases	%	Cases	%
Mixed type of hypertension	83	71.6	97	63.8
Isolated systolic hypertension	3	2.6	13	8.6
Isolated diastolic hypertension	12	10.3	11	7.2
Normal blood pressure	18	15.5	31	20.4

Table 13. Comparison of Positive Findings of the Atherosclerosis of Extracranial Portion of CAS of TIA, Cerebral Infarction and Healthy Groups

Age group (years)	Healthy group			TIA			Cerebral infarction		
	No.	Positive cases	%	No.	Positive cases	%	No.	Positive cases	%
20–	24	–	–	1	–	–	–	–	–
30–	24	–	–	–	–	–	–	–	–
40–	24	1	4.2	10	4	40.0	7	3	42.9
50–	21	1	4.8	40	17	42.5	54	37	68.5
60–	49	13	26.5	20	11	55.0	47	31	66.0
70–	32	9	28.1	20	13	65.0	22	20	90.9
80–	12	5	41.7	3	3	100.0	3	3	100.0
	186	29	15.6	94	48	51.1	133	94	70.7

and cerebral infarction in the aged. The isolated systolic hypertension was resulted from the decrease of arterial wall elasticity and compatibility caused by atherosclerosis.[4]

6. THE INCIDENCE OF CEREBRAL STROKE AND MYOCARDIAC INFARCTION IN BEIJING SHIJINGSHAN DISTRICT

Based on the epidemiological study of the population with cerebral stroke and myocardiac in Beijing Shijingshan district, the incidence was significantly different. In the population of 72,638 studied in eight years an average of 133 and 23 occurrences of cerebral stroke and myocardiac infarction in every year were found. The former was four times of the latter. It indicated that the incidence of cerebral stroke was much more frequent than myocardiac infarction in China.[5]

7. THE STUDY ON ATHEROSCLEROSIS OF EXTRACRANIAL PORTION OF CAROTID ARTERY SYSTEM IN HEALTHY PERSONS, TRANSIENT ICHEMIC ATTACKS AND CEREBRAL INFARCTION

The atherosclerosis in extracranial portion of the carotid artery system was an important cause of cerebral stroke, such as transient ichemic attacks (TIA) and thrombosis. The European and American scientists reported that the incidence of atheroslerotic lesions was higher in extracranial portion of carotid artery system. In China, there has been no systematic study on it, but there were more cases with atherosclerotic lesions located intracranially. Hundred and eighty two healthy persons (119 males and 63 females), 94 cases of TIA and 133 cases of cerebral infarction were examined. The results revealed that diameter of the carotid artery of the healthy persons was increased with the increase of age. The positive findings of extracranial atherosclerosis were 4.2–4.8% in the presenile stage and 26.5–41.7% in the senile stage. Compared with the data reported by McRae LP, our data was significantly lower.[6] It explained that the prevalence rate of CVD in China was increased with the increase of age and incidence of CVD caused by extracranial atherosclerosis in the carotid artery system of TIA and cerebral infarction were 51.1% and 70.7%, apparently higher than in 15.9% of the healthy persons (Table 13).[7] In the presenile stage the positive findings in the patients with TIA, cerebral infarction and healthy person were 40.0–42.5%, 42.9–68.5% and 4.2–4.8%, respectively. In the senile stage the positive findings in the cases with TIA cerebral infarction and healthy persons were 55.0–100%, 66–100% and 26.5–41.7% respectively. It showed that the positive findings of the atherosclerotic lesions were markedly higher in the patients with TIA and cerebral infarction than in the healthy persons.

REFERENCES

1. Feng E.J. (1990) Epidemiological study on nervous diseases in the elderly. in: Geriatric Neurology pp. 36 X.D. Wang, ed., People's Health Publisher, Beijing.
2. Wang Z.C.et al (1984) Epidemiologic study of neurological disorders in an urban community of Beijing. Chinese Journal of Neurology and Psychiatry 17:72.
3. Zhang W.J., Wang X.D. (1988) Cerebral stroke and myocardiac infarction. Natl Chinese Medical Journal 27(6):345.
4. Li K.Y. et al (1990) Correlation between the types of hypertension and occurrence of stroke. Chinese Journal of Geriatrics 9(3):149.
5. Tao S.C. et al (1982) Incidence of acute myocardial infarction and stroke in the capital and steel complex region of Beijing, China. Magnesium 26(1):144.
6. Wang X.D. (1989) Study on atherosclerosis of extracranial portion of carotid artery system in healthy persons. Chinese Medical Journal 89:718.
7. Wang X.D. (1990) Chinese patients with transient ischemic attacks. Chinese Medical Journal 103(9):699.

CELLULAR PHYSIOLOGY OF EPILEPTOGENIC PHENOMENA

A Hint for Future Therapy by a Herbal Mixture

Eiichi Sugaya,[1,2] Aiko Sugaya,[3] Kagemasa Kajiwara,[1,2] Tadashi Tsuda,[3] Noriyo Kubota,[3] Noriyuki Yuyama,[1] Masahiro Motoki,[3] Tamaki Takagi,[1] Hisaaki Takagi,[1] Tamiko Ookura,[2] and Hideko Nagasawa[2]

[1] Department of Physiology
Kanagawa Dental College
Yokosuka, Japan
[2] Laboratory of Neurobiology
Tsumura Institute for Oriental Medicine
Keio University, School of Medicine, Tokyo, Japan
[3] Faculty of Pharmaceutical Sciences
Josai University
Saitama, Japan

Epilepsy is one of the most frequently occurring nervous diseases. One out of 200 persons suffers from epilepsy. Nevertheless, there is still no definite radical therapy against epilepsy. Actual therapy against epilepsy is merely the expectation of a spontaneous cure by long-term inhibition of convulsions that facilitate the next convulsions using so-called anticonvulsant drugs, even when patients are tortured with side effects. To achieve a reasonable radical therapy, it is dispensable to elucidate the process of intracellular change of normal regular firing neurons into seizure-evoking neurons; and the differences between neurons manifesting bursting activity and those showing normal regular firing. We have hitherto investigated the intracellular events during bursting activity in a single neuron. I would like to talk about our experiments on such cellular mechanisms of epileptogenic phenomena, especially intracellular events during seizure activity. I will also describe anticonvulsant effects of a herbal mixture,. TJ-960, which shows a different mechanism of action from pure chemical anticonvulsants; and finally I will give some hints for future therapy against epilepsy.

1. PENTYLENETETRAZOL-INDUCED BURSTING ACTIVITY

When seizure activity is observed on the mammalian cerebral cortex, the intracellular membrane potential shows a characteristic potential change named bursting activity or paroxysmal depolarization shift, i.e., huge depolarization with grouped spike discharges

Neurochemistry in Clinical Application,
Edited by Lily Tang and Steven Tang, Plenum Press, New York, 1994

Figure 1. Intracellular calcium increase during PTZ treatment. A: Control primary cultured neuron from the cerebral cortex of the mouse. B: 5 min after extracellular application of PTZ (10 mM). C: 10 min after rinsing with normal medium. A confocal laser microscopic observation with an ultraviolet laser source (335nm) using indo 1 calcium sensitive dye. Red indicates the most dense and blue, the lowest concentration of calcium.

Figure 2. Intracellular calcium increase during PTZ application under the condition of calcium channel blockade with cobalt chloride in the medium. A: Control cultured neuron from the cerebral cortex of the mouse. B: 5 min after extracellular application of PTZ with cobalt chloride. C: 10 min after rinsing with normal medium. A confocal laser microscopic observation with an ultraviolet laser source (335 nm) using indo 1 calcium sensitive dye. Red indicates the most dense and blue, the lowest concentration of calcium.

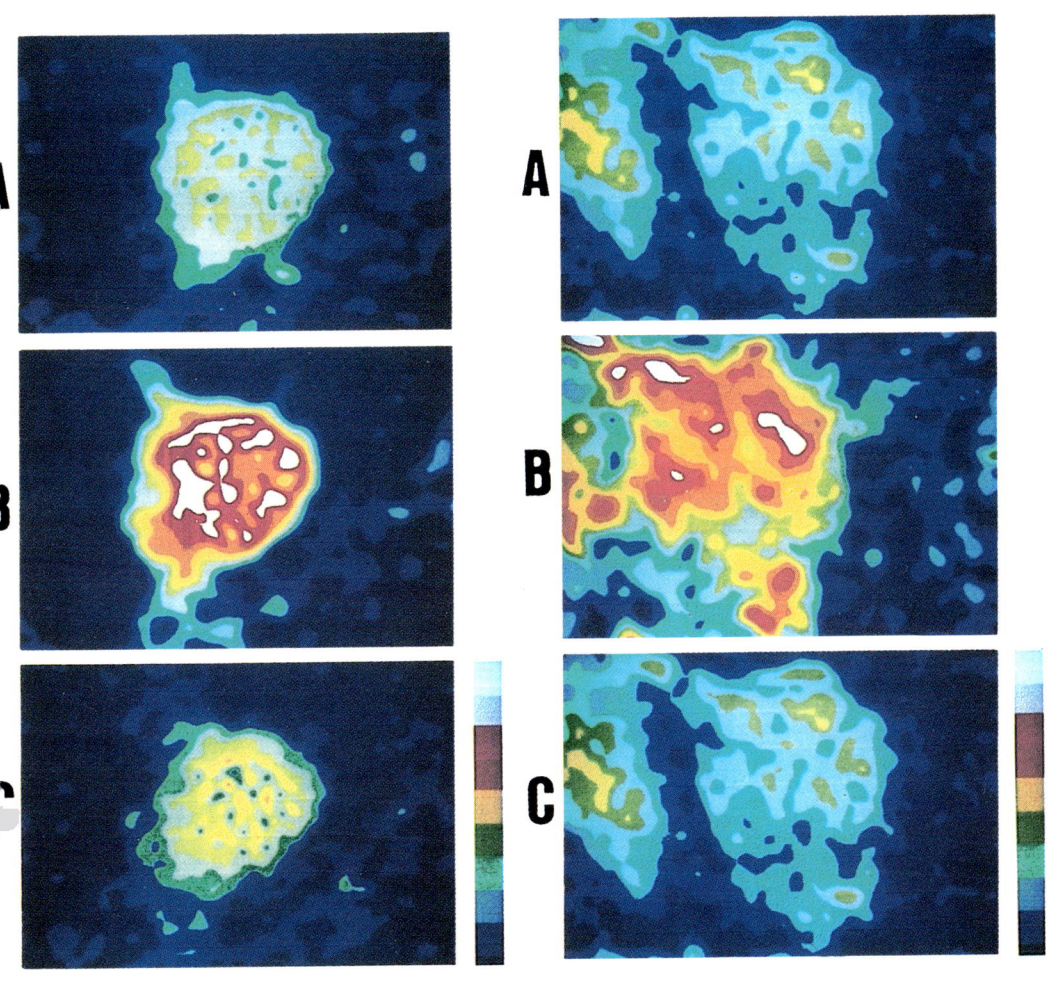

Figure 1

Figure 2

Figure 6

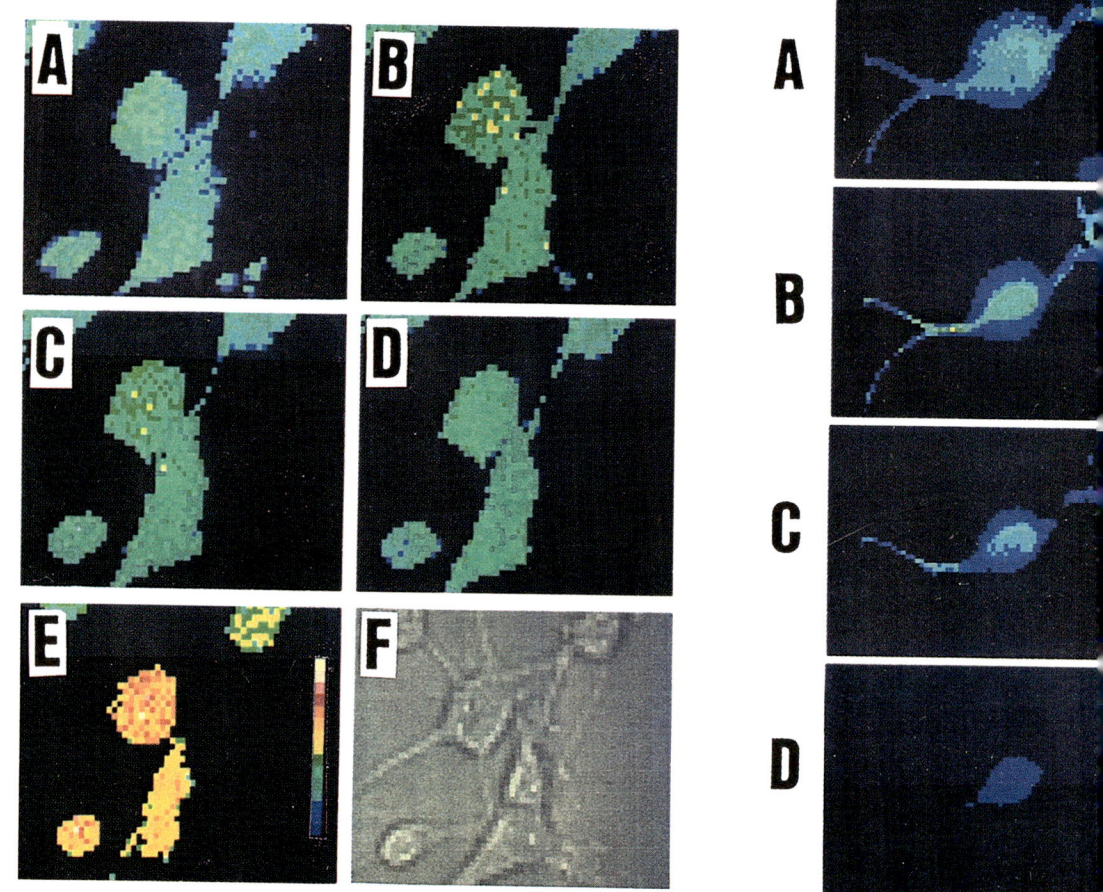

Figure 9

Figure 6. Normalizing effect of TJ-960 on PTZ-induced intracellular calcium increase. A: Normal neuron. B: Treatment of PTZ (10 mM) with TJ-960 (75 µg/ml). C: 10 min After rinsing with normal medium D: 20 min after rinsing with normal medium. E: PTZ without TJ-960. Intracellular calcium increased markedly. F: Phase contrast photograph of the observed neuron. Observation using ARGUS-50 (Hamamatsu Photonics) with calcium sensitive dye fura 2. Primary cultured neuron from the cerebral cortex of the mouse.

Figure 9. Inhibitory effect of PHT on PTZ-induced intracellular calcium increase. A: Calcium distribution in a normal cultured neuron. B: 7 min after PTZ (5 mM) and PHT (0.05mM) application. C: 12 min after PTZ and PHT. D: 16 min after PTZ and PHT. Observation using ARGUS-50 (Hamamatsu Photonics) with calcium sensitive dye fura 2.

on its rising phase and sometimes also on its falling phase (Sugaya, E. et al., 1964). Such characteristic bursting activity can be seen not only in the cerebral cortex of mammals, but also in special neurons in the ganglion of the snail, *Euhadra peliomphala*, by extracellular application of pentylenetetrazol (PTZ) (Sugaya, A. et al., 1973). Moreover, in the case of *Euhadra* ganglion, besides PTZ-sensitive neurons, there are PTZ-non-sensitive neurons that never show bursting activity even by the concentrated and long-term application of PTZ (Sugaya, A. et al., 1973). The changes during seizure activity concerning transmembrane ion movement can be investigated in detail using advanced electrophysiological techniques; but the intracellular events during seizure activity were not elucidated because of technical difficulties. Fifteen years ago, we used an electron probe X-ray microanalyzer (EPXMA) to make a calcium distribution map of a single neuron during bursting activity (Sugaya, E. et al., 1978d). We performed various experiments, always comparing the PTZ-sensitive neurons to PTZ-non-sensitive neurons, and elucidated the differences of the intracellular events during bursting activity and the normal state (Sugaya, E. et al., 1991b).

2. INTRACELLULAR EVENTS DURING BURSTING ACTIVITY

To apply EPXMA to biological specimens, we must devise a method for preparation of the specimens. After various electrophysiological experiments, a ganglion was frozen in freon 12 cooled with liquid nitrogen. Then the specimen was sliced, and the identified neurons were picked up and arranged on a pure carbon disk. The carbon disk was placed on a measuring stage and the calcium content was successively measured spot by spot and stored in a computer. Finally, the stored results were printed out and a calcium distribution map of a single neuron was obtained.

This intracellular calcium distribution map of a single neuron showed the following results. In the normal state, the intracellular calcium was scattered evenly in the cytoplasm; however, after PTZ application, the intracellular calcium in the PTZ-sensitive neurons gathered near the cell membrane area (Sugaya, E. at al., 1978a). In the PTZ-non-sensitive neurons, these PTZ-induced changes were never observed (Sugaya, E. et al., 1978a). These changes were also observed under conditions of blockade of calcium current by cobalt or lanthanum, which means that the PTZ-induced intracellular change occurred regardless of transmembrane calcium current change (Sugaya, E. et al., 1978a). The source of intracellular calcium release by PTZ was lysosome-like granules which showed calcium release with morphological changes by an examination with an energy disperse type EPXMA on frozen ultra thin sections of specimens (Sugaya, E. et al., 1978b). The site of the calcium gathered near the cell membrane was the inner surface of the cell membrane (Sugaya, E. et al., 1978c) according to an examination using stepwise etching of a frozen single neuron with an ion shower milling machine and measurement of the calcium content with a flameless atomic analyzer using a tungsten boat combined with weighing of a single neuron by a quartz fiber fishpole balance (Lowry et al., 1972). The calcium binding state near the cell membrane of the PTZ-treated neuron was different from that of the normal neuron, i.e., the substance bound to calcium near the cell membrane of the PTZ-treated neuron was different from that of the normal neuron (Sugaya, E. et al., 1981).

The concentration change of the intracellular free calcium was measured by a calcium ion sensitive microelectrode (Tien et al., 1980) inserted in the PTZ-sensitive neurons and PTZ-non-sensitive neurons. In the PTZ-sensitive neurons, intracellular calcium concentration increased gradually and when it reached a certain level higher than normal but lower than that which evokes the Meech effects (Meech et al., 1972), bursting activity occurred (Sugaya, E. et al., 1987c). In the PTZ-non-sensitive neurons, no remarkable change was observed (Sugaya et al., 1987c). These types of intracellular calcium concentration changes were confirmed using calcium sensitive dye fura 2

(Grynkiewicz et al., 1985) in the cultured neurons from the cerebral cortex of the mouse (Sugaya, E. et al., 1993).

We also examined the intracellular calcium concentration level using a confocal microscope with an excitation laser source of 335 nm wave length and calcium sensitive dye indo 1 (Grynkiewicz et al., 1985). Using this confocal microscope, the intracellular calcium concentration changes at the slice level of a cultured neuron can be observed. The results showed that the intracellular free calcium was unevenly increased by extracellular application of PTZ (Fig. 1). This change was the same as that when the calcium current was blocked by cobalt chloride (Fig. 2).

Thus, during bursting activity, intracellular stored calcium was released and intracellular calcium concentration increased in both *Euhadra* neurons and primary cultured neurons from the cerebral cortex of mice.

Intracellular proteins of approximately 5 kDa and 15 kDa in molecular weight increased during bursting activity in *Euhadra* PTZ-sensitive neurons. Incorporation of calcium into these two proteins increased during bursting activity (Sugaya, E. et al., 1982). We also confirmed the increase in 15 kDa acidic proteins during PTZ treatment in primary cultured neurons from the cerebral cortex of rats (Uyemura et al., 1986).

3. ANTIEPILEPTIC HERBAL MIXTURE FORMULATION TJ-960

We became aware of a formulation of a herbal mixture, SK or TJ-960, which had frequently been prescribed for epilepsy by specialists of herbal medicine who reported dramatic effects on intractable epilepsy (Sugaya, E., 1987a). TJ-960 is a mixture of nine herbal drugs, i.e., root of *Bupleurum falcatum*, bark of *Cinnamomum cassia*, root of *Panax ginseng*, root of *Glycyrrhiza uralensis*, root of *Paeonia lactiflora*, root of *Scutellaria baicalensis*, tuber of *Pinellia ternata*, root of *Zingiber officinale* and fruit of *Zizyphus jujuba*. This prescription has been called "Saiko-keishi-to-ka-shakuyaku" (SK) or "Shosaiko-to-go-keishi-ka-shakuyaku-to" (TJ-960). Both prescriptions are slightly different in the doses of the component herbal drugs, but the experimental and clinical effects were the same.

A Japanese doctor, Saburo Aimi, a specialist in herbal medicine, wrote a new interpretation of "Shokanron," a famous Chinese classical medical book. He applied this formulation to more than 300 epilepsy patients and obtained remarkable results. He first reported these results in the Journal of Oriental Medicine in Japanese in 1956 (Aimi, 1956). We did not believe him at first since we thought an intractable disease such as epilepsy could not be cured by a mixture of roots of various herbs. Therefore, I asked my colleague Dr. Narita, chief of the psychiatric clinic of The Second National Tokyo Hospital to apply this herbal mixture. He selected 23 patients with intractable epilepsy whose convulsions had never been controlled with various pure chemical anticonvulsants. To our surprise, 19 out of 22 patients --one case dropped out---were improved (Narita et al., 1982).

Then, we performed general pharmacological screening tests. However, the results were almost all negative except for a slight prolongation in the hexobarbital sleep test and inhibition of convulsions in the DBA mouse, an audiogenic epilepsy animal model (Takato et al., 1981). What does it mean when a drug is clinically very effective on intractable epileptics but negative in pharmacological screening tests?

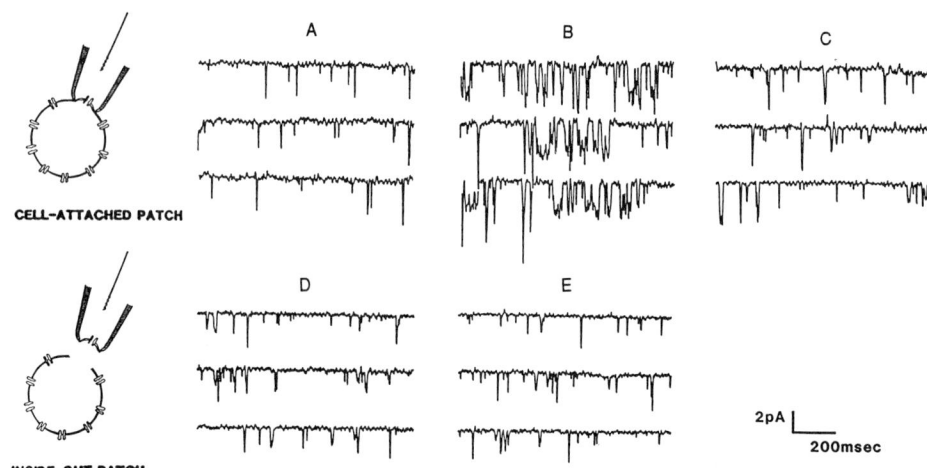

Figure 3. PTZ-induced changes of the single potassium channel. A: Normal open-close state. Cell-attached configuration. B: PTZ-induced frequent and burst type open-close state. C: After rinsing. D: After switching to the inside-out configuration. E: PTZ application from inside the cell membrane. No change was observed. Primary cultured neuron from the cerebral cortex of the mouse. 3-day-old culture.

4. NORMALIZATION OF SEIZURE-RELATED CELLULAR PHENOMENA BY TJ-960

Then, we examined the effects on the various above-mentioned intracellular pathological phenomena during seizure activity. PTZ-induced bursting activity in *Euhadra* neurons was completely normalized by this herbal mixture (Sugaya, A. et al., 1978, 1979).

Intracellular calcium shift to the cell membrane area was completely normalized by extracellular administration of this herbal mixture (Sugaya, A. et al., 1985a). The abnormal calcium binding state change during bursting activity was also normalized by this herbal mixture (Sugaya, A. et al., 1985a). The abnormal increase in 5 kDa and 15 kDa intracellular proteins was also completely inhibited. The increased calcium incorporation into the proteins was also normalized (Sugaya, A. et al., 1985a).

It is said that neurons which tend to easily evoke seizure discharge show so-called negative resistance characteristics. This herbal mixture completely inhibits these negative resistance characteristics (Sugaya, A. et al., 1985b).

Research to clarify the principle is usually performed using adequate materials for experiments such as a snail's neuron. However, some people ask if snails have epilepsy. Therefore, we performed the experiments on mammalian cerebral cortical neurons. Primary cultured neurons from the cerebral cortex of mice or rats were used.

Using a patch clamp technique (Hamil et al., 1981), we examined PTZ effects of primary cultured neurons. They showed clear bursting activity by PTZ (Sugaya, E. et al., 1989). Using a cell-attached patch clamp technique, we examined the behavior of a single potassium channel. In the normal state, a single potassium channel showed a random open-close state with short open time and long close time. When PTZ was applied extracellularly, the open-close state became frequent and finally a burst type open-close state was observed (Sugaya, E. et al., 1989). After the PTZ effect was confirmed and specimen was rinsed, the same channel was switched to the inside-out patch configuration from the cell-attached patch. In this condition,

Cellular Physiology of Epileptogenic Phenomena

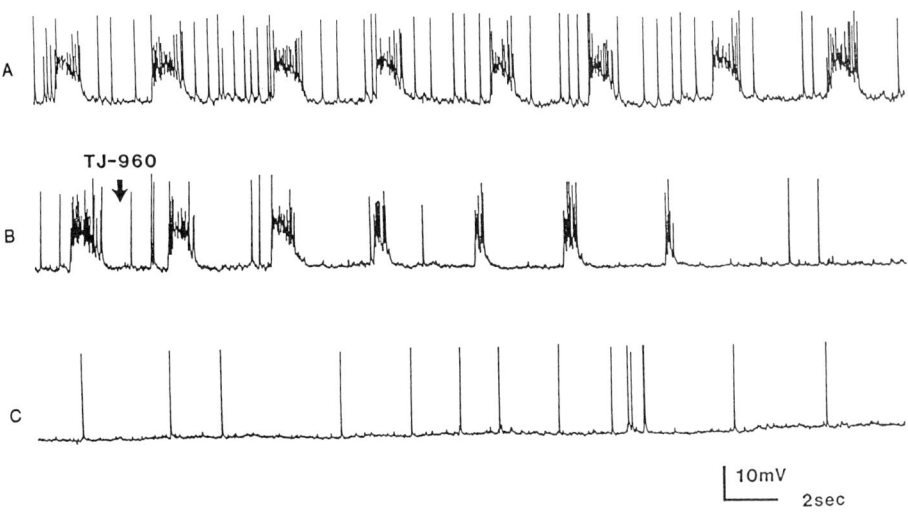

Figure 4. Normalizing effects of TJ-960 on PTZ-induced bursting activity in primary cultured neuron. A: PTZ-induced bursting activity. Whole-cell patch recording. B: Normalization of bursting activity by extracellular application of TJ-960 (75 µg/ml).

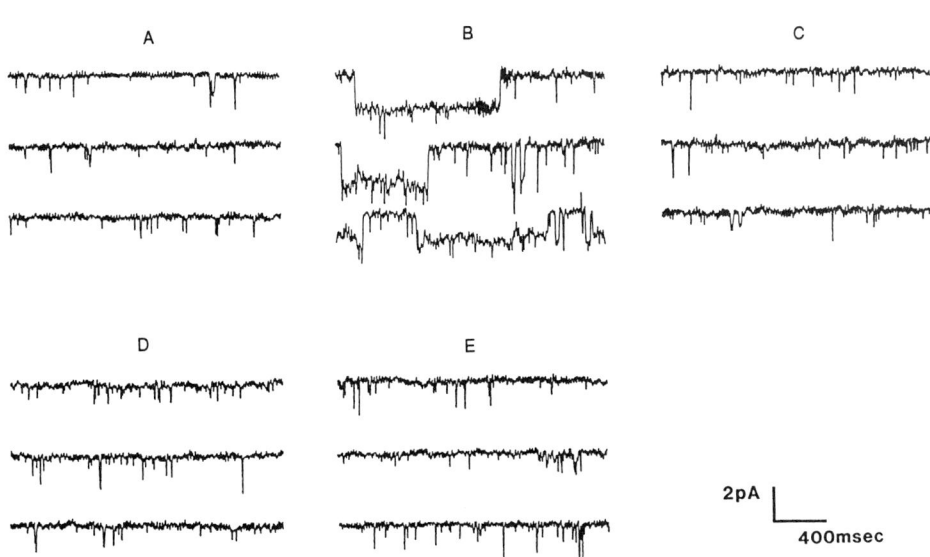

Figure 5. Normalizing effects of TJ-960 on PTZ-induced burst type open-close state of the potassium channel of a cultured neuron. A: Random open-close state in a normal neuron. B: PTZ-induced frequent burst type open-close state. C: After rinsing with normal medium. D: TJ-960 (75 µg/ml) added medium. E: PTZ and TJ-960 added medium. No PTZ effect was observed. Primary cultured neuron from the cerebral cortex of the mouse. Cell-attached patch configuration.

PTZ application from inside the membrane no longer evoked the burst type open-close state as shown in the E part of Fig. 3. This result suggests that to provoke PTZ effects, an intracellular change is necessary prior to the changes in the ionic channel. The intracellular calcium shift in the primary cultured neurons from the cerebral cortex of the mouse was also confirmed (Sugaya, E. et al., 1991b). By PTZ application, the intracellular calcium shifted to unevenly near the cell membrane. The intracellular increase of free calcium measured by the calcium sensitive dye fura 2 was also confirmed (Sugaya, E. et al., 1993).

We examined the effect of TJ-960 on cultured neurons. PTZ-induced bursting activity in primary cultured neurons was completely inhibited by TJ-960 application (Fig. 4). The PTZ-induced burst type open-close state was completely normalized by TJ-960 (Fig. 5). PTZ-induced intracellular calcium shift in the primary cultured neuron was normalized by TJ-960 (Sugaya, E. et al., 1988c). PTZ-induced intracellular free calcium increase was also inhibited by TJ-960 (Fig. 6). Fig. 6B shows PTZ-with TJ-960 and Fig. 6E, PTZ application after rinsing. Thus, TJ-960 normalized PTZ-induced intracellular seizure-related changes in primary cultured neurons. Such inhibition was also observed by phenytoin (Sugaya, E. et al., 1985a, b). Ionic channel changes induced by PTZ were also inhibited by PHT (Figs. 7, 8). Intracellular free calcium changes were also inhibited by PHT (Fig. 9).

5. EPILEPSY ANIMAL MODEL, THE EL MOUSE AND AMELIORATION OF DEVELOPMENTAL DEFECTS by TJ-960

There is a famous epilepsy animal model, the El mouse (Imaizumi, 1964). The El mouse shows very high susceptibility to convulsions (Suzuki, 1971). Only 18 mg/kg of PTZ is sufficient to evoke convulsions although the normal mouse needs at least 48 mg/kg (Sugaya, E. et al., 1986). TJ-960 shows clear inhibitory effects on convulsions of the El mouse. Oral administration of 1 mg/kg of TJ-960 showed 100 % inhibition of PTZ-induced convulsions of El mice (Sugaya, E. et al., 1988b).

In addition to high susceptibility to convulsions, the behavior of the El mice is different from that of normal mice. The normal mouse is active at night and silent in the day time, but the El mouse does not show any clear circadian rhythm, and the total movement is greater than that of normal mice (Sugaya, E. et al., 1990). However, administration of TJ-960 at a dose of 2 g per kg a day for two weeks ameliorated the abnormal circadian rhythm and high motor activity of the El mouse (Sugaya, E. et al., 1990).

When we performed a primary culture from the embryonic cerebral cortex of the El mouse, neuron and glial cell development of the El mouse showed defects in development compared with those of the normal mouse (Sugaya, E. et al., 1987b). When TJ-960 was added to the culture medium at 75 µg/ml, the El mouse neurons proliferated as well as the normal mouse neurons. The number of neurons and neurite length of neurons in the TJ-960 treated groups were about twice those in the group without TJ-960 (Sugaya, E., et al., 1993).

When astrocytes of the El mouse were subcultured several times, they became so-called type 1 astrocytes that are inactive glial cells. However, by addition of TJ-960 for 48 h, all of the type 1 astrocytes changed into type 2, the active type (Sugaya, E. et al., 1993). We also performed the following experiment. First, astrocytes of the El mouse were activated in the TJ-960 containing medium. Then, TJ-960 was washed out. When El mouse neurons were placed on the TJ-960 pretreated and washed astrocytes, the El mouse neurons proliferated markedly (Sugaya, E. et al., 1993).

For neuron development and neurite extension, the ganglioside is said to be important (Tsuji et al., 1985). We measured the ganglioside contents of 3 day primary cultured neurons from 15 to 18 day-old embryonic cerebral cortex. The content of gangliosides was markedly

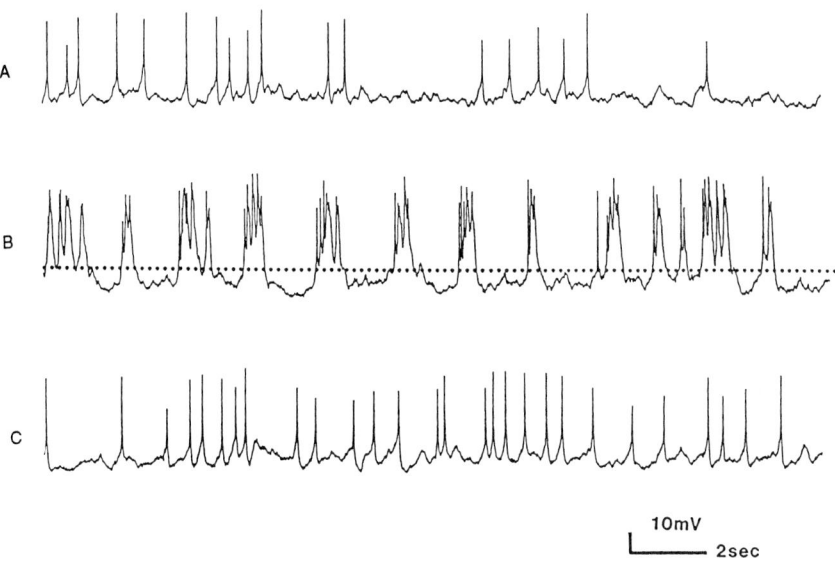

Figure 7. Inhibitory effect of PHT on PTZ-induced bursting activity in a cultured neuron from the cerebral cortex of the mouse. A: Normal regular firing. B: PTZ-induced bursting activity. C: Inhibitory effect of PHT (0,025 mM) on PTZ-induced bursting activity.

lower, especially in the primary cultured neurons from an 18 gestation day embryo. However, contents of the ganglioside in the cultured neurons of the El mouse were also markedly increased by addition of TJ-960 to the culture medium for 96 hours (Sugaya, E. et al., 1993).

In the case of El mouse primary cultured neurons, spontaneous bursting activity was observed without PTZ (Sugaya, E. et al., 1993). The spontaneous bursting activity was completely normalized by extracellular addition of TJ-960 to the culture medium (Sugaya, E. et al., 1993). In the case of the El mouse neurons, a frequent open-close state was observed without PTZ and also a spontaneous burst type open-close state was observed (Sugaya, E. et al., 1993). This spontaneous burst-type open-close state was completely normalized by addition of TJ-960 to the culture medium (Sugaya, E. et al., 1993).

The bursting activity of the El mouse neuron was incompletely inhibited by phenytoin (Fig. 10). However, most of the frequent and spontaneous burst-type open-close state of a single potassium channel of the El mouse neurons was not improved by phenytoin (Fig. 11).

6. PROTECTIVE EFFECTS OF TJ-960 ON NEURON DAMAGE

When normal neurons were cultured in cytochalasin B (5µg/ml) containing medium, the neurite extension was distorted and showed the so-called looping phenomenon, and the growth cone disappeared. However, when neurons were cultured in medium containing cytochalasin B and TJ-960 (75 µg/ml), this abnormal phenomenon was almost completely prevented (Sugaya, A. et al., 1987, 1988). When the culture was continued with cytochalasin B containing medium, most of the neurons ceased to extend their neurites and died. However, when TJ-960 was mixed with the cytochalasin B containing medium, neurons survived with neurite extension although they were not completely sound (Sugaya, A. et al., 1987, 1988).

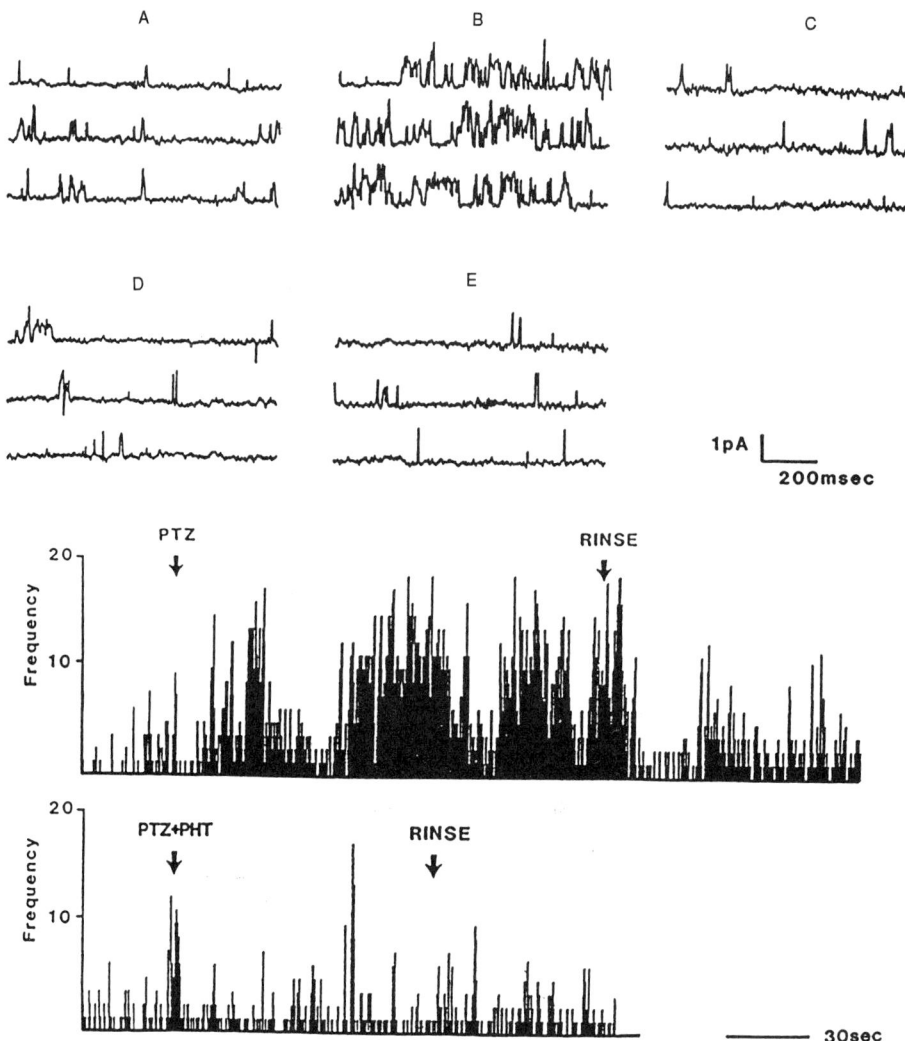

Figure 8. Inhibitory effect of PHT on the PTZ-induced burst type open-close state of a single potassium channel. A: Normal random open-close state. B: PTZ-induced burst type open-close state. C: Rinsing with normal medium. D: Application of PTZ (10 mM) and PHT (0,025mM)g. E: After rinsing with normal medium. The lowest two panels are continuous records and show sequential frequency histograms. Primary cultured neuron from the cerebral cortex of the mouse. Cell-attached configuration.

There is a famous local epilepsy experimental model called the cobalt focus epilepsy model (Dow et al., 1962). When a small quantity of metallic cobalt is placed on the cerebral cortex, the contralateral cortex shows severe epileptic spike discharges. Moreover, from two days after cobalt application, neuron degeneration began, and approximately twenty days after cobalt application, most of the neurons in the CA1 area of the hippocampus had degenerated and finally showed neuron loss (Sugaya, E. et al., 1988a). However, when TJ-960 was administered one month prior to the cobalt application, this neuron loss in the hippocampus was completely prevented (Sugaya, E. t al., 1991). One month prior application of phenytoin at 100 mg/kg could not prevent but rather accelerated neuron degeneration (Sugaya, E. et al., 1991).

TJ-960 also showed almost complete protective effects on neuron loss in the hippocampal CA1 area caused by ischemia by administration one week prior to clamping of both carotid and vertebrate arteries of rats (Sugimoto et al., 1986)

7. ANTITERATOGENIC ACTION OF TJ-960

Valproic acid is one of the most frequently used anticonvulsants. This drug is also known as a strong teratogenic agent. However, when TJ-960 was administered from the beginning of pregnancy, a significant decrease in malformations was observed (Minematsu et al., 1990).

8. CONCLUSION AND A HINT FOR FUTURE THERAPY

During bursting activity, intracellular stored calcium was released and moved toward the inner surface of the cell membrane. During bursting activity, the calcium binding state near the cell membrane was different from that of the normal cell membrane. During bursting activity, intracellular proteins of approximately 5 kDa and 15 kDa increased and calcium was incorporated into these proteins. During bursting activity, intracellular free calcium increased more than that in the normal state. Therefore, when we can inhibit all of these intracellular phenomena, seizure activity will be controlled.

TJ-960 as well as phenytoin completely inhibited the above-mentioned intracellular seizure-related pathological phenomena. In addition to the anticonvulsant effects, TJ-960 showed proliferating effects on the underdeveloped neurite extension of El mouse cultured neurons, normalizing effects on abnormal behavior of the El mouse, protective effects on

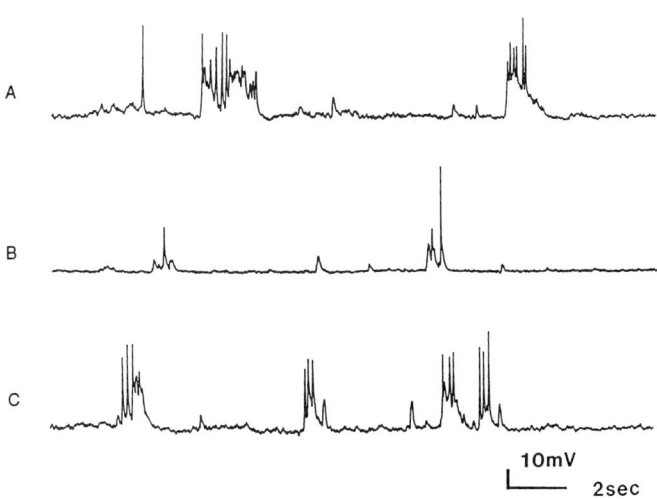

Figure 10. Incomplete inhibitory effect of PHT on spontaneous bursting activity of the El mouse cultured neuron. A: Spontaneous bursting activity. B: Application of PHT (0,025mM). C: After rinsing. Whole-cell patch recording.

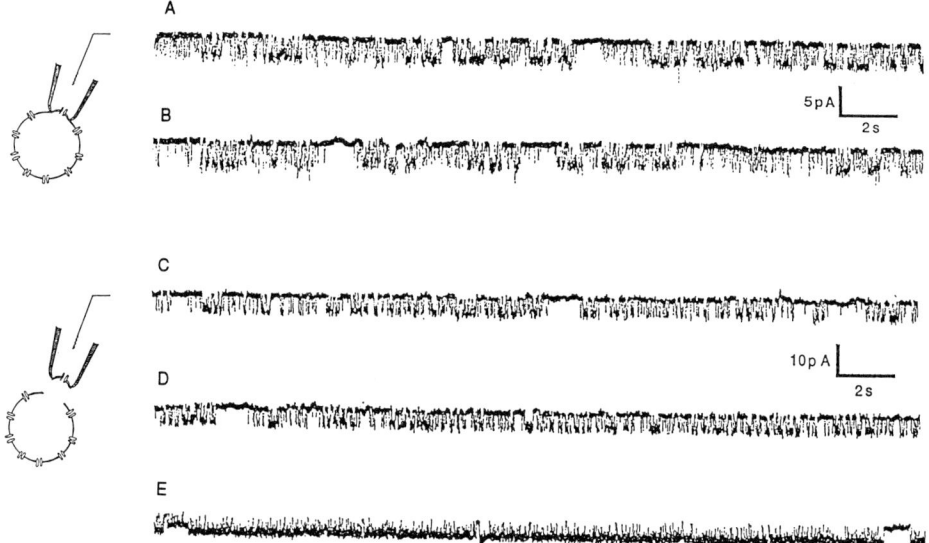

Figure 11. Loss of inhibitory effect of PHT on spontaneous frequent burst type open-close state of a single potassium channel of a cultured neuron from the cerebral cortex of the El mouse. A: Spontaneous frequent burst type open-close state of a single potassium channel. Cell-attached configuration. B: 5 min after extracellular application of PHT (0.025 mM). No inhibitory effect was observed. C: Spontaneous burst type open-close state of a single potassium channel after long-term rinsing (more than 30 min) and switching to the inside-out configuration of the same channel. D: 5 min after application of PHT (0.025 mM) from the inside of the cell membrane. No inhibitory effect was observed. E: 20 min after PHT. No inhibitory effect was observed and the open time was prolonged.

hippocampal neuron damage in cobalt focus epilepsy and protective effects on the appearance of the looping phenomena in cytochalasin-B treated neurons. TJ-960 also showed decreasing effects on the teratogenic action of valproic acid. These favorable actions on the nervous system were never observed with pure chemical anticonvulsants.

Table 1 shows a comparison of effects between phenytoin and TJ-960. Anticonvulsant action is almost the same in both drugs, but PHT showed no protective effect against neuron damage, no ameliorative effects on underdevelopment of neurons, no antiteratgenic effect and no normalizing effect on the hereditary abnormal cell membrane.

The above-mentioned results indicate that a herbal mixture, TJ-960, has actions to maintain optimal nervous function even during severe environmental changes around the nerve cells in addition to its anticonvulsant effects. Such effects will ameliorate intractable nervous symptoms caused by extraordinarily severe changes in the environment of the nerve cells. For better therapy of epilepsy, it is preferable to use this herbal mixture combined with pure chemical anticonvulsants.

ACKNOWLEDGMENT

These studies were performed in collaboration with the following laboratories: Department of Physiology, Kanagawa Dental College; Department of Pharmacognosy, Faculty of Pharmaceutical Sciences, Josai University; Central Laboratory for Experimental Animals;

Table 1. Comparison between Phenytoin Effects and TJ-960 Effects

	Phenytoin	TJ-960 (SK)
Bursting activity	inhibited	inhibited
Intracellular calcium shift	inhibited	inhibited
Calcium binding state change	inhibited	inhibited
Intracellular protein changes	inhibited	inhibited
El mouse convulsions	inhibited	inhibited
Cobalt focus EEG	unchanged or worsened	normalized
Cobalt focus—hippocampal neuron loss	unchanged or worsened	protected
Neuron damage by cytochalasin-B	damage accelerated	protected
Cultured neuron development	damaged—death	romoted
Developmental defect (neuron and glia, El mouse)	—	ameliorated
Abnormal behavior (El mouse)	—	normalized
Teratogenicity (valproic acid)	—	ameliorated
Single channel abnormality (El mouse)	unchanged	normalized

Tsumura Institute for Oriental Medicine, Keio University, School of Medicine; Tsumura Research Institute for Pharmacology, Tsumura Co., Ltd.; Research Institute, Chugai Pharmaceutical Co., Ltd.; Department of Physiology, Saitama Medical College; Department of Physiology, School of Medicine, Toho University; Department of Physiology, School of Medicine, Keio University; Section of Psychiatry, The Second National Tokyo Hospital; Epilepsy Section, Asaka Kosei Hospital and Section of Epilepsy and Herbal Medicine, Roppongi Hospital; I am very grateful to the staff of these laboratories. Part of these studies was supported by grants from the Ministry of Education, Science and Culture, the Ministry of Health and Welfare, the Genichi Kato Foundation for Physiology and Medicine and the Epilepsy Research Foundation.

REFERENCES

Aimi, S. (1956) Treatment of epilepsy by "Sho-saiko-to," *J. Orient. Med.* 7, 23–25 (in Japanese)
Dow, R.S., Fernandez-Guardiola, A. and Malik, V. (1962) The production of cobalt experimental epilepsy in the rat. *Electroencephalogr. Clin. Neurophysiol.* 14, 399–407.
Grynkiewicz, G., Poenie, M. and Tien,R.Y. (1985) A new generation of Ca^{2+} indicators with greatly improved luorescence properties. *J. Biol. Chem.* 260, 3440–3450.
Hamil, O.P., Marty, A., Neher, E., Sakmann, B. and Sigworth, F.J. (1981) Improved patch clamp techniques for high-resolution current recording from cells and cell-free membrane patches. *Pflügers Arch.* 391, 85–100.
Imaizumi, K. and Nakano, T. (1964) Mutant stocks, Strain El. *Mouse News Letter* 31, 57.
Lowry, O.H. and Passonneau, J.V. (1972) The quartz fiber fishpole balance. A Flexible System of Enzymatic Analysis. pp. 236–249. Academic Press. New York.
Meech, R.W. (1972) Intracellular calcium injection causes increased potassium conductance *Aplysia* nerve cells. *Comp. Biochem. Physiol.* 42A, 493–499.
Minematsu, S., Taki, M., Watanabe, M., Takahashi, M., Wakui, Y., Ishihara, K., Takeda, S. and Fujii, Y. (1990) Effects of Shosaiko-to-go-keishi-ka-shakuyaku-to (TJ-960) on the valproic acid induced anomalies of rat fetuses. *Folia Pharmacol. japon.* 96, 265–273. (in Japanese).
Narita, Y., Satowa, H., Kokubu, T. and Sugaya, E. (1982) Treatment of epileptic patients with the Chinese herbal medicine "Saiko-keishi-to" (SK). *IRCS Med. Sci.* 10, 88–89.
Sugaya, A., Sugaya, E. and Tsujitani, M. (1973) Pentylenetetrazol-induced intracellular potential changes of the neuron of the Japanese land snail *Euhadra peliomphala*. *Jpn. J. Physiol.* 23, 261–274.
Sugaya, A., Tsuda, T., Sugaya, E., Takato, M. and Takamura, K.(1978) Effect of Chinese medicine "Saiko-keishi-to" on the abnormal bursting activity of snail neurons. *Planta Medica* 34, 294–298.
Sugaya, A., Tsuda, T., Sugaya, E., Usami, M. and Takamura, K. (1979) Local anaesthetic action of the Chinese medicine "Saiko-keishi-to," *Planta Medica* 37, 274–276.

Sugaya, A., Tsuda, T., Yasuda, K., Sugaya, E. and Onozuka, M. (1985a) Effect of Chinese herbal medicine "Saiko-keishi-to" on intracellular calcium and protein behavior during pentylenetetrazole-induced bursting activity in snail neurons. *Planta Medica* 51, 2–6.

Sugaya, A., Tsuda, T., Yasuda, K. and Sugaya, E. (1985b) Effect of Chinese herbal medicine "Saiko-keishi-to" on transmembrane ionic current of snail neurons. *Planta Medica* 51, 60–61.

Sugaya, A., Yuzurihara, M., Tsuda, T., Yasuda, K., Sugaya, E., Kajiwara, K., Takagi, T. and Komatsubara, J. (1987) Normalizing effect of Saiko-keishi-to commercial formula on cytochalasin-B distorted neurites using primary cultured neurons of rat cerebral cortex. *J. Ethnopharmacol.* 21, 193–199.

Sugaya, A., Yuzurihara, M., Tsuda, T., Yasuda, K., Sugaya, E., Kajiwara, K., Takagi, T., Komatsubara, J., Asou, H. and Hirano, S. (1988) Normalizing effect of Sho-saiko-to-go-keishi-ka-shakuyaku-to (TJ-960) on cytochalasin B distortion using primary cultured neurons of the rat cerebral cortex. Recent *Advances in the Pharmacology of Kampo (Japanese Herbal) Medicines*, Proc. Satellite Meeting on Kampo Medicines of the 10th International Congress of Pharmacology (Auckland) (Eds, Hosoya E. & Yamamura Y.), pp. 105–111, Excerpta Medica, Amsterdam.

Sugaya, E., Goldring, S. and O'Leary, J.L. (1964) Intracellular potentials associated with direct cortical response and seizure discharge in cat. *Electroencephalogr. Clin. Neurophysiol.* 17, 661–669.

Sugaya, E. and Onozuka, M. (1978a) Intracellular calcium: its movement during pentylenetetrazole-induced bursting activity. *Science* 200, 797–799.

Sugaya, E. and Onozuka, M. (1978b). Intracellular calcium: its release from granules during bursting activity in snail neurons. *Science* 202, 1195–1197.

Sugaya, E. and Onozuka, M. (1978c) Ion shower milling: its application to cell membrane removal. *Science* 202, 1197–1198.

Sugaya, E., Onozuka, M., Usami, M. and Sugaya ,A. (1978d) Ionic distribution changes during bursting activity induced by pentylenetetrazol in a single isolated snail neuron: tentative application of electron probe X-ray microanalyzer to a single isolated neuron. *Abnormal Neuronal Discharges*. (Eds, Chalazonitis, N. & Boisson, M.). pp. 407–418. Raven Press. New York.

Sugaya, E., Onozuka, M., Furuichi, H. and Sugaya, A. (1981) Cellular calcium binding state change during pentylenetetrazole-induced bursting activity in snail neurons. *Experientia* 37, 1080–1081.

Sugaya, E.,Onozuka, M., Kishii, K., Sugaya, A. and Tsuda, T. (1982) Intracellular protein changes during pentylenetetrazole induced bursting activity in snail neurons. *Brain Research* 253, 271–279.

Sugaya, E., Onozuka, M., Furuichi, H., Kishii, K., Imai, S. and Sugaya, A. (1985a) Effect of phenytoin on intracellular calcium and intracellular protein changes during pentylenetetrazole-induced bursting activity in snail neurons. *Brain Research* 327, 161–168.

Sugaya, E., Kishii, K. and Onozuka, M. (1985b) Inhibitory effect of phenytoin on intracellular cyclic nucleotide and calcium changes during pentylenetetrazole-induced bursting activity in snail neurons. *Brain Research* 341, 313–319.

Sugaya, E., Ishige, A., Sekiguchi, K., Iizuka, S., Ito, K., Sugimoto, A., Aburada, M. and Hosoya, E. (1986) Pentylenetetrazol-induced convulsion and effect of anticonvulsants in mutant inbred strain El mice. *Epilepsia* 27, 354–358.

Sugaya, E. (1987a) SK (TJ-960). *Drugs of the Future* 12, 360–363.

Sugaya, E., Asou, H., Itoh, K., Ishige, A., Sekiguchi, K., Iizuka, S., Sugimoto, A., Aburada, M., Hosoya, E., Takagi, T., Kajiwara, K., Komatsubara, J. and Hirano, S. (1987b) Characteristics of primary cultured neurons from embryonic mutant El mouse cerebral cortex. *Brain Research* 406, 270–274.

Sugaya, E., Furuichi, H., Takagi, T., Kajiwara, K. and Komatsubara, J. (1987c) Intracellular calcium concentration during pentylenetetrazol-induced bursting activity in snail neurons. *Brain Research* 416, 183–186.

Sugaya, E., Ishige, A., Sekiguchi, K., Yuzurihara, M., Iizuka, S., Sugimoto, A. and Hosoya, E., (1988a) Damage of hippocampal neurons caused by cobalt focus in the cerebral cortex of rats. *Brain Research* 459, 196–199.

Sugaya, E., Ishige, A., Sekiguchi, K., Iizuka, S., Sugimoto, A., Yuzurihara, M. and Hosoya, E. Inhibitory effect of a mixture of herbal drugs (TJ-960, SK) on pentylenetetrazol-induced convulsions in El mice. (1988b) *Epilepsy Res.* 2, 337–339.

Sugaya, E., Takagi, T., Kajiwara, K., Komatsubara, J., Sugaya, A., Tsuda, T., Yasuda, K., Sekiguchi, K., Ishige, A. and Iizuka, S. (1988c) Pentylenetetrazol-induced calcium-related intracellular abnormal phenomena and Shosaiko-to-go-keishi-ka-shakuyaku-to (TJ-960). *Recent Advances in the Pharmacology of Kampo (Japanese Herbal) Medicine*, Proc. Satellite Meeting on Kampo Medicine of the 10th International Congress of Pharmacology, Auckland, (Eds., Hosoya, E. & Yamamura, Y.), pp. 87–96, Excerpta Medica, Amsterdam.

Sugaya, E., Sugaya, A., Takagi, T., Tsuda, T., Kajiwara, K., Yasuda, K. and Komatsubara, J. (1989) Pentylenetetrazole-induced changes of the single potassium channel in primary cultured cerebral cortical neurons. *Brain Research* 497, 239–244.

Sugaya, E., Sugaya, A., Takagi, T., Tsuda, T., Kajiwara, K., Yasuda, K., Yuyama, N., and Takagi, H. (1990) Seizure-related pathological changes in the El mouse neurons and their normalization by a new antiepileptic drug, TJ-960. *Neurosciences* 16, 95–104.

Sugaya, E., Ishige, A., Sekiguchi, K., Yuzurihara, T., Iizuka, S., Sugimoto, A., Takeda, S., Waku, Y., Ishihara, K., and Aburada, M. (1991a) Protective effects of TJ-960 herbal mixture on hippocampal neuron damage induced by cobalt focus in the cerebral cortex of rats. *J. Ethnopharmacolgy* 34, 13–19.

Sugaya, E. and Sugaya, A. (1991b) Cellular physiology of epileptogenic phenomena and its application to therapy against intractable epilepsy. *Comp. Biochem. Physiol.* 98C, 249–270.

Sugaya, E., Sugaya, A., Yuyama, N., Tsuda, T., Kajiwara, K., Kubota, K., Katoh, K., Hosoya, S., Takagi, T., and Motoki, M. (1993) Developmental defects of primary cultured neurons from the cerebral cortex of the El mouse and their amelioration with a herbal formulation, TJ-960. Neurosciences (in press).

Sugimoto, A., Ishige, A., Sudo, K., Sekiguchi, K., Iizuka, S., Itoh, K., Yuzurihara, M., Aburada, M., Hosoya, E. and Sugaya, E. (1988) Protective effect of Sho-saiko-to-go-keishi-ka-shakuyaku-to (TJ-960) against cerebral ischemia. *Recent Advances in the Pharmacology of Kampo (Japanese Herbal) Medicine*, Proc. Satellite Meeting on Kampo Medicine of the 10th International Congress of Pharmacology, Auckland, (Eds., Hosoya, E. & Yamamura, Y.), pp. 112–119, Excerpta Medica, Amsterdam.

Suzuki, J., (1976) Paroxysmal discharges in the electroencephalogram of the El mouse. *Experientia* 32, 336–338.

Takato, M., Takamura, K., Sugaya, A., Tsuda, T. and Sugaya, E. (1982) Effect of the Chinese medicine "Saiko-keishi-to" on audiogenic seizure mice, kindling animals and conventional pharmacological screening procedures. *IRCS Med. Sci.* 10, 86–87.

Tsien, R.Y. and Rink, T.J. (1980) Neutral carrier ion-selective microelectrodes for measurement of intracellular free calcium. *Biochem. Biophys. Acta* 599, 623–638.

Tsuji, S., Nakajima, J., Sakai, T. and Nagai, Y. (1985) Bioactive gangliosides. IV: Ganglioside GQ_{1b}/Ca dependent protein kinase activity exists in the plasma membrane fraction of neuroblastoma cell line, GOTO. *J. Biochem.* 97, 969–972.

Uyemura, K., Tanaka, S., Wang ,Y., Sakamoto, Y., Tsuda, T., Yasuda, K., Sugaya, A., Takagi, T., Kajiwara, K. and Sugaya, E. (1988) Effects of Sho-saiko-to-go-keishi-ka-shakuyaku-to (TJ-960) on pentylenetetrazol-induced protein changes in primary cultured cerebral cortical neurons. *Recent Advances in the Pharmacology of Kampo (Japanese Herbal) medicines.* Proc. Satellite Meeting on Kampo Medicines of the 10th International Congress of Pharmacology. (Auckland) (Eds. Hosoya, E. & Yamamura, Y.). pp. 97–101. Excerpta Medica. Amsterdam.

THE CLINICAL PHARMACOLOGY OF ANTIEPILEPTIC DRUGS

Qu Zhi-ping

WHO Collaborating Center
Institute of Neurology
Shanghai Medical University
Shanghai, China

In China, work on clinical pharmacology, blood level monitoring of antiepileptic drugs are carried on in many laboratories of medical schools and teaching hospitals, in laboratories of hospitals in large cities as well. Phenytoin (PHT), carbamazepine (CBZ), sodium valproate (VPA-Na) and phenobarbital (Pb) are the main antiepileptic drugs under investigation.

In addition to blood level monitoring, we did the pharmacokinetics of phenyhydantoin, carbamazepine, entericcoated and sustain released sodium valproate the interactions of PHT and CBZ, the saturation kinetic phenomena of PHT, the relationship between blood level and therapeutic effects.

The pharmacokinetics of PHT and CBZ were studied in 16 healthy volunteers by taking PHT and CBZ and then both. The kinetic course could be described by a one compartment model and one order elimination kinetics. Thus the pharmacokinetic parameters Co, Cm, AUC, T1/2, Tm, Td and clearance (CL) under the interaction between PHT and CBZ in Chinese healthy volunteers were obtained for the time. When CBZ at dose of 400 mg/day was added on the basis of 200 mg PHT daily, serum PHT levels rised (7.01 ± 4.35 µg/ml to 15.54 ± 5.04 µg/ml). $P < 0.05$. The results were similar to those of some reports in literatures. As PHT at dose of 200 mg/d was added on 400 mg CBZ daily, serum CBZ levels fell (from 5.96 ± 1.02 µg/ml to 4.89 ± 1.21 µg/ml), $P < 0.01$, and T1/2 shortened (from 18.71 ± 3.62 h to 13.97 ± 3.31 h), $P < 0.0001$. Therefore, from pharmacokinetic viewpoint we conclude that there is interaction between PHT and CBZ. The possible mechanisms are that CBZ inhibits PHT from converting to metabolites and on the contrary, PHT induces metabolic enzymes which catabolize CBZ to metabolites.

The clinical drug level monitoring was exercised in 26 epileptic patients taking PHT and CBZ. Serum CBZ levels of 10 patients fell significantly with increasing PHT dose. When CBZ dose was increased, serum PHT levels in 6 of 16 patients lowered while in the others rised. The results that CBZ caused serum PHT levels either a fall or a rise in particular individuals differed from those in volunteers. Drug concentration determining and clinical observation in 2 patients showed that interaction between PHT and CBZ could change the therapeutic effects. It is suggested that when PHT and CBZ are taken together, drug concentration monitoring and individualized dosage tailoring be carried on to increase therapeutic efficacy and avoid toxic effects.

Phenytoin exhibits saturation kinetic, which should be bear in mind of clinicians when the dose will be adjusted. If the blood level is below the effective therapeutic range, it is better to increase 50 mg each time, but to increase 25 mg while the level is within the effective therapeutic range. After increasing the dose, it needs to follow up 3 to 4 weeks for avoiding the pseudosteady state phenomena. We measured more than 150 blood samples of 30 patients who took phenytoin only and analyzed 71 values (some are the values of 3 consecutive measurements). 19 patients were older than 14 years of age. The blood levels were higher than 26 µg/ml in 9 of 19 patients (47.6%), while 6 patients were within the effective range (10~20 µg/ml). Only in 4 cases were the blood levels lower than 10 µg/ml. This showed that about 50% of patients above 14 years old reached toxic level and suggested that the dose should be started from 200 mg/d and gradually increase 25 to 50 mg when necessary. The dose in children should be started from 150 mg daily and increase gradually to maximum 250 mg daily. The drug time curve of saturation kinetics has the tendency of shifting to left in comparison to the data in literature.

Individualizing of PHT and other antiepileptic drugs is important and example is presented.

HEMODYNAMIC ACTIONS OF HUATUO RECONSTRUCTION PILL ON ANESTHETIZED ANIMALS

Huang Shou Jian

Pharmacology Department
Sun Yat-sen University of Medical Sciences
Guanzhou, P. R. China

ABSTRACT

The Huatuo reconstruction pill (HTRP) is a Chinese traditional herbal preparation which has been used for hundreds of years with remarkable effect on the cerebral palsy. The significant increase in both the common and internal carotid blood flow and the positive inotropic action on the heart were observed in anesthetized cat, dog and rabbit experiments following intravenous (iv) injection of HTRP extract 0.125 to 1.0 ml per kilogram body weight. 1.0 nl of which is equivalent to 0.2067 gm crude drugs. The carotid blood flow was selectively increased without change in vascular resistance of the hind limb. The internal carotid blood flow reached the maximal efficacy as high as 173% of the control level. Neither the action of potassium ion in the drug nor the histamine release was excluded. The positive inotropic action was demonstrated by the increase in LVP, dP/dtmax, Vmax, CO, TTI and dP/dt-LVP vectogram. Yet HR, BP, EEG, ECG and respiration remained unchanged. All these results may provide a theoretical explanation to HTRP therapeutic effect on neurological sequelae of cerebral vascular accident.

INTRODUCTION

The Huatuo reconstruction pill (HTRP) is a famous Chinese traditional herbal preparation which has been used for hundreds of years with remarkable effect on sequlae of cerebral vascular accidents. Data of a clinical trail on 644 cases with cerebral or coronary vascular diseases were collected in several Chinese famous hospitals illustrates that HTRP can significantly improve the clinical symptoms of cerebral vascular diseases including headache, dizziness, extremity weakness and paralysis, asphasia, memory diminution, sensitive disturbance, hemoplegia and palsy of cranial nerves with total effective rate up to 94.24%. Its outstanding effects were also proven by both CT scan and hemodynamic examinations. This paper reports the effect of HTRP on the common/internal carotid blood flow and the performance function of heart to reconfirm its action characteristic, to estimate its safety margin, and to provide scientific basis for its clinical use.

MATERIALS AND METHODS

HTRP : Its extract was obtained from Qi-Xin pharmaceutical factory. The aqueous solution for this study is equivalent to 0.2067 gm crude drug per millimeter. The LD_{50} of administering iv or per oral to mice were 7.06 (95% confidence limits 6.46 to 7.71) and 9.34 (95% confidence limits 8.13 to 10.7) gm/kg respectively.

Animals : Cats, weighing 1.8 to 2.9 Kg; dogs, weighing 10.5 to 11 Kg and rabbits, weighing 2.3 to 2.8 Kg, either male or female were used for this study.

Determination of common/internal carotid blood flow : The common carotid blood flow (CCBF)/internal carotid blood flow (ICBF) was recorded on a physiograph (RM-6000, Japan) through an electromagnetic blood flow meter, of which a probe was put on left common/internal carotid artery.

Determination of vascular resistance of hind limb : The vascular resistances of hind limb were performed on anesthetized cats by Mcleod's method.[1] Vascular resistance was represented as change in femoral artery pressure at constant blood flow, and was recorded in a potential meter along with CCBF synchronously.

Experiments on cardiac hemodynamic study : These experiments were performed in cats anesthetized by intraperitoneal (ip) injection of pentobarbital. The trachea was separated. A canula connected to a pulmoter for artificial ventilation was inserted to the trachea. The chest cavity was opened to reveal the heart. The ascending aorta was separated by blunt dissection for attaching a suitable probe of electromagnetic flow meter to record aorta blood flow (ABF). Left ventricular pressure (LVP) was detected by a pressure transducer through a catheter set in ventricular cavity. Left femoral artery was inserted with a canular connected to another pressure transducer for recording artery blood flow pressure. Needle electrodes were inserted through the subcutaneous tissue in the extremities for recording the ECG. All tracings, namely, LVP, dP/dt, LVEDP, ECG, ABF and BP were simultaneously recorded on an eight channel physiograph. LVP-dP/dt vector loops were revealed on a memory oscilloscope (VC-10. Nihon Kohden, Japan) and were plotted with a X-Y recorder.[2]

HTRP with total volume of 1.0 ml were given to groups of animals of 6, intravenously through the catheter of femoral vein by cumulative method.[3] Another dosage was given while the ABF has restored to the baseline for 2 minutes. Control animals were administered normal saline. Another 3 groups of 6 animals each were given sodium nitrite or phenergan, an agent that inhibit histamine release, before and after administration of HTRP, for the study of ICBF in comparison with HTRP treated ones.

The hemodynamic parameters were calculated based on the following method : Heart rate (HR), from the P-R interval; maximal contractile element shorting velocity (Vmax), from the dP/dt and LVP tracings at the interval from 0 to dP/dtmax; ejection time (ET) was the period from dP/dtmax to -dP/dtmax; cardiac index (CI), stroke volume (SV) and stroke index (SI) are calculated according to ABF and the body weight; time tension index (TTI) was the area under the curve of LVP tracing in the ejection period multiplying HR.

Experiments on cardiac hemodynamic study in intact dogs : The instruments used and the recording methods were the same as above except recording LVP was by a catheter entering left ventricular cavity through the common carotid artery.

RESULTS

Effect on CBF : The increased CCBF can be seen after given HTRP 0.125 - 1.0 mg/kg intravenously in anesthetized animals including cats, dogs, and rabbits. The maximal effect was reached up to 137.4% of the baseline value. The increased CCBF by HTRP in dogs and rabbits were more significant than that in cats. Simultaneously recording CCBF and vascular resistance in hind limb, performed in anesthetized cats revealed that increased CCBF did not accompany the change in vascular resistance. This meant that HTRP increased the blood flow into the head organ selectively.

Effect on ICBF : In anesthetized cats, HYRP 0.125 - 1.0 ml/kg increased the ICBF. The maximal effect was reached up to $173 \pm 12\%$ of the baseline values, which was much more significant than that in CCBF. The hypotensive effect can also be seen, but there were no significant changes in ECG, EEG and respiratory rate.

HTRP had higher efficacy of increasing ICBF than sodium nitrite. HTRP 0.5 mg/kg increased ICBF up to 113.4 ± 4.7 ml/min. (mean $\pm$ SE, n=6), while sodium nitrite at the maximal efficacy dose only increased up to 106.7 ± 3.3 ml/min. (mean $\pm$ SE, n=6). There was a significant difference between the data obtained from HTRP treated and sodium nitrite treated animals ($p<0.05$).

The increasing effect of HTRP on ICBF was unrelated to its potassium salt content. HTRP 0.5 ml/Kg increased ICBF by 13.4 ± 1.9 ml/min. ($16.4 \pm 1.5\%$ in percentage, mean $\pm$ SE, n=6). 0.5 ml of HTRP only contained 0.46 mg of potassium chloride. Equal volume of normal saline containing the same amount of potassium chloride as HTRP. The ICBF of the normal saline treated animals only increased 3.18 ± 0.52 ml/min ($4.8 \pm 0.42\%$, in percentage, mean $\pm$ SE, n=6). HTRP, as compared with the effect of equal volume of normal saline, the difference of the former was more significant, $p<0.01$, while p value of the latter was merely 0.05.

Phenergan 5 mg/Kg has no effect and did not antagonize the increased effect of HTRP 0.5 ml/Kg on ICBF. The increased ICBF in percentage were 39.7 ± 0.6 and 30.5 ± 0.7 % before and after given phenergan respectively. There was no tachyphylaxis observed, suggesting that the effect of HTRP on ICBF is not relative to the release of histamine.

Hemodynamic effect on anesthetized cats : The effect of HTRP on hemodynamic parameter is presented in Table 1.

The fact that LVEDP was unchanged by HTRP 0.125 - 1.0 ml/kg but blood pressure was decreased with a dose dependent manner, illustrated that HTRP did not change the preload but decrease the afterload. HTRP increased the myocardial contractility represented by the increased LVP, dP/dtmax and Vmax, enlarged the area of dP/dt-LVP vector loops and steep slope of dP/dt-LVP loops, and shortened the time to dP/dtmax. HTRP can improve the cardiac performance function, which includes increased ABF, CI, SV, and SI, unchanged HR, less increased TTI than ABF and shortened ejection time.

The experimental results performed in anesthetized dogs were similar to those in cats (Table 2). As compared with isoprenaline 0.5 µg/Kg, HTRP increased CCBF more significantly, but augmented contractility, decrease HR and BP insignificantly.

DISCUSSION

The characteristics of hemodynamic effect of HTRP on intact animals are as follows:

1. HTRP increased CCBF and ICBF with dose dependent manner. The increase is more significant than sodium nitrite. This is a combined results of selectively vasodialation in the head and from increased ABF. The selectivity of HTRP on vessels in the head was manifested as follows : firstly, increased CCBF but without changing the vascular resistance in the hind limb resulting in redistribution of blood to the head from other parts of the body; secondly, the percentage of increased ICBF or CCBF is greater than that of decreased BP; thirdly, greater increased ICBF in percentage than CCBF. The phenomenon above suggested they are induced by expanded cerebral vessels. Although sodium nitrite can also expand cerebral vessels, but its action was attenuated by decreased cardiac output from decreased preload and by stolen

Table 1. Influence of HTRP on Hemodynamic Parameters in Intact Cats

			HTRP ml/KG			
	BASELINE	NS	0.125	0.25	0.5	1.0
SBP mmHg	120.00± 5.00	113.00± 5.00	104.00±05.00	101.00±5.00*	86.00± 5.00*	83.00± 6.00*
DBP mmHG	77.00± 3.00	74.00± 4.00	68.00±10.00	61.00± 5.00	53.00± 5.00*	50.00± 5.00*
HR bpm	207.00±14.00	204.00±15.00	205.00±13.00	205.00±11.00	205.00±12.00	203.00±11.00
LVEDP mmHg	0.20± 1.10	1.00± 1.40	0.70± 1.40	0.20± 1.30	1.20± 1.20	1.70± 1.60
LVP mmHg	135.00±07.00	143.00± 9.0	153.00± 9.00	155.00± 9.00	167.00± 7.00*	143.00± 4.00*
dP/dtmax 0.1mmH	349.00±40.00	350.00+39.00	393.00±39.00	416.00±41.00	444.00±44.00	526.00±54.00*
Vmax/s	63.40±04.70	64.10± 4.60	72.00± 3.10	79.20± 3.10	35.10± 5.40*	97.10±31.00*
t-dt/dtmax msec	40.40± 3.10	40.40± 3.20	37.50± 2.50	35.00± 2.50	32.10± 3.00	31.30± 2.40
ET msec	133.00±10.00	131.00±10.00	129.00±10.00	127.00±11.00	118.00± 7.00	113.00± 6.00
ABF m//min	142.00±11.00	152.00±10.00	161.00±10.00	1.75±12.00	193.00±15.00	219.00±11.00*
CI 2 L/min/M	0.84± 0.06	0.91± 0.07	0.97± 0.08	1.10± 0.10	1.17± 0.13	1.32± 0.10*
SV ml/beat	0.70± 0.07	0.75± 0.08	0.84± 0.08	0.88± 0.08	0.97± 0.12*	1.09± 0.08*
SI ml/beat/M	24.60± 0.50	4.50± 0.60	4.90± 0.70	5.30± 0.70	6.00± 1.00	6.47± 0.84*
TTI X10 mmHgXm	215.00±27.00	242.00±18.00	242.00±18.00	258.00±19.00	272.00±13.00	296.00±17.00

n+6 $*p < 0.05$

Table 2. Influence of HTRP on Hemodynamic Parameters in Dogs

			HTRP ml/kg		
	BASELINE	NS	0.125	0.25	0.5
CCBF%		5.00± 1.70	14.80± 2.70**	19.40± 3.10**	6.00± 3.90**
DBP mmHg	102.00± 7.000	99.00± 7.00	94.00± 9.000	92.00± 8.00*	86.00±10.000*
LVP mmHg	188.00± 7.00	188.00± 7.00	195.00± 7.000*	202.00± 6.0**	206.00± 5.000
dp/dtmax 0.1mmHg/s	803.00±79.00	806.00±78.00	830.00±79.000	860.00±76.00**	800.00±74.00**
LVEDP mmHg	-5.80± 1.80	-6.00± 1.70	-6.80± 1.600	-6.30± 1.400	-6.80± 1.300
HR bpm	197.00± 3.00	197.00± 3.00	201.00± 4.000	202.00± 5.000	200.00± 6.000
RESP /MIN	56.00±13.00	60.00±13.00	59.00±14.000	60.00±13.000	62.00±14.000

n-8, $*p < 0.05$, $**p < 0.01$.

blood from expanded other vessels. The effect of HTRP may not be related to its potassium salt content and cannot be antagonized by antihistaminic drugs. Thus one can exclude these nonspecific factors.

2. HTRP increased myocardial contractility, decreased afterload and improved the heart performance function. There are many parameters to assess the cardiac contractility, but none of them is satisfactory because they can not fully exclude the influence of preload or afterload. Therefore, many parameters were chosen to analyse the action of HTRP. The experimental results including increased LVP, dP/dtmax and Vmax with dose dependent manner, enlarged dP/dt-LVP loops and steep slope in isovolume contraction phase, improved contractility, demonstrated that HTRP had cardiotonic action. Due to the fact that there is no change in HR, this action is not induced by released catecholamines from hypotension. Increased contractility can improve the cardiac performance function. It is due to increased ABF, CI, SV, and SI. More significantly increased in ABF than TTI means HTRP improved efficiency of performance. The experimental results including HTRP do not induce arrhythmia, insignificantly shortened ET and t-dP/dtmax, illustrated toward positive inotropic action. The iv LD_{50} of HTRP in mice is more than 30 times as the maximal dosage use in these experiments.

Synchronously recording EEG, ECG and respiratory exhibit insignificant changes, suggested wide margin of the safety of HTRP.

The cerebral vascular accidents can be divided into hemarrhagic and ischemic. The drugs that increase cerebral blood flow help the blood clot to reabsorbed after the acute phase has subsided and to improve the blood supply to ischemia focus. With the high efficacy of increasing CCBF and ICBF and improving heart performance functions, HTRP, which is superior to sodium nitrite and isoprenaline is expected to be beneficial to cerebral vascular diseases therapeutically.

REFERENCES

1. McLeod, L.J. (1970) Pharmacological experiments on intact preparations. Churchill Livingstone pp 70.
2. Parmley, W. et al (1971) Cardiac muscle function studies. in W. Parmeys Methods in Pharmacology 1:105, Meredith Corporation.
3. Kenneth, C. B. et al. (1979) Various techniques for evaluating cardiodynamic function. Pharmacol Ther 5:105.

TREATMENT OF AFFECTIVE DISORDERS

Clinical and Basic Aspects

Nobutaka Motohashi

Department of Neurology and Psychiatry
Hiroshima University School of Medicine
1-2-3 Kasumi, Minami-ku, Hiroshima 734
Japan

INTRODUCTION

The neurological separation of affective disorders into unipolar and bipolar was first propounded by Kleist and Leonhard[1,2]. Their proposal was examined by Angst and Perris[3], who demonstrated the difference in the premorbid personality and clinical course in these two entities. First, unipolar depression is more frequent in females than in males. As for bipolar disorder, there is no difference in frequency between males and females. Second, the genetic penetrance is more pronounced in bipolar disorder than in unipolar depression. Third, the premorbid personality of bipolar disorder is syntonic, while that of unipolar depression is asthenic. Lastly, the periodicity is more rapid in bipolar disorder than in unipolar depression. In their follow-up study in 1978, Angst and colleagues[4] demonstrated the change of diagnosis. In about 10% of unipolar depression patients, the diagnosis was changed into bipolar disorder. The unipolar patients with three or more episodes were prone to change their diagnosis.

In the course of the treatment with antidepressants, some bipolar patients become manic and a few of them may become rapid cyclers[5,6]. Furthermore, neuroleptic treatment induces post-manic depression[5]. In this respect, the introduction of lithium is a new strategy because of its acute and prophylactic effects on both manic and depressive phases. Lithium has been classified into a mood stabilizer. Lithium therapy, however, has its limitations. Some bipolar patients, especially rapid cyclers, do not respond to lithium[7]. Moreover, some side effects on thyroid and renal function do exist, although they may be exaggerated[8].

Tricyclic and non-tricyclic antidepressants have been widely used for the treatment of major (unipolar) depression. However, not a small number of patients are refractory to antidepressant drugs. Protracted major depression is one of the major problems in the treatment of affective disorders.

I am going to consider the treatment of refractory bipolar disorder and major depression in clinical and basic aspects.

TREATMENT OF REFRACTORY BIPOLAR DISORDER

Recently, anticonvulsants such as carbamazepine and valproate have been introduced for the acute and prophylactic treatment of bipolar disorder[9]. Like lithium, these anticonvulsants are regarded as mood stabilizers. These anticonvulsants, alone or in additon to lithium, have some efficacy in the treatment of lithium-refractory bipolar disorder, especially rapid-cycling one. Although many cases under carbamazepine therapy have been reported[10], the efficacy of valproate must be determined.

A prodrug of valproate, valproic acid amide, has been used alone or in combination with other psychotropic drugs for the treatment of affective disorders in Europe[11–13]. Antimanic effect of valproate was first reported by Emrich and colleagues[14] in 1980. They also showed the prophylactic treatment of bipolar disorder with the combination of lithium and valproate. More recently, antimanic effct was recognized in the United States[15,16]. Furthermore, the efficacy of valproate alone or in combination with lithium or other psychotropic drugs, in treating rapid-cycling bipolar disorder, has been reported[17,18].

I will demonstrate two cases of rapid-cycling bipolar disorder successfully treated with lithium and valproate.

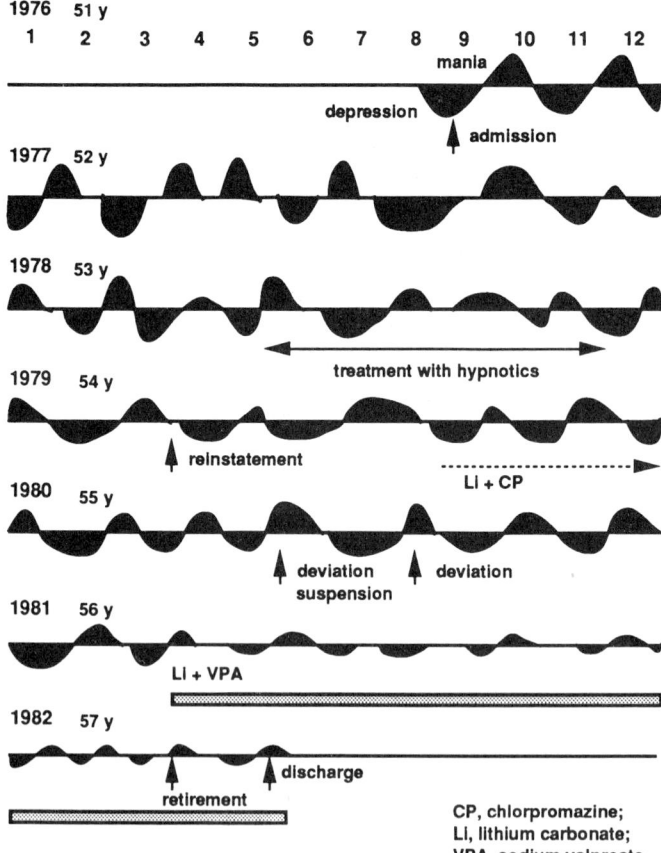

Figure 1. Clinical course of Case 1.

Case 1 is a male ex-schoolteacher. He became depressed for the first time at the age of 37. After treatment with antidepressant drugs, he became euthymic but soon he became manic. Thereafter, the manic and depressive phases appeared continuously at a interval of 2 months. He was admitted to a university hospital in 1976. Lithium, antidepressants, antipsychotics, carbamazepine, acetazolamide, reserpine and thyroid hormone did not change the circular course of his illness. Because his affective states improved slightly with lithium and chlorpromazine, he was reinstated in his school. But he showed deviated behavior at school and he was again suspended. From March, 1981, combined therapy of lithium and valproate has been introduced. His state improved moderately and he was discharged from hospital (Fig. 1). Although he shows subdepressive and hypomanic states at a regular interval, he leads a carefree life after retirement under combined therapy of lithium (lithium carbonate, 600 mg/day) and sodium valproate (1000 mg/day).

Case 2 is a housewife. Completely tired from care of her sick husband, she became hallucinatory-paranoid state at the age of 37. She was successfully treated with antipsychotic drugs but she gradually became depressed. Because she attempted suicide, amoxapine was prescribed. She soon recovered completely and had taken amoxapine for several months. Then she suddenly became hypomanic. Although the amount of amoxapine was reduced, hypomanic and subdepressive phases appeared continuously. Because she became manic, lithium and haloperidol were prescribed instead of amoxapine. Her manic state subsided soon, but she became depressed and attempted suicide again. She was admitted to the hospital and had been taken care of without medication. As she soon became manic, lithium and haloperidol had been used for the treatment. Carbamazepine, added in expectation of its mood-stabilizing effects, had to be stopped because of its side effects (skin rash and high fever). Her manic state subsided in 40 days. She, however, became a little depressed under lithium therapy. After sodium valproate (600 mg/day) was added to lithium carbonate (600 mg/day), she became euthymic in a week. She has remained euthymic up to now (Fig. 2).

From these two cases, the combination of lithium and sodium valproate may be effective in the treatment of rapid-cycling bipolar disorder, especially that of continuous circular type. Furthermore, this combination may be safe using for a long time.

MECHANISMS OF ACTION OF MOOD STABILIZERS

From the clinical observation of the longitudinal course of affective illness, Post and colleagues[19] proposed the kindling model of affective illness. Kindling phenomenon was first reported by Goddard[20]. He demonstrated that an intermittent electrical stimulation of brain at low intensity develops epileptic seizures. Like kindling, the frequency of relapse increases with the number of previous episodes, with later episodes having a more sudden and spontaneous onset. Furthermore, recurrent depression may become rapid-cycling bipolar disorder[19]. It is interesting in this regard that anticonvulsants such as carbamazepine and sodium valproate are effective in the treatment of bipolar disorder, especially rapid-cycling one.

Because γ-aminobutyric acid (GABA) mechanisms are thought to be involved in the pathophysiology of epilepsy and affective disorders[21,22], the effects of mood stabilizers on GABA receptor subtypes have been examined. Chronic, but not single, treatment with lithium chloride, carbamazepine or sodium valproate increased $GABA_B$ receptors in the rat hippocampus. $GABA_A$ receptors did not change in several brain regions after chronic administration of these three drugs[23,24]. Because a selective $GABA_B$ agonist, baclofen, which is devoid of mood-stabilizing effect[10], did not induce such changes, a common mechanism of action of mood stabilizers may be related to $GABA_B$ receptors in the hippocampus (Fig. 3, 4).

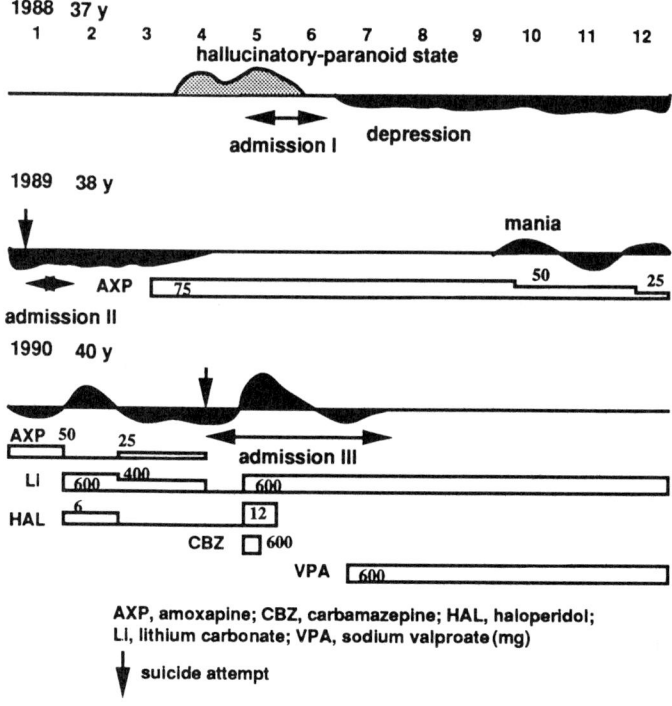

Figure 2. Clinical course of Case 2.

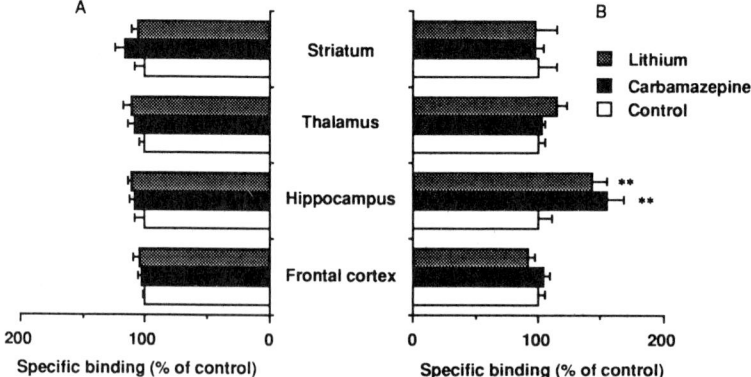

Figure 3. Effects of chronic treatment with lithium and carbamazepine on $GABA_A$ (A) and $GABA_B$ (B) receptors in the rat brain (Motohashi, 1992). Lithium chloride (1.5 mEq/kg) or carbamazepine (50 mg/kg) was injected i.p. once daily for 14 days. The rats were decapitated 24 hrs after the last injection. **$p < 0.01$ as compared to Contral.

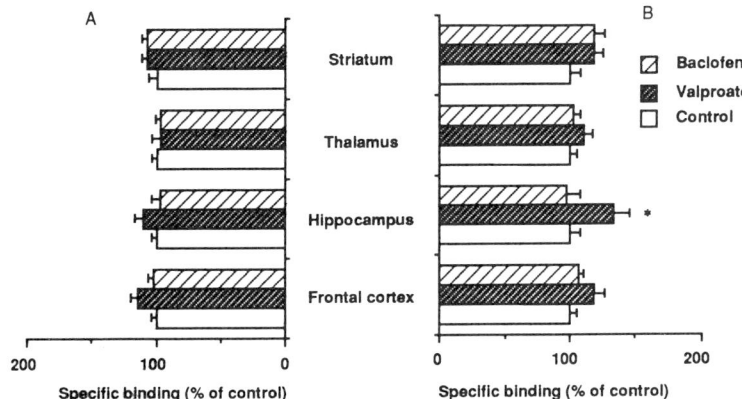

Figure 4. Effects of chronic treatment with sodium valproate and baclofen on $GABA_A$ (A) and $GABA_B$ (B) receptors in the rat brain (Motohashi, 1992). Sodium valproate (150 mg/kg) or baclofen (10 mg/kg) was injected once daily for 14 days. The rats were decapitated 24 hrs after the last injection. *p < 0.05 as compared to Control.

Accumulating evidence suggests that the adenosine system may be related to epilepsy as well as affective disorders[25]. Because carbamazepine directly acts at adenosine A_1 receptors[26], the influence of mood stabilizers on adenosine A_1 receptors and adenosine uptake sites has been examined[27]. Carbamazepine directry acted at adenosine A_1 receptors at concentrations within the human therapeutic ranges. Both adenosine A_1 receptors and adenosine uptake sites did not change in the cerebral cortex after chronic treatment with lithium or carbamazepine. Chronic treatment with sodium valproate, however, increased adenosine A_1 receptors and adenosine uptake sites in the cerebral cortex (Fig. 5). These results failed to demonstrate that the adenosine system is related to a common mechanism of action of mood stabilizers, although part of action of valproate may be mediated via the adenosine systems.

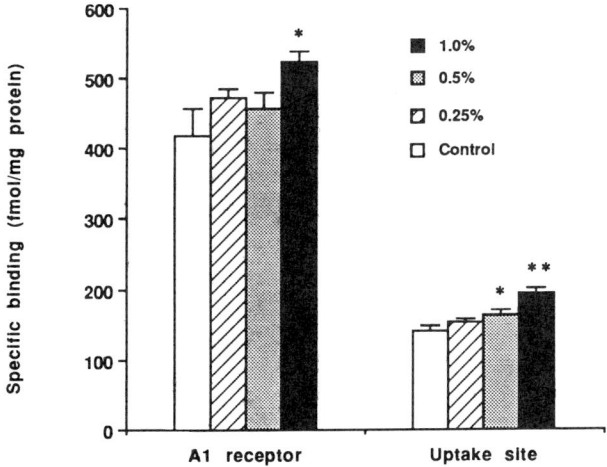

Figure 5. Effects of chronic treatment with sodium valproate on adenosine A1 and adenosine uptake sites in the rat cerebral cortex. The rats were fed with food containing sodium valproate (0.25–1.0%) for 40 days. *p < 0.05, **p 0.01 as compared to Control.

TREATMENT OF REFRACTORY MAJOR DEPRESSION

Approximately 30% of patients with major depression fail to improve satisfactorily while receiving treatment with tricyclic antidepressant drugs. Several pharmacologic and nonpharmacologic strategies exist for refractory depression[28]. Among pharmacological strategies, the addition of lithium to tricyclic antidepressant medication has been effective in some patients[29].

MECHANISMS OF LITHIUM AUGMENTATION OF ANTIDEPRESSANT ACTION

Although de Montigny and colleagues[30] proposed that this augmentation is due to enhancing effect of lithium on serotonin neurons, the precise mechanism of lithium potentiation is not known. We have examined the effects of chronic clomipramine and citalopram administration with short-term (5 days) lithium treatment on monoamines and their metabolites in the rat brain (Okamoto et al., in preparation). Short-term lithium treatment increased the turnover rate of serotonin (Fig. 6), while that of dopamine did not change after the same treatment. These results suggest that part of mechanisms of lithium augmentation may be related to the increased turnover of serotonin.

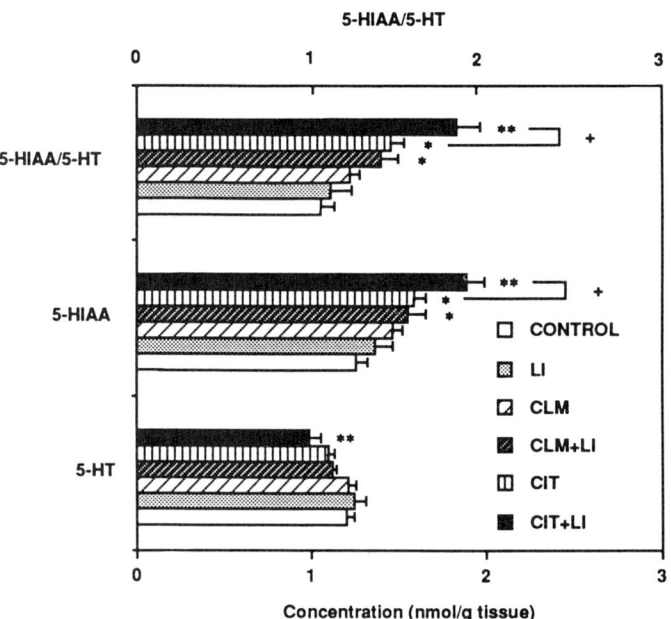

Figure 6. Effects of lithium augmentation of antidepressant treatment on serotonin (5-HT) systems in the frontal cortex of the rat. CONTROL, 0.9% saline i.p. for 19 days; LI, 0.9% saline for 14 days and LiCl (2 mEq/kg, i.p.) for the last 5 days; CLM, clomipramine HCl (20 mg/kg, i.p.) for 19 days; CLM+LI, CLM for 14 days and CLM+LI for 5 days; CIT, citalopram HBr (20 mg/kg, i.p.) for 19 days; CIT+LI, CIT for 14 days and CIT+LI for 5 days. The rats were decapitated 24 hrs after the last injection. 5-HIAA, 5-hydroxy indole acetic acid. *$p < 0.05$, **$p < 0.05$ as compared to Control. +$p < 0.05$.

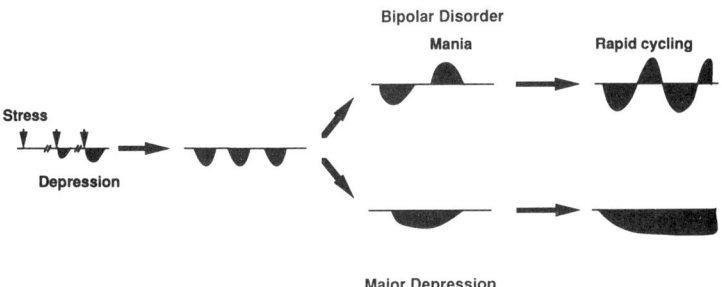

Figure 7. Modified kindling model of affective disorder.

CONCLUSIONS

I propose the modified kindling model of affective disorders (Fig. 7). In the course of affective disorder, some patients may become manic and rapid-cyclers, while others may show protracted depressive state. Treatment must be changed according to the stage of their illness. Mood stabilizers may be effective in the treatment of refractory affective disorder.

ACKNOWLEDGEMENT

The author thanks Drs. M. Nankai, K. Koyama, Y. Atsumi (Tokyo Medical and Dental University), Prof. T. Kariya, Dr. K. Shioe (Yamanashi Medical College), Prof. S. Yamawaki, Dr. Y. Okamoto (Hiroshima University) and Dr. H. Mitsushio (National Musashi Hospital) for their collaboration to perform this study. This work was supported in part by grants from the Ministry of Education, Science and Culture and from the National Center of Neurology and Psychiatry (NCNP) of the Ministry of Health and Welfare, Japan

REFERENCES

1. Kleist, K. (1947) Froteschritte der Psychiatrie. Frankfurt a. M., Kramer.
2. Leonhard, K. (1957) Aufeiling der Endogenen Psychosen, Berlin, Academie Verlag.
3. Angst, J. and Perris, C. (1968) Zur Nosologie endogener Depressionen. Vergleich der Ergebnisse zeier Untersuchungen. Arch. Psychiat. Z. ges. Neurol. 210: 373–386.
4. Angst, J., Felder, W., Frey, R. and Stassen, H.H. (1978) The course of affective disorders. I. Change of diagnosis of monopolar, unipolar, and bipolar illness. Arch. Psychiat. Nervenkr. 226: 57–64.
5. Kukopulos, A., Reginaldi, D., Laddomada, P., Floris, G., Serra, G. and Tondo, L. (1980) Course of the manic-depressive cycle and changes caused by treatments. Pharmacopsychiatry 13: 156–167.
6. Wehr, T.A. and Goodwin, F.K. (1987) Can antidepressants cause mania and worsen the course of affective illness? Am. J. Psychiatry, 144: 1403–1411.
7. Dunner, D.L. and Fieve, R.R. (1974) Clinical factors in lithium carbonate prophylaxis failure. Arch. Gen. Psychiatry 30: 229–233.
8. Schou, M. (1989) Lithium prophylaxis: myths and realities. Am. J. Psychiatry, 146: 573–576.
9. Motohashi, N., Shioe, K. and Kariya, T. (1992) Mechanisms of action of antidepressants in affective disorders: comparison with lithium. Lithium 3: 7–14.
10. Post, R.M. (1990) Sensitization and kindling perspectives for the course of affective illness: toward a new treatment with the anticonvulsant carbamazepine. Pharmacopsychiatry, 23: 3–17.

11. Lambert, P.-A., Borselli, S., Marcou, G., Bouchardy, M., Cabrol, G. (1971) Action thymo-régulatrice à long terme du Dépamide dans la psychose maniaco-dépressive. Ann. Med. Psychol. 2: 442–448.
12. Semadeni, G.W. (1976) Etude clinique de l'effet normothymique du dipropylacétamide. Acta Psychiat. Belg. 76: 458–466.
13. Puzynski, S. and Klosiewicz, L. (1984) Valproic acid amide in the treatment of affective and schizoaffective disorders. J. Affect. Disord. 6: 115–121.
14. Emrich, H.M., v Zerssen, D., Kissling, W., Möller, H.J. and Windorfer, A. (1980) Effect of sodium valproate on mania: the GABA hypothesis of acute mania. Arch. Psychiatr. Nervenkr. 229: 1–16.
15. Pope, H.G.Jr., McElroy, S.L., Keck, P.E. and Hudson, J.I. (1991) Valproate in the treatment of acute mania. Arch. Gen. Psychiatry 48: 62–68.
16. Freeman, T.W., Clothier, J.L., Pazzaglia, P., Lesem, M.D. and Swann, A.C. (1992) A double-blind comparison of valproate and lithium in the treatment of acute mania. Am. J. Psychiatry 149: 108–111.
17. McElroy, S.L., Keck, P.E.Jr., Pope, H.G.Jr. and Hudson, J.I. (1988) Valproate in the treatment of rapid-cycling bipolar disorder. J. Clin. Psychopharmacol. 8: 275–279.
18. Calabrese, J.R. and Delucchi, G.A. (1990) Spectrum of efficacy of valproate in 55 patients with rapid-cycling bipolar disorders. Am. J. Psychiatry 147: 431–434.
19. Post, R.M., Uhde, T.W., Putnam, F.W., Ballanger, J.C. and Berrettini, W.H. (1982) Kindling and carbamazepine in affective illness. J. Nerv. Ment. Dis., 170: 717–731.
20. Goddard, G.V. (1967) Development of epileptic seizures through brain stimulation at low intensity. Nature 214: 1020–1021.
21. Lloyd, K.G. (1983) Role of GABAergic systems in the mechanism of action of antiepileptic drugs. Thérapie 38: 355–362.
22. Lloyd, K.G., Morselli, P.L., Bartholini, G. (1987) GABA and affective disorders. Med. Biol. 65: 159–165.
23. Motohashi, N., Ikawa, K. and Kariya, T. (1989) $GABA_B$ receptors are up-regulated by chronic treatment with lithium or carbamazepine. GABA hypothesis of affective disorders? Eur. J. Pharmacol. 166: 95–99.
24. Motohashi, N. (1992) GABA receptor alterations after chronic lithium administration. Comparison with carbamazepine and sodium valproate. Prog. Neuro-Psychopharmacol. & Biol. Psychiatry 16: 571–579.
25. Durcan, M.J. and Morgan, P.F. (1990) Prospective role for adenosine and adenosinergic systems in psychiatric disorders. Psychol. Med. 20: 475–486.
26. Clark, M. and Post, R.M. (1989) Carbamazepine, but not caffein, is highly selective for adenosine A_1 binding sites. Eur. J. Pharmacol. 164: 399–401.
27. Motohashi, N., Shioe, K., Mitsushio, H. and Kariya, T. (1992) Effects of mood stabilizers on adenosine systems in rat brain. Jpn. J. Psychiat. Neurol. 46, 559.
28. Nierenberg, A.A. and White, K. (1990) What next?: a review of pharmacologic strategies for treatment resistant depression. Psychopharmacol. Bull. 26: 429–460.
29. Austin, M.-P. V., Souza, F.G.M. and Goodwin, G.M. (1991) Lithium augmentation in antidepressant-resistant patients. A quantitative analysis. Br. J. Psychiatry 159: 510–514.
30. de Montigny, C., Cournoyer, G., Morissette, R., Langlois, R. and Caille, G. (1983) Lithium carbonate addition in tricyclic antidepressant-resistant unipolar depression. Arch. Gen. Psychiatry 40: 1327–1334.

STUDIES OF DIAGNOSIS AND PATHOGENESIS OF WILSON'S DISEASE

Xiuling Leung, Rong Chen, Zhoulin Liu and Yinru Zhang

Department of Neurology
1st Affiliated Hospital
Sun Yat-sen University of Medical Sciences
Guangzhou, People's Republic of China

ABSTRACT

We have studied the copper and metallothionein (MT) in culture skin fibroblasts of patients and heterozygotes with Wilson's disease (WD) and controls (5 cases each) after incubation in mediums containing various concentrations of copper (C_1: 15.74 µmol/L, C_2: 78.70 µmol/L, C_3: 157.38 µmol/L, C_4: 314.76 µmol/L). The results are as follows:

1. In each of the three groups, the copper/protein ratio (Cu/P) in cytosols of 1–5 passages is significantly higher than that of 6–10, 11–15, 16–20 passages. There are no significant differences among 6–10, 11–15 and 16–20 passages.

2. In standard medium, Cu/P in cytosols of three groups are not significantly different, but Cu/P of patients is significantly higher than that of the other two groups after incubation in C_4 medium for 12 or 24 hours.

3. After incubation in C_1, C_2, C_3 and C_4 mediums respectively, the Cu/P in cytosols of the three groups only increased in C_4 medium with time (within 72 hours). It is higher in the patient group than the other two groups.

4. In the three groups, cytosol copper distributed similarly in two peaks after Sephadex G-75 chromatography, which are on high molecular weight (HMW) proteins and MT fractions respectively. It remained the same after incubation in various concentrations of copper. The copper content found in MT fractions in WD patients is much higher than that of controls and heterozygotes. It is even higher after incubation in various concentrations of copper but no changes was found in controls and heterozygotes.

5. The MT contents were not significantly different in cytosols of the three groups when cultured in standard medium. However, they were all elevated in the three groups after incubation in C_4 medium but not in C_1, C_2 and C_3 mediums. WD patients showed higher MT contents than the other two groups.

The results of our experiments suggest that :

1. The copper content is stabled in 6–20 passages of cultured skin fibroblasts in the three groups. The skin culture fibroblasts are ideal for further studies of copper metabolism of WD.

2. The copper content measurement in cytosol of culture skin fibroblasts in C_4 medium for 12 or 24 hours can be used in distinguishing WD patients from heterozygoates and normals. Yet the copper contents in cytosols of the WD patients differ from those of the heterozygotes and normals only when culture in C_4 medium but not the standard ones.

3. The deposition of copper in cytosol of WD patient fibroblasts is highly related to MT.

4. The MT in WD patients may have an abnormal high inductivity by copper, which might also play an important role in the copper deposition in WD cells.

INTRODUCTION

Hepatolenticular degeneration, also called Wilson disease (WD), is an autosomal recessive inherited disease with metabolic dysfunction of copper. Its main features consist of liver cirrhosis, basal ganglia degeneration, corneal K-F rings and disturbance of renal functions. The frequency of Wilson's disease gene in population reported around the world is among 1/100 - 1/200. The prevalence of WD is rated one per 20,000 - 200,000. It seems likely that the prevalence of WD in China is much higher than that in the western countries.[1]

Enormous progress has been achieved in the study of WD. The most outstanding pathological change of WD is the excessive deposition of copper in various organs, particularly in the liver, brain, kidney and cornea. Yet the pathogenesis of WD still remains unknown. Some investigators have shown that in the liver, brain and kidney of patients, there are elevated content of metallothionein, which binds mostly cytosol copper.[2,3] Therefore, it is suggested that metallothionein abnormality may play an important role in the copper deposition in WD. Clinically, rational copper biochemical examination (including serum ceruloplasmin, serum copper and urine copper measurements) has made the diagnosis of the patients with typical manifestation easy. The diagnosis of patients of early stage (including presymptomatic ones) remains difficult because there is lack of an unambiguous biochemical examination.[4] However, since the application of skin fibroblast culture on WD in 1980,[5] it has provided an in vitro model for further study of copper metabolism in WD. The copper content measurement in cytosol of cultured skin fibroblasts of patient has been recommended by Chan et al[5] in the diagnosis of WD. In this paper, we studied the copper and metallothionein metabolism in culture fibroblasts of patients and heterozygotes with WD and controls after incubated in mediums containing various concentrations of copper. This study is to established an effective diagnostic method for the patients of the early stage and attempt to understand the pathogenesis of this disease.

MATERIALS AND METHODS

Five WD patients (8–25 years old) were studied. All had low serum ceruloplasmin and copper concentrations, with increased urinary copper excretion. Five WD heterozygotes (34–54 years old) were the parents of the WD patients. They all had a normal phenotype. Five individual served as controls (9–55 years old)
were excluded the diagnosis of WD and other genetic diseases by physical and copper biochemical examinations.

Skin fibroblast cultures were established from biopsies of the above three groups. Cells were grown in Eagle's minimal essential medium with fetal calf serum (10 percent) and penicillin-streptomycin (1 percent) and adjusted to pH 7.2–7.3 with 7.5 percent sodium bicarbonate solution.

Copper concentrations in the above standard medium were 1.89–3.15 μmol/L and 12.24–15.30 μmol/L respectively. Cells were also incubated in mediums that contained C_1:15.75 μmol/L, C_2:78.70 μmol/L, C_3:158.38 μmol/L and C_4:314.76 μmol/L copper for 3, 6, 12, 24, 36, 48, 60, and 72 hours.

Confluent cultures were harvested with a rubber policeman. Cells were suspended with demineralized 0.05 mol/L tris-HCl (pH 8.6) and lysed by ultrasonic method (80W, 60 seconds). The lysate was centrifuged at 108,000 X g for 60 minutes to obtain cytosol (supernatant).

To examine the distribution of cytosol copper in protein with different molecular weight, the cytosols of patients, heterozygotes and controls were fractionated by gel filtration on a Sephadex G-75 column (80 by 1.5 cm) equilibrated with 0.05 mol/L tris-HCl (pH 8.6). Proteins were eluted from the column with the equilibrated buffer at a flow rate of 0.5 ml/min (1.5 ml per fraction). The fractionation procedures were carried out at 4°C.

Copper concentrations of cytosol and column fractions were determined by flameless atomic absorption spectroscopy equipped with a carbon furnace. Protein concentration in cytosol of cultured skin fibroblasts was assayed by Bradford's dye-binding method,[6] using recrystalized bovine albumin as a standard.

Metallothionein content in cytosol of fibroblasts was measured by the Ag/Hemolysate method.[7]

RESULTS

1. Cytosol copper content of fibroblasts of different passages

Being cultured in standard medium, cells of four passage groups were collected (1–5, 6–10, 11–15, 16–20 passages). Their cytosol copper/protein ratios were determined and compared to each other (see Table 1). The results showed, in each of the patient, heterozygote and control groups, the copper/protein ratio in 1–5 passage group was significantly higher than those of 6–10, 11–15, and 16–20 passage groups. There were no significant difference among 6–10, 11–15, and 16–20 passages. In each passage group, there were also no significant difference among patients, heterozygotes and controls.

Since the fibroblasts of 6–20 passages from patients, heterozygotes and controls all had stable cytosol copper contents, they were collected for further experiments.

2. Copper contents in cytosol of fibroblasts after copper incubation

After incubation in C1, C2, C3 and C4 mediums respectively, it was found that the copper/protein ratios in cytosols of WD patients, heterozygotes and controls (two cases each) were not influenced by C1 and C2 mediums. They were similarly elevated after C3 medium incubation, but in C4 medium, the copper/protein ratio of patients was elevated much more than that of heterozygotes and controls. (Figure 1). Therefore, the copper/protein ratios of all patients, heterozygotes and controls (5 cases each) were determined and compared after incubation in C4 medium for 12 and 24 hours. It was found that the copper/protein ratio of the patient group was significantly higher than those of the other two groups ($p<0.01$). There was no overlapse of the data obtained from the patients and the other two groups. The heterozygotes and controls had no significant difference in cytosol copper/protein ratio. (Table 2)

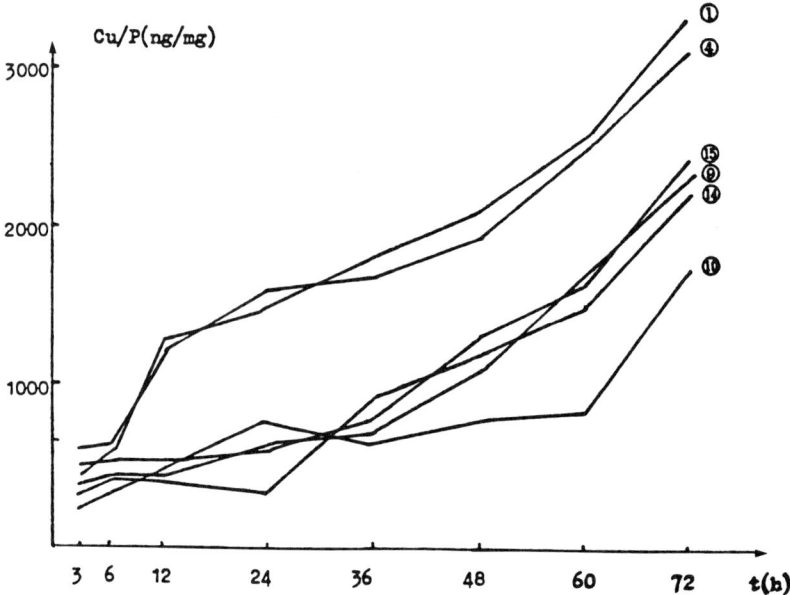

Figure 1. Copper/protein ration(Cu/P) in cytosols of patients, heterozygotes and controls after incubation in C4 medium ① ④ patients, ⑨ ⑩ heterozygotes, ⑭ ⑮ controls

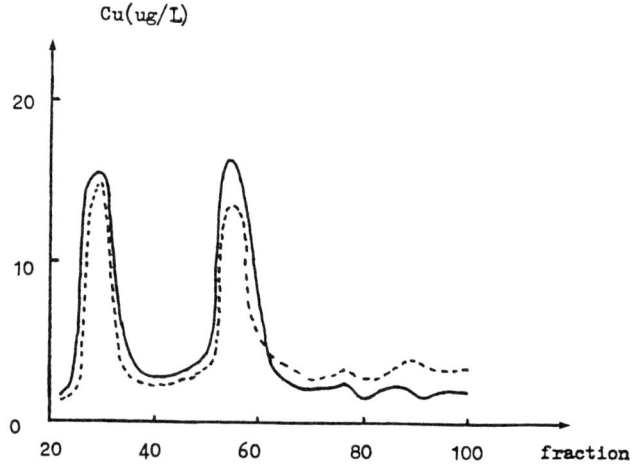

Figure 2. Copper distribution after Sephadex G-75 chromatography of cytosol from control fibroblasts (two cases)

3. Copper distribution in different molecular weight proteins in cytosol of fibroblasts cultured in standard medium (Figure 2,3 4)

In each of WD patient, heterozygote and control groups, the copper in cytosol of cultured fibroblasts were distributed up two peaks after Sephadex G-75 chromatography. The two peaks were on high molecular weight (HMW) protein fractions and low molecular weight (LMW) protein fractions respectively. The LMW protein fractions had a Ve/Vo about 1.8. Its high

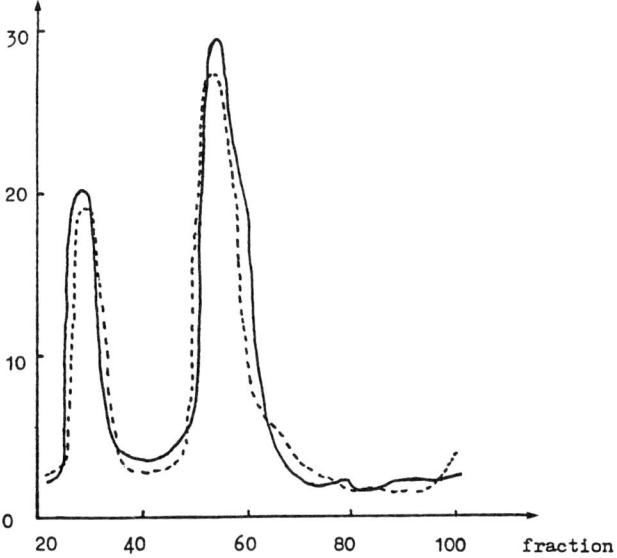

Figure 3. Copper distribution after Sephadex G-75 chromatography of cytosol from patient fibroblasts (two cases)

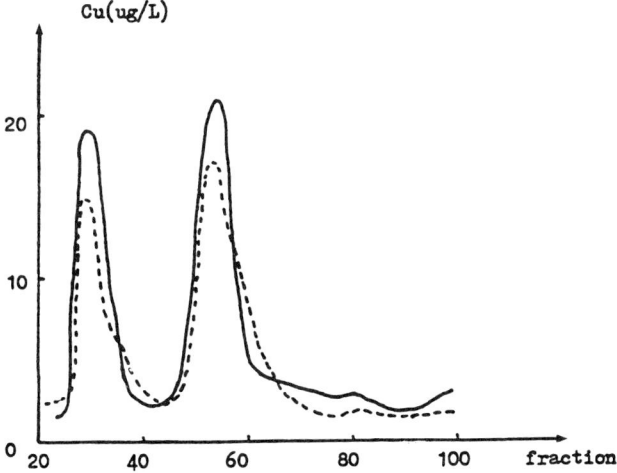

Figure 4. Copper distribution after Sephadex G-75 chromatography of cytosol from heterozygote fibroblasts (two cases)

copper content at A220 and A280 in ultraviolet spectrophotometer supported the presence of metallothionein in this peak.

In controls, the copper contents found in two peaks were nearly equal. Yet there were much more copper found in the metallothionein fractions of patients than heterozygotes and controls. It was only slightly higher in heterozygotes than in controls.

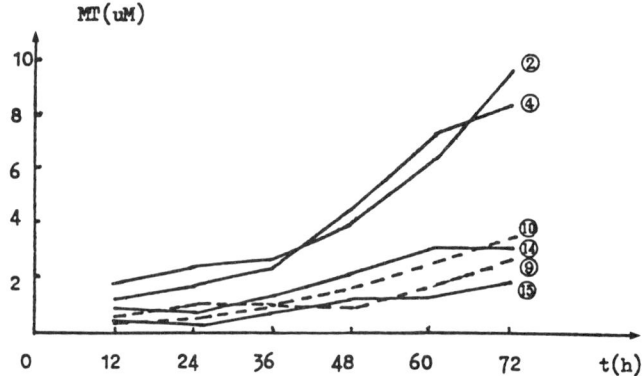

Figure 5. Metallothionein content in cytosols of fibroblasts after incubation in C4 medium. ② ④ patients, ⑨ ⑩ heterozygotes, ⑭ ⑮ control

Table 1. Copper/protein ratio in cytosol of patient, heterozygote and control fibro

	PASSAGES			
	I (1 - 5)	II (6 - 10)	III (11 - 15)	IV (16 - 20)
(1) PATIENTS	438.9±141.1	151.0±65.5	161.3±81.2	167.1±34.3
(2) HETEROZYGOTES	357.5±171.6	161.2±65.7	124.7±33.0	158.6±45.5
(3) CONTROL	489.6± 73.6	222.5±66.4	173.3±82.4	176.8±73.5

	p VALUE				
	I to II	I to III	I to IV	II to III	III to IV
(1) PATIENTS	<0.01	<0.01	<0.01	>0.05	>0.05
(2) HETEROZYGOTE	<0.01	<0.01	<0.01	>0.05	>0.05
(3) CONTROLS	<0.01	<0.01	<0.01	>0.05	>0.05

	PASSAGES			
	I (1-5)	II (6-10)	III (11-15)	IV (16-20)
	p VALUE			
(1) TO (2)	>0.05	>0.05	>0.05	>0.05
(1) TO (3)	>0.05	>0.05	>0.05	>0.05
(2) TO (3)	>0.05	>0.05	>0.05	>0.05

* Values are $\bar{X} \pm SD$

Table 2. Comparison of copper/protein ratio (Cu/P) in cytosol of patients, heterozogote and control cells after incubation C4 medium for 12 or 24 hours (ng/mg)

		Cu/P (X+SD)	p VALUE		
			(1) to (2)	(1) to (3)	(2) to (3)
12 h	(1) PATIENTS	1369.4±182.8			
	(2) HETEROZYGOTES	424.6±187.2	<0.01	<0.01	<0.05
	(3) CONTROLS	489.7±244.0			
24 h	(1) PATIENTS	1540.7±184.7			
	(2) HETEROZYGOTES	569.8±120.0	<0.01	<0.01	<0.05
	(3) CONTROLS	498.7±294.1			

Table 3. Comparison of metallothionein content (uM) in cytosols of patient, heterozygote and

(1) PATIENTS	(2) HETEROZYGOTES	(3) CONTROLS	P VALUE		
			(1) TO (2)	(1) TO (3)	(2) TO (3)
0.2487	0.2254	0.2432			
0.2521	0.2851	0.2700			
0.1689	0.2677	0.2833	>0.05	>0.05	>0.05
0.1771	0.2473	0.2612			
0.2477	0.2745	0.2554			

4. Metallothionein concentration in sytosol of fibroblasts

There were no significant difference of metallothionein contents in cytosols of fibroblasts among WD patients, heterozygotes and controls both cultured in standard medium (Table 3) and incubated in C1, C2, and C3 medium (Figure 5)

DISCUSSION

Since the first application of skin fibroblasts culture in the patients with WD in 1980[5], it has provided an effective in vitro model for the study of diagnosis and pathogenesis of this disease. The cultured fibroblasts of 6–13 passages were used in all such studies, according only to an incomplete result of Goka et al.[8] In the study about the cultured skin fibroblasts of Menkes' kinky hair disease, Goka et al found that the copper content in cells was relatively stable within 1–5 passages. However, the exact cytosol copper content in fibroblasts of different passages both in normal and WD patients have still not been studied. In this study, it is shown that the copper contents in cytosols of fibroblasts in WD patients, heterozygotes and controls were stable from 6–20 passages. It is ideal for further investigation of copper metabolism in WD, and can provide enough material for study. It is expected that the copper contents might remain stable in some more passages.

Effective treatment has been confirmed for WD. Yet it has to start before irreversible damage of the organs. Therefore, early detection and diagnosis is very important.[1] Despite the low serum ceruloplasmin and serum copper concentration, increased urinary copper excretion are the characteristic changes in WD patients. It has been reported that about 10 percent of WD patients had a normal value of copper biochemical examination. On the other hand, abnormal serum copper and ceruloplasmin concentration may be seen in some chronic liver disease, for example, in the primary cirrhosis and chronic active hepatitis patients. The diagnosis of early stage and atypical patients with WD remains difficult.[4]

Chan et al[5] first reported that the copper content in cultured skin fibroblasts of WD patients was about three times of controls and suggested that this method may be used in diagnosis of WD. Unfortunately, other investigators obtained negative results from this method.[9,10] The usefulness of this diagnostic method remain controversial. Our results showed that when cultured in standard medium, the copper/protein ratios among WD patients, heterozygotes and controls are not significantly different. This suggests that the copper content measurement of cultured skin fibroblasts cannot be used as the diagnosis of either WD in patients or heterozygotes.

In consideration of the low copper concentration in standard medium (much lower than the normal serum copper concentration), we took further examination of the copper content in cytosol of cultured cells after incubation in various copper concentrations. The results showed, the copper/protein ratios of WD patients were significantly higher than those of the other two groups after incubation in C4 medium for 12 or 24 hours. The data has no overlap between WD patients and the other two groups. It suggests that the copper/protein ratio determination in cytosol of cultured skin fibroblasts after incubation in C4 medium for 12 or 24 hours can be used in distinguishing the WD patients from heterozygotes and controls. It is suitable for the diagnosis of the early and atypical patients and perhaps might be used for presymptomatic and prenatal diagnosis of WD patients.

The basic biochemical defect of WD is still unclear. It was first suggested by Uzman[11] (1953) that the liver of WD patient may contain an abnormal protein with high copper-capacity, which may prevent the release of copper in the hepatocytes. Soon after that, other investigators had shown that the metallothionein from the liver of WD patient may have an abnormally high copper affinity.[12] It has also been shown that the metallothionein contents in the liver and kidney of patients with WD were all elevated, which binds most intracellular copper.[13] Additionally, two abnormal metallothionein-like proteins have been isolated from the brain of a WD patient.[14] A study showed that the deposition of copper in the corneal Descenet's membrane of WD patient might be related to metallothionein.[15] All these results suggest that metallothionein may play an important role in the pathogenesis of WD.

Our results showed, even though the copper/protein ratios of WD patients were not significantly higher than that of heterozygotes and controls when cultured in standard medium, the copper contents found in metallothionein fractions were much higher than those of the other two groups. It suggests that the metallothionein in WD patients may have a primary high copper affinity, which could cause the accumulation of copper in WD cells. Otherwise, despite the content of metallothionein in cytosol of WD patient fibroblasts is normal when cultured in standard medium. It was much higher in patients than in heterozygotes and controls after incubation C4 medium. This suggests that the metallothionein in patients may have an abnormal high inducitity by copper. The excessive metallothionein in WD may result in the further deposition of copper within cells.

It has been reported that the metallothionein in hepatocyte may be collected and degraded by lysosome, copper than be discharged into bile. Schechiger et al[16] has suggested that the copper in metallothionein in hepatocyte may be transported to apoceruloplasmin to form ceruloplasmin. If these would be confirmed, the low serum ceruloplasmin concentration of the patients might be the consequence of the excessive combination of copper into metallothionein.

REFERENCES

1. Liu, ZL. and Liang, XL, (1988) Genetics in Neurology. Publishing House, Beijing, China p.86.
2. Barty, NO et al : (1987) Fepatic copper and metallothionein distribution in Wilson's disease. Lab Invest. 57:397.
3. Evans, J., et al: Wilson's disease : (1973) Identification of an abnormal copper-binding protein. Science 181:75.
4. Sternlieb, I. et al: (1968) Prevention of Wilson's disease in asymptomatic patients. New Eng. J. Med. 278:352.
5. Chan, WY et al: (1980) Genetic expression of Wilson's disease in cell culture : A diagnosis marker. Science 208:299.
6. Bradford, MM.: (1976) A rapid and sensitive method for the quantitation of microgram quantities of protein utilizing the principle of protein-dye binding. Anal. Biochem. 72:248.
7. Lee, DY et al: (1989) treatment of wilson's disease with zinc: VII. Protection of the liver from copper toxicity by zinc-induced metallothionein in a rat model. J. Lab. Clin. Med. 114:639.
8. Goka, JJ et al: (1976) Menkes' disease: a biochemical abnormality in cultue human fibroblasts. Proc. Batl. Acad. Sci. USA 73:604.
9. Favier A et al: (1983) Menkes' and Wilson's diseases; study of copper in skin fibroblasts. J. Inhe. Metab. Dis. 6 suppl 2:86.
10. Van Den Berg, GJ et al: (1989) Cultured skin fibroblasts: useful for diagnosis of wilson's disease? J. Inher. Metab. Dis. 12:64.
11. Uzman, LL: (1953) On the relationship of urinary copper excretion to the aminoaciduria in Wilson's disease. Am. J. Med. Sci. 226:645.
12. Shapiro, J. et al.: (1961) A copper-protein of human live. J. Clin. Invest. 40:1081.
13. Sato, M.: (1986) copper accumulation and copper binding proteins in tissues and cultured fibroblasts from patients with Wilson's disease. J. Med. Soc. Toho. Japan 33:20.
14. /Sato, M. st al: (1986) Copper-binding proteins in liver, kidney, and brain tissues from Wilson's disease patient. Brain and Berve 38:933.
15. Johnson, RE et al: (1982) Wilson's disease : Electron microscopic, X-ray energy spectroscopic, and atomic absorption spectroscopic studies of corneal copper deposition and distribution. Lab. Invest. 46:564.
16. Schechinger, J. et al.: (1986) Copper transport from Cu(I)-thionein into apoceruloplasmin mediated activated leucocyte. Biochem. J. 240:281.

EVIDENCE FOR PRESYNAPTIC DAMAGE IN MYASTHENIA GRAVIS

Chuan-Zhen Lu

WHO Collaborating Center for Training and Research in Neurosciences
Shanghai Medical University
Shanghai, 200040
People's Republic of China

INTRODUCTION

Myasthenia Gravis (MG) has been considered as an autoimmune disease at neuromuscular junction caused by circulating antibodies to acetylcholine receptors (AchR) since 1973. Antibodies to AchR can be detected in 65–95 % sera from patients with MG. In the mean time, no correlation between the antibodies titer and severity of symptoms of the disease was found. Experiemntal myasthenia gravis (EAMG) not only can be induced by injection of antibodies to AchR positive sera from MG patients, but also can be induced by injecting sera with undetectable antibody to AchR. The last few years, we looked into other factors in the pathogenesis of myasthenia Gravis.

ANTIBODIES TO BETA-BUNGAROTOXIN (β-BUTX) BINDING PROTEIN

Utilizing β-BuTx to capture its corresponding protein from crude receptor fraction or affinity purified β-BuTx binding protein, we found that antibodies to β-BuTx are present in two third of patients with MG. Among the patients studied, 44% had antibodies to both AchR and β-BuTx binding protein (PrsmR), 12% has antibodies to PrsmR and 11% had antibodies to AchR only. Antibodies to both AchR and PrsmR can be absorbed by alpha-BuTx and beta-BuTx affinity columns.

ANTIBODIES TO OTHER NEUROTOXIN BINDING PROTEIN AND MYELIN COMPONENTS

A neurotoxin of Shanghai snake and a Chinese herb toxin (chuen dong xu) bind specifically to presynaptic membrane. The later is bound on the active zone at presynaptic membrane. Utilizing these two neurotoxins to identify their corresponding protein, we also found that antibodies to these two neurotoxin binding proteins are present in 50% of patients with MG. They do not completely overlap with the antibodies to PrsmR. It is well known that PO, MBP, MAG, MOG are of components of myelin, which have been widely considered as autoantigen in demyelinating diseases of nervous system. Using these myelin components as antigen, we detected the antibodies and the antibodies secreting cells to those myelin components with ABC-ELISA and ELISPOT in sera and peripheral blood lymphocytes in patients with MG respectively. We also found 40–60% MG patients had such antibodies and antibodies secreting cells. The antibodies titer and number of antibodies secreting cells were less than those to AchR.

ANIMAL EXPERIMENTS

Two kinds of experiments have been conducted. 20 samples of sera from patients with MG with various antibodies to AchR, PrsmR only or to both AchR and PrsmR positive or to both AchR and PrsmR negative were studied in rats diaphragm muscle in vitro. The results showed that sera with antibodies to PrsmR only and to both AchR and PrsmR induced Mepps frequency increase or decrease, particularly, in two samples of sera with antibodies to PrsmR only induced the Mepps decrease after short term of elevation. No membrane potential changes was found.

Passive transfer of EAMG model induced by injection of plasma from one patient with antibodies to both AchR and PrsmR and one patient with antibodies to PrsmR only have been done. Thirty male BLBC/C mice were used in the first experiment. Thirteen mice were injected with MG plasma, 10 injected with normal plasma and 7 injected with normal saline. Two mice died of weakness during the period of the experiment. The rest of the mice were killed on day 69th. MP, the Mepps and EPP were recorded during the experiment. No changes of the amplitude, and frequency of Mepps were found in the experiment with normal Kreb's solution. However, the frequency of Pepps was greatly altered when K^+ concentration in Kreb's solution increased to 17.5 mM. The mean frequency of Mepps in MG, normal plasma and normal saline groups were 10.1 ± 2.2, 139.0 ± 4.3 and 124.9 ± 8.7 respectively. These results show that high concentration of K^+ can not enhance the release of Ach from presynaptic membrane in species from MG plasma immunized animals. The EPP induced by 6 strain stimulations was significantly changed in MG group. The rates of first to six EPP of the three groups were 3.0 ± 1.9 (n = 27), 0.9 ± 0.3 (n = 23) and 0.9 ± 0.3 Hz respectively. These results indicated a typical neurophysiological findings of myasthenia gravis, not an Eaton-Lambert syndrome. In another experiment, EAMG was produced by injection of plasma with antibodies to PrsmR only. The results also showed that Mepps frequency in high concentration of K^+ in Kreb solution was significantly different in MG, normal plasma and normal saline groups. The mean values were 2.2 ± 0.2, 35.8 ± 4.5 and 41.2 ± 3.3 respectively. The rates of first to fifth EPP induced by 50 Hz strain stimulation were 2.18 ± 0.03, 1.05 ± 0.06 and 1.04 ± 0.02 respectively. The quantum contents were 0.45 ± 0.07, 0.66 ± 0.07 and $0.0).65 \pm 0.05$ respectively. These results again indicated that sera or plasma from patients with MG with antibodies to PrsmR only or to both AchR and PrsmR can produce a presynaptic damaged EZMG model.

PURIFICATION OF β-BUTX BINDING PROTEIN

Beta-BuTx binding protein was purified from bovine diaphragm muscle by detergent extraction with 1.5% Triton X-100 followed by affinity chromatography on (1). Lentillectin sepharose 4B (2). beta-BuTx Affi-Gell 15 and (3). Wheat germ agglutinin sepharose 6 MB. The SDS-PAGE og β-BuTx binding protein revealed two major bands with molecular weight of about 69 KD an 41 KD. Its isoelectric point (PI) was about 7.7. The AchR, alpha-BuTx binding protein was also purified from bovine diaphragm muscle. In compaing these two proteins, we found that they have different PI, molecular weight, fractions and specific binding abilities to alpha or beta bungarotoxin. Beta-BuTx bound on beta-BuTx binding protein not AchR, and vice versa. Their contamination was less than 3%.

CONCLUSION

We do consider that myasthenia gravis is an autoimmune disease at neuromuscular junction, not caused by circulating antibodies to AchR only, but to the presynaptic membrane as well.

ACKNOWLEDGEMENT

This work was supported by National Nature Scientific Foundation of China and Shanghai Medical University.

INDEX

Acetylcholine
  in Alzheimer's disease, 1, 5–7
  neurons, 4
  in Parkinson's disease, 1
  receptors, 4, 8, 65, 75, 207–209
    density, 5
    junction, 6
    membrane, 6
    subtypes
      muscarinic, 4
      nicotinic, 66–73
        conformation states
          resting state, 67
          putative activated, 68
          putative desensitized, 68, 69
  subtypes
    muscarinic, 4, 5
    nicotinic, 6, 7
  synthesis, 4, 6
Acetylcholinesterase, 4
  in Alzheimer's disease, 5
    anticholinesterase agents, 5, 6, 65
    release, 5
  inhibitors
    action
      effect on cognition, 51
      effect on cfos gene expression, 51
Activity monitor, 32
Acupuncture, 2, 109
Adenylate cyclase, 7
Alzheimer's disease, 1
  diagnostic scales
    DSMIII, Reisburg Global Determination Scale, 57
    NINCDS-ADRDA Critiria, 57
  eteology, 4, 6
  pathology, 6
    hisopathological changes, 6, 57, 58
    Neurochemcal changes, 1, 4–6, 47, 48, 57
      cholinergic hypothesis of memory dsyfunction, 5
      beta amyloid plague, 58

Alzheimer's disease
  therapy strategies
    basic research
      chemicals investigated for efficacies
        brain gangliosides, 59–61
        genetic risk, 6
        MAO inhibitors on cognitive behavior, 15
        neuron rescue, 16, 18
        neurotransmitters release, 48
        putative drugs tests on monkey, 5
      surgical invasive reseach in animals
        embryonic septo-hippocampus graft
    clinical
      correct neurochemical dysfunctions
        anticholinesterase, 5, 47
        muscarinic receptor agonists, 5
        precursor therapy (choline), 5, 47
        muscarinic agonists, 5, 47
Amantidine, 3, 25
Aminoethyl-chloro-benzamide, 25
Aminoethyl-chloropyridine, 25
Amitriptyline, 79
AMPA, 123–128
Amphatamine, 17–19
Angiotensin II, 5
Apolipoprotein E (APOE), 6
Apomorphine, 3, 25
APP, 6
Arecoline, 5
Arginine, 3
Asiogliosides, 57, 60, 61

Baclofen progabide, 3, 191
Behavioral tests on learning and memory, 50
Blood brain barrier, 2, 19
Botulinum A, 6
Brain-derived neurotrophic factor, 4
Brain ischemia
  animal model, 133
  study on occlusion time and neurotransmitters, 139–141

Brain ischemia (cont.)
  symptoms, 124
    antioxidant enzyme activities, 141
    delay neuronal death, 124
    electrical activity arrest, 124
    glutamate release, 125
    loss of consciousness, 124
    monoamine and metabloties levels changed, 133–136
    transmitters release, 125
  types, 124
    focal, 133
    global, 133
Bromocriptine, 3, 25
Budipine, 25

Carbidopa, 3
Cerebral palsy
  clinical symptom, 183
  CT scan, 183
  treatment
    Chinese herbs, 183
Cerebrovascular disease
  cerebral hemorage, 158
  cerebral ischemic lesion, 158, 187
  cerebral infarction, 161, 162
    with myocardiac, 162
  epidemiological survey, 156
  prevalance rate, 157–159
  risk factors, 160
Cereloplasmin, 198, 204
Chinese tradional medicine, 6, 27, 28, 143, 169, 184
Choline, 5, 47
Chromosome, 6, 21
Clomipramine, 78
Clonazapam, 3
Copper in cytosol of fibroblasts
  concentration, 199
  distribution, 200
Cobalt, 168
Corydalis, 27
Curare, 6, 66
Cyclic adenine, 3'5' monophosphate cAMP, 7, 143, 151
Cyclic guanine, 3'5' monophosphate cGMP, 7, 151

Dementia, 14, 53
Depression
  clinical assessment
    Clinical Global Impression Chart (CGI), 112–119
    Grading Scale, 112
    Rating Scale of Side Effects (ASBERG), 112–119
  diagnostis criteria
    Criteria at Huangshan China Symposium, 1981 110
    Hamilton Deprssive Rating Scales, 110

Depression (cont.)
  diagnostis criteria (cont.)
    Handbook of Epidemiologic Survey of Mental Disease in China, 110
    Present State Examination, 110
    WHO Standardized Assessment of Depression, 110
  types
    depressive psychosis, 112
    manic depression, 112
    neurosis, 112
    schizophrenia, 79, 112
  treatment
    acetazolamide, 191
    adenosine, 193
    amoxapine, 191
    carbamazepine, 190
    chlorpromazine, 191
    citalopram, 194
    clomipramine, 78, 194
    haloperidol, 191
    lithium, 189
      mechanism, 194
    thyroid hormone, 191
    valproate, 190
    antimanic, 190, 191
Depressive disorders
  basic research
    development of putative drugs
      serotonin reuptake inhibitors, 79, 80, 84
      serotonin inhibitors efficacies, 89
      interaction of chemicals with neurotransmitter receptors, 88
  clinical
    classification
      unipolar
        asthenic, 189
      bipolar
        genetic penetrant, 189
        syntonic, 189
  therapy
    chemicals
      mood stablizer, 191
        mechanism, 191–193
      norepinephrine uptake inhibitors in neuron, 78
      serotonin inhibitors, 78
      tryptophan, 79
      5HTP + MAO, 79
    others
      acupuncture, 109
      changes in neurochemicals, 79, 120
      computer control electro-acupuncture, 110
      electro-acupuncture, 109
Desimipramine (DMI), 28, 31, 78
Dexetimide, 25
Diabetes mellitus, 160

Index

Dopa
  D,L-dopa, 3
  L-dopa, 3, 4, 18, 25
Dopac, 18, 21, 134, 135, 137
Dopamine, 30, 133, 134
  in neurons, 27, 30
  in Parkinson's disease, 1
  in Psychopharmacological study and psychiatric patients, 1, 2
  levels in brain, 1, 21, 25, 29, 31, 194
  receptors
    D_1, 27, 28
    D_2, 27, 28
  release, 48
  structure, 2
  uptake in brain, 21, 77, 79
DSP-4 18, 21

Eaton-Lambert Syndrome, 208
Epidermal growth factor (EGF), 4
Electron probe x-ray microanalysis (EPXMA), 168
Epilepsy, 193
  animal model
    pentylenetetrazol-induced, 165
    cobalt focus epilepsy model, 174
  basic research
    techniques
      astrocytes culture, 172
      patch clamp, 170
      primary neuron culture, 170
  symptoms
    bursting activity in cerebral cortex, 165, 168
    change in intracellular calcium, 168
    siezure activity, 168
    transmembrane ion movement, 168
  therapy
    herbal mixture, 168
      effect on intractable epilepsy, 169
      inhibit convulsion, 169
      normalized
        calcium incorporate to protein, 170
        calcium shift, 170
        protein level, 170
        seizure-related cellular phenomena, 170
        protect neuron damage, 170
    phenytoin PHT, 172–176
    valproic acid, 175

Fluoxetine (prozac), 82, 83
  functions, 86, 87
  pharmacological responses, 88
Fura, 2 168

Galanin, 4
Gama amino-butyric acid (GABA), 3, 137, 138, 191
Ganglioside
  assay, 58
  content, 59, 61, 172

Ganglioside (cont.)
  synthesis, 57, 58
GM1 ganglioside, 4
Glutamate, 125
  excitatory hypothesis, 125
  effects on
    cellular calcium influx, 125
    plasma glucose level, 125
    temperature
  revive excito toxic, 126
L-glutamic acid, 123
G-protein, 6, 123
Graphic computer (3-D), 8
Growth hormone, 4
Guanylate cyclase, 7, 27

5HIAA, 18, 21, 134, 136, 137
Homovanilic acid (HVA), 18, 21, 134, 135, 137
Huatuo reconstruction pill (HTRP), 183
  actions
    determining techniques, 184
    effects on
      blood supply to ischemia focus, 187
      carotid blood flow, 184, 185
      heart performance, 187
      histamine, 185
      myocardial contractility, 186
6-hydroxydopamine (6-OHDA), 25, 27, 28
5-hydroxytryptamine, 5HT (serotonin), 3, 18, 31, 77, 97, 133–137, 194
  receptor subtypes, 97–105
    antagonists, 97
    binding assay, 98
    intra-spieces heterogeneity, 105
    intra-tissue heterogeneity, 105
Hypertension, 160

Ilexonin A
  source - Chinese herb, 143
  therapy for cerebrovascular disease, 143
    actions
      affect action potential of muscle cell, 147
      beta-receptor blocker, 151
      effects on Hiss bundle electrograph, 143, 146
    influence
      arterial blood pressure, 145, 150
      calcium flux, 148, 149
      carotid sinus reflex, 145
      cyclic AMP, 148, 150
      myocardial contractility, 144
Ion channels, 6, 97, 125
  calcium, 1, 2, 53, 123, 125, 148
  potassium, 52, 125, 170
  sodium, 2, 7, 8
Isoprenaline, 151

Lanthanum, 168
Lecithin, 5

Linopiridine
  actions on
    cerebral glucose metabolism, 51
    cerebral cortical electroencephalographic, 51
    cfos gene expression, 51
    cognitive behavior, 50
    transmitters release, 48, 49
  mechanism, 52, 53
  pharmacological activity, 53
Lisuride, 3, 25

Metal toxin, 6
Metallothionein, 198, 200, 203, 204
Methylene blue dye, 32
Milacemide, 19
Molecular receptor modeling, 8
  electro interaction, 8
  steric interaction, 8
Monoamine oxidase (MAO) inhibitors, 3, 15, 19
  function, 3
  in Alzheimer's disease, 15–18
  in depression, 17
  in Parkinson's disease, 3, 15–18, 25
  sub-types
    MAO-A, 20
    MAO-B
      deamination, 15, 19
      neurochemical effects, 15, 16, 21
      neuroprotective effects, 15, 16, 21
      with amphetamine moiety, 15–17
      with aliphatic amines, 17–21
      analogues, 19–22
MPP^+ 15, 18, 19, 29–44
  actions
    on behavior, 37, 38
    on dopamine nerve terminal, 30
    on free radicals, 30
    on monoamine, 37
  ^3H dopamine uptake, 39, 41
  ^3H NE uptake, 40, 42
  ^3H 5HT uptake, 41, 42
MPTP, 3, 15, 18, 19, 21, 22, 29–44
  metabolites (DMI,NOH)
    on behavior, 35
    on monoamine, 35
    on NE neurons, 30
    on tremor frequency, 38
Myasthenia Gravis, 6, 207
  alpha-bungarotoxin, 209
  beta-bungarotoxin, 207–209
  antibody, 207
  cGMP content in brain, 7
  neurological change, 208
Myelin, 208

NBQX
  block synaptic transmission, 126
Neostigmine, 151

Neurotoxin, 6, 7, 30, 207
Neurotransmitters
  imbalance, 3–6
  interaction with receptors, 88
  structures, 2
  uptakes, 77
Nicotinamide adenine dinucleotide (NADH), 25
Nicotine, 3, 6
Nifedipine, 149, 151
NMDA, 52, 123, 125
Nomifensin, 29, 31
Nortriptyline, 79
Norepinephrine (NE), 2, 18, 21, 29–31, 77, 78, 133, 148

Parkinson's disease
  basic research
    animal models for Parkinson's disease
      reserpine treated, 1
      MPTP model, 3, 15, 18, 21, 29
        effect on behavior, 37, 38
        effects on monoamines, 37
        effects on tremor frequency, 3
        MPTP metabolites, 15, 18, 21, 29–44
        neurotoxicity, 30
    perspective therapeutic strategies
      antioxidants, 4
      Chinese herbs extract, 27, 28
      fetal and adrenal transplants with MRI, 4
      gama-aminobutyric acid, 3
      glutamate inhibitors, 3
      GM1 ganglioside, 4
      growth factors, 4
      novel dopamine agonists, 4, 21, 27
      neuroleptics, 4,
  clinical
    symptoms, 30
    side effects after treatment, 3, 18, 25
    treatment
      chemicals
        anticholinergic agents, 3, 25
        COMT inhibitors, 3, 25
        decarboxylase + l-dopa, 2, 3, 25
        dopaminergic agonists, 2, 3, 25
        dopamine release enhancers, 3
        dopamine synthesis stimulants, 3
        MAO inhibitors, 3, 15, 18, 25
        melanocytes stimulating hormone antagonist, 25
      surgery
        adrenal medulla autografts, 4
        fetal substantia transplantation, 4
        stereotactic surgery, 4
Pentyenetetrazol, 168–176
Pergoline, 3, 25
Phenoxyphenyl propylamine
  structure-activity relationship, 82
Phentolamine, 149

Index

Phosphotidylcholine, 47
Physostigmine, 5, 47
Piracetam, 5
PLG (MIF), 25
Propranololi, 151
Potassium, 1, 52

Reserpine, 1, 27, 78, 191

Selegiline (L-deprenyl)
  Actions
    on dopamine neurotransmission, 15
    on MAO inhibition, 15, 18, 19
    on production of toxic oxidative radical, 15, 19
    trophic-like, 15
Signal transduction, 8
Sinemet, 3, 25
Sodium nitrite, 185
Somatostatin, 5
Synaptosome, 32

Tetrabiopterine, 25
Tetracyclic, 109
Tetrahydroaminoacridine, 47
Tetrahydroprotoberberines (THPBs), 27
Thioridazine, 7
ThreoDOPs, 25

Toxins
  alpha-bungarotoxin, 6, 65, 75, 207, 209
  alpha-cobrotoxin, 66, 75
  beta-bungarotoxin, 207–209
  Kappa-bungarotoxin, 75, 76
    action, 76
    binding site, 76
  organophosphates, 65
  types
    phencyclidine, 65
    sarin, 65
    soman, 65
    tabun, 65
Tremor monitor, 32
Tricyclic, 3, 78, 109, 189
Tyrosine hydroxylase, 30
Trihexyphenidyl, 25

Wilson disease, 197
  clinical diagnosis, 198, 204
    biochemical examination, 198
    copper/protein ratio, 199, 204
    skin fibroblast culture, 198, 204
  copper metabolism, 198
  in vitro model, 203
  pathological symptom, 198
  prevalence, 198